Magnetic Structures of
2D and 3D Nanoparticles

Magnetic Structures of 2D and 3D Nanoparticles

Properties and Applications

edited by

Jean-Claude Levy

Published by

Pan Stanford Publishing Pte. Ltd.
Penthouse Level, Suntec Tower 3
8 Temasek Boulevard
Singapore 038988

Email: editorial@panstanford.com
Web: www.panstanford.com

British Library Cataloguing-in-Publication Data
A catalogue record for this book is available from the British Library.

Magnetic Structures of 2D and 3D Nanoparticles: Properties and Applications

Copyright © 2016 Pan Stanford Publishing Pte. Ltd.

All rights reserved. This book, or parts thereof, may not be reproduced in any form or by any means, electronic or mechanical, including photocopying, recording or any information storage and retrieval system now known or to be invented, without written permission from the publisher.

For photocopying of material in this volume, please pay a copying fee through the Copyright Clearance Center, Inc., 222 Rosewood Drive, Danvers, MA 01923, USA. In this case permission to photocopy is not required from the publisher.

ISBN 978-981-4613-67-5 (Hardcover)
ISBN 978-981-4613-68-2 (eBook)

Printed in the USA

Contents

Preface   xiii

1   Magnetic Structures of 2D and 3D Nanoparticles   **1**
Jean-Claude Serge Levy
1.1   Introduction   1
1.2   Experimental Status   7
     1.2.1   Nanoparticle Preparation   7
         1.2.1.1   Magnetic nanoparticles from grinding   8
         1.2.1.2   Ionic or atomic jet beam clustering   9
         1.2.1.3   Sputtering on convenient surfaces or within voids close to the surface   10
         1.2.1.4   Molecular beam epitaxy on convenient surfaces   14
         1.2.1.5   Chemical preparation of magnetic nanoparticles   16
         1.2.1.6   Electrochemical preparation of magnetic nanoparticles   17
     1.2.2   Simulated Nanoparticles: Structures   17
         1.2.2.1   Simulated supported nanoparticles   19
         1.2.2.2   Simulated free 2D nanoparticles   19
         1.2.2.3   Simulated three-dimensional nanoparticles   23
         1.2.2.4   Simulated aggregation of magnetic nanoparticles   29
     1.2.3   Experimental Evidence of the Magnetic Structure of Nanoparticles   31
         1.2.3.1   Bitter powder and other tracking techniques   32

| | 1.2.3.2 | Faraday and Kerr effect: Magneto-optical Kerr effect (MOKE) | 35 |

1.2.3.2 Faraday and Kerr effect: Magneto-optical Kerr effect (MOKE) — 35

1.2.3.3 Scanning electron microscopy with polarized analysis (SEMPA) — 36

1.2.3.4 Photoemission electron microscopy (PEEM) — 37

1.2.3.5 X-ray magnetic circular dichroism (XMCD) — 39

1.2.3.6 Magnetic force microscopy (MFM) — 40

1.2.3.7 X-ray holography — 41

1.2.4 Numerical Evidence of the Magnetic Structure of Nanoparticles — 42

1.2.4.1 Numerical methods for determining the magnetic ground state of a nanoparticle — 42

1.2.4.2 Monte Carlo studies of 2D magnetic nanoparticles — 43

1.2.4.3 Langevin's simulations of 2D magnetic nanoparticles — 44

1.2.4.4 Monte Carlo studies of 3D magnetic nanoparticles — 48

1.2.4.5 Micromagnetic and OOMMF studies of magnetic nanoparticles — 50

1.2.5 Experimental Evidence of the Magnetic Excitations of Nanoparticles — 50

1.2.5.1 Gyrotropic vortex (antivortex) motion — 52

1.2.5.2 Spin wave motion within a topological defect — 53

1.2.6 Numerical Evidence of the Magnetic Excitations of Nanoparticles — 53

1.2.6.1 Gyrotropic vortex and antivortex motion simulated — 54

1.2.6.2 Spin wave simulations in presence of topological defects — 57

1.3 Theory — 60

1.3.1 Magnetic Carriers — 60

1.3.2 Magnetic Interactions — 60

1.3.2.1 Exchange and superexchange — 61

		1.3.2.2	Anisotropy	61
		1.3.2.3	Dipolar interactions	62
		1.3.2.4	RKKY interactions	62
		1.3.2.5	Zeeman effect	62
	1.3.3	Determination of the ground state		62
		1.3.3.1	Ground state of 2D nanoparticles	66
		1.3.3.2	Ground state of 3D nanoparticles	72
	1.3.4	Excited states		78
		1.3.4.1	Effective "1D" excited states	79
1.4	Conclusions			82

2 Vortex Lines in Three-Dimensional Magnetic Nanodots by Langevin Simulation

101

Ph. Depondt and J.-C. S. Lévy

2.1	Introduction			102
2.2	The Simulation			103
	2.2.1	Model		103
	2.2.2	Method		104
		2.2.2.1	Numerical integration of the precession motion	104
		2.2.2.2	The dipole–dipole interaction	106
		2.2.2.3	Finite temperature simulations	107
		2.2.2.4	Seeking vortices	108
2.3	Results			109
	2.3.1	Initial Conditions: One Central Vortex Line, Normal to the (xOy) Plane		110
	2.3.2	Ground State?		117
2.4	Concluding Remarks			120

3 In-Plane Magnetic Vortices in Two-Dimensional Nanodots

123

Sławomir Mamica

3.1	Introduction	123
3.2	Dynamical Matrix Method	127
3.3	Normal Modes in Circular Dots	133
3.4	Spin-Wave Spectrum	135
3.5	Spin Waves vs. d and L	140
3.6	Stability of the In-Plane Vortex	149
3.7	Final Remarks	152

viii | Contents

4 Magnetic Properties of Nanostructures in Non-Integer Dimensions   **159**
Pascal Monceau

4.1 Introduction	159
4.2 Fractality and Scale Invariance	160
4.2.1 Discrete Deterministic Scale Invariance	161
4.2.2 Discrete Random Scale Invariance	161
4.2.3 Random Scale Invariance	162
4.2.4 Hausdorff Fractal Dimension	162
4.2.5 Configuration Entropy and Disorder Quantification	163
4.2.6 Geometrical Properties of Deterministic Sierpinski Fractals	164
4.2.6.1 Order of ramification	165
4.2.6.2 Connectivity and mean number of links per site	166
4.3 Effects of Deterministic and Random Discrete Scale Invariance on Spin Wave Spectra	166
4.3.1 Spin Waves Eigenmodes Problem, Integrated Density of States and Spectral Dimensionality	166
4.3.2 Discrete Deterministic Fractals	168
4.3.3 Random Scale Invariance	172
4.4 Phase Transitions on Fractals	173
4.4.1 Critical Behavior of the Ising Model on Deterministic Fractals	173
4.4.1.1 Theoretical background	173
4.4.1.2 Finite-size scaling and Monte Carlo simulations methods	175
4.4.1.3 Overall review of the numerical simulation results: weak universality	178
4.4.2 First-Order Magnetic Transitions on Deterministic Fractals: The Potts Model	181
4.4.2.1 Theoretical background	181
4.4.2.2 Multicanonical simulation of first-order transitions: the Wang Landau algorithm	182
4.4.2.3 Canonical simulations of first-order transitions: the Meyer-Ortmanns and Reisz criterion	183

| | | 4.4.2.4 Phase diagram of the Potts model | 184 |
| | 4.4.3 | Critical Behavior of the Ising and Potts Model on Random Fractals | 186 |

5 Magnetic Anisotropy and Magnetization Reversal in Self-Organized Two-Dimensional Nanomagnets — 193

Vincent Repain

5.1	Introduction	193
5.2	Self-Organized Magnetic Nanocrystals Supported on Crystalline Surfaces	194
5.3	A Model System for the Study of Ultra-High Density Magnetic Recording Media	198
5.4	The Thermal Stability of Nanomagnets: Beyond the Néel–Brown Model	200
5.5	Magnetic Anisotropy in Core–Shell and Alloys Nanomagnets	202
5.6	Conclusion	206

6 High-Aspect-Ratio Nanoparticles: Growth, Assembly, and Magnetic Properties — 213

Frédéric Ott, Jean-Yves Piquemal, and Guillaume Viau

6.1	Introduction		213
6.2	Elaboration of High-Aspect-Ratio Magnetic Nanoparticles: Relation between Growth and Structural Properties		214
	6.2.1	Chemical Synthesis	215
		6.2.1.1 Polyol process	216
		6.2.1.2 Organometallic chemistry	219
	6.2.2	Electrochemical Synthesis in Solid Templates	222
	6.2.3	Pulsed Laser Deposition	225
	6.2.4	Summary	226
6.3	Magnetism of High-Aspect-Ratio Nanoparticles		227
	6.3.1	Description of the Magnetism of Nanometric and Micrometric Scaled Objects	227
	6.3.2	Description of the Magnetization Reversal Processes of Magnetic Nanoparticles within the Stoner–Wohlfarth Model	230
	6.3.3	Non-Uniform Magnetization Reversal Modes	232

x | Contents

6.3.4 Description of the Magnetization Reversal
Processes of Magnetic Nanoparticles within a
Micromagnetic Modelling   233
6.3.4.1 Effects of the shape of elongated
magnetic particles on the coercive
field   234
6.3.4.2 Coherent and incoherent
magnetization processes   235
6.3.5 Comparison of Experimental Data with
Calculations   238
6.4 Magnetic Properties of Assemblies of Nanoparticles   240
6.4.1 Effect of the Particles' Orientation   240
6.4.2 Effect of the Macroscopic Demagnetizing Field   242
6.4.3 Effect of the Dipolar Interactions between
Wires   244
6.5 Applications of Magnetic Nanowires   246
6.6 Conclusion and Outlook   249

**7 Magnetoferritin Nanoparticles as a Promising Building
Blocks for Three-Dimensional Magnonic Crystals   257**
Sławomir Mamica
7.1 Introduction   257
7.2 Magnetoferritin Nanoparticles and Their 3D
Assemblies   261
7.3 Theoretical Approach: 3D PWM   264
7.4 Spin-Wave Spectra of Magnetoferritin-Based
Magnonic Crystals   266
7.5 Final Remarks   274

8 Magnonic Crystals: From Simple Models toward Applications   283
Jarosław W. Kłos and Maciej Krawczyk
8.1 Introduction   283
8.2 Spin Wave Dynamics in Homogeneous Ferromagnetic
Film   287
8.2.1 Exchange and Dipole-Exchange Spin Waves   290
8.3 Theory of Spin Waves in Planar Magnonic Crystals   293
8.3.1 Spin Waves in Inhomogeneous Medium   293

	8.3.1.1	Exchange field in inhomogeneous media	296
	8.3.1.2	Demagnetizing field in inhomogeneous media	297
	8.3.1.3	Boundary conditions for magnetization at interface with nonmagnetic material	300
8.3.2	Bloch Theorem		300
8.3.3	Plane Wave Method		301
8.4	Spin Wave Spectra in Magnonic Crystals		305
8.4.1	Periodicity		306
8.4.2	Dipole-Exchange Spin Waves		310
8.4.3	Exchange Waves		316
8.4.4	Magnonic Waveguides		318
8.5	Concluding Remarks		320

9 Physical Properties of 2D Spin-Crossover Solids from an Electro-Elastic Description: Effect of Shape, Size, and Spin-Distortion Interactions — **333**

Kamel Boukheddaden, Ahmed Slimani, Mouhamadou Sy, François Varret, Hassane Oubouchou, and Rachid Traiche

9.1	Introduction		333
9.2	Some Recent Experimental Results of Optical Microscopy: Visualization of the Spin Transition		339
9.2.1	A Front Propagation Slower Than the Snail		342
9.2.2	The HS–LS Interface Deduced from Optical Microscopy Measurements		347
9.2.3	Photo-Induced Effects and Relaxation of the Metastable HS State at Low-Temperature		348
9.3	Theoretical Description of the Spin-Crossover Problem		351
9.3.1	The Hamiltonian		353
9.3.2	Structure of the Hamiltonian and the Analogy with the Ising-Like Model		355
9.3.3	The Monte Carlo Procedure		358
9.3.4	The Parameter Values		359
9.4	Results and Analysis		361
9.4.1	The Thermally Induced First-Order Transition		361

	9.4.2	Square-Shaped Lattice	362
9.5		Lattice Configurations upon the SC Transition of the Square Lattice	365
	9.5.1	Derivation of the Elastic Stress from the Displacement Field	367
9.6		Spatio-Temporal Aspects of the HS to LS Relaxation Process: Case of the Square Lattice	370
	9.6.1	The Front Shape and the Interface Velocity	375
9.7		Lattice Configurations upon the SC Transition of the Rectangular Lattice	379
	9.7.1	Relaxation of the Metastable HS Fraction at Low-Temperature	381
	9.7.2	On the Interface Propagation	382
9.8		Relation between the Interface Orientation and Lattice Symmetry	388
	9.8.1	Thermal Properties and Isothermal Relaxation at Low Temperature	390
	9.8.2	Isothermal Relaxation of the Metastable HS Fraction at Low Temperature	392
9.9		Thermal Behavior of Spin-Crossover Nanoparticles and Their Theoretical Description	399
	9.9.1	Size and Shape Effects	404
	9.9.2	Role of the Elastic Misfit between the Matrix and the SC Nanoparticle	407
		9.9.2.1 Effect of a soft shell	409
		9.9.2.2 Effect of a rigid shell	411
9.10		Conclusion and Perspectives	412

Index 435

Preface

Magnetic nanoparticles already have a long story since they appear naturally in rock magnetism together with a large distribution of magnetic particle sizes and shapes, as studied more than 60 years ago by Louis Néel in several famous papers. The recent interest in magnetic nanoparticles comes from the new ability in producing artificial nanoparticles of arbitrary shape and size in a controlled way with the opportunity to observe their properties, which offer many applications from nano-size magnetic memories to metamaterials for electromagnetic waves as well as biological applications such as nanosurgery with minimal traumatism.

It can be said that two steps can be distinguished in nanoparticle progress as well as in magnetic nanoparticle progress. First, the improvement of surface deposition by molecular beam epitaxy, which was soon followed by the accurate observation of surface tunneling microscopy and scanning microscopy, which enabled analysis of these samples and gave physicists and chemists the ability to deal with 2D nanodots, and of course 2D magnetic nanodots. The possibility of making accurate 2D maps both from observation and from numerical simulation was used in order to obtain numerous results within a quite short time.

A natural extension of this work has been the production of ultrathin films and multilayers, with the success of giant magnetoresistance of multilayers and its applications. So the next step consisted in producing, observing, and simulating 3D magnetic nanoparticles, an already very active topic. A curious thing during this process is that the effort in solving the magnetic structure of 3D nanoparticles leads to a new approach of the general magnetic structure of 3D samples of any size, a still unclear point. These remarks define this book's goals, which are to introduce the reader to the many aspects of 2D and 3D magnetic nanoparticles. Nine chapters are written by different authors. This partition ensures both the existence of complementary views and also the occurrence of several connections between these chapters. So we will review briefly the contents of these chapters and underline the connections between these chapters which all mix experimental, theoretical, and numerical observations of magnetic structures and their dynamic properties.

The first chapter introduces in the first part the quite numerous experimental ways which have been used to produce magnetic nanoparticles and compares these results with basic numerical simulations of their atomic structures. A very large amount of new structures can be reached in such a way. The second part is devoted to experimental observations and numerical simulations of the magnetic structures of these nanoparticles, with the result that while 2D magnetic structures are rather well known, 3D magnetic structures are not easy to observe. The third part deals with a theoretical modeling of 3D magnetic structures of very small nanoparticles which shows the stability of S-shaped lines of vortex cores when dipolar interaction is strong enough. This model leads to further steps of complex magnetic structures when either the sample size is increased or the dipolar interaction is increased for a given exchange. An introduction to spin waves in such structures is also given.

This new picture with S-shaped lines of vortex cores is confirmed by the numerical simulation with Langevin spin dynamics done in the second chapter about samples of various thicknesses which mimic the transition from a flat sample toward a cubic or spherical one. The Langevin dynamics results confirm the occurrence of twisted vortex core lines, which for strong enough dipolar interaction self-organize themselves in networks, and for stronger dipolar interaction organize themselves in a more complex way in order to satisfy to general symmetry requirements in the 3D space. These results extend to 3D the basic properties observed for 2D samples by means of these twisted cortex core lines.

The third chapter deals with the stability of an in-plane vortex in 2D models as deduced from an analysis of the dynamical matrix in presence of both exchange and dipolar interactions. Here the ground state must be assumed before and its stability is calculated as well as its first changes. Basically, this method is quite complementary of the numerous methods which can be used to derive the ground state, such as Monte Carlo simulations, Langevin dynamics, or other solutions of the Landau–Lifshitz equations. The good agreement which is obtained here with Langevin simulations is interesting as a validity test for these two independent methods. Another advantage of the dynamical matrix method is to deduce the spin wave modes which can be observed by magnetic resonance, and computing their profiles enables to deduce their intensities. It opens a clear comparison with experimental data of magnetic resonance to come.

The fourth chapter deals with dynamical properties and also with critical properties, but now for a very large and rich class of samples, the ones which are scale invariant with non-integer dimensions. Here such calculations neglect dipolar contributions. Such samples have a final, very low density and can be produced with quite different levels of randomness. An interesting point is that dynamical properties as well as critical properties are shown to be quite sensitive to the topology of such scale-invariant samples. Samples with very low randomness exhibit specific singular continuous spin wave spectra, which mean the occurrence of gaps in the excitation spectra. This gap property can be used for metamaterials as discussed in chapter 7 and 8. And these gaps were recently observed for such materials. Moreover, when the level of randomness of the sample is increased, the spin wave excitation spectrum becomes continuous and the exponent which characterizes the relative increase of the frequency as a function of mode number is shown to be dependent on this level of randomness. In other words, dynamical properties are quite sensitive to the level of complexity and disorder of the underlying structure. About critical properties, a similar sensitivity is observed even with the occurrence of first-order magnetic transition.

The fifth chapter is based upon experimental observations which are done both on deposit structures and on their magnetic properties. These 2D deposits are mainly produced over gold reconstructed surfaces. This surface reconstruction is due to elastic forces and thus the magnetic deposits upon this surface are also submitted to strong elastic forces and so to distortions. This complex elastic field induces a magnetic effect which in the case of cobalt, a good candidate for magnetic applications because of its high magnetization, also leads to magnetic anisotropy. Another feature of this experimental work is the occurrence of simultaneous numerous magnetic dots, of which the main separation distance can be selected by a convenient choice of the substrate surface orientation. Such a set of dots sees a distribution of elastic contributions as well as a specific dipolar effect. Thus the distribution of local anisotropies competes with dipolar effects. This realistic situation induces specific magnetic features, for instance, in the magnetization reversal process, which is a basic one for writing and reading a magnetic memory. And such magnetization reversals strongly depend upon the observation temperature.

Chapter 6 deals with another specific experimental situation with magnetic nanoparticles, that of nanorods and nanowires, which are interesting because of their specific, unusual shape. And dipolar effects are known to be sensitive to shape effects. As a matter of fact, there are different ways used to produce such samples with different specific morphology—a set of dumbbells, for instance. As observed in chapter 5, the elastic properties which lead to obtain such elongated samples also induce magnetic anisotropy in a non-uniform way within the sample and from sample to sample there is also a distribution of magnetic anisotropy. Thus dipolar effects are here too competing with this distribution of local anisotropy. Thus magnetization reversal cycles are also strongly influenced by this distribution of anisotropy, as it was in the case of rock magnetism.

Chapter 7 has also an experimental origin since its deals with the magnetic properties of magnetoferritin blocks which occur in biology where these blocks are encapsulated within proteins and which can be obtained artificially in a similar way. One advantage of this method is the overall symmetry of these nanoparticles even if elastic interactions occur at the protein surface. Here the basic feature is the simulation of the spin wave spectrum which can be observed for such a set of magnetoferritin blocks. As in the case of chapter 4, gaps occur in the spin wave spectrum; this is a magnonic effect, as developed in chapter 8. Here, dipolar interactions are taken into account but in presence of an external field which is strong enough to align all spins. A quite interesting thing is that such gaps of the excitation spectrum are sensitive to the magnetization orientation, i.e., to the external field orientation. This effect could be useful for magnonic applications and could even explain the detection of the orientation of the Earth's magnetic field by birds, since their brain contains magnetoferritin blocks, and these gaps in the excitation spectrum can induce some coupling with mechanic oscillations which appear during flight. Another aspect of this careful study is the interest in the spin wave profiles which show localization effects within magnetoferritin blocks.

Chapter 8 is devoted to the general study of magnonic effects when considering different sets of blocks submitted to an external saturating field. As in the rather symmetric samples of chapter 4, gaps occur in the excitation spectrum, now even when taking into account dipolar interactions. This fact can be used for applications with metamaterials. An interesting point is the nature of the spin wave profiles which show unusual localization effects. These localization effects can be compared to that observed in chapter 4, where they were due to the geometric complexity, while here they can be due either to the geometric complexity or to specific long-ranged dipolar effects. A quite large amount of samples can be conceived in such a way for metamaterials.

Chapter 9 reports on experimental data on molecules with two spin states which can be occupied at rather low temperatures. Basically, these two spin states correspond to two slightly different molecule geometries. Thus, when dealing with a network of such molecules, strong elastic distortions occur when these molecules change their spin states. Such a situation can be compared with the experimental situations of chapters 5 and 6, where substratum elastic forces were responsible for a distribution of magnetic anisotropies. Here these strong distortions are intrinsic to the spin system and thus lead to unavoidable effects, even if in some molecules the morphology change is weaker than for others. This very interesting feature can be activated by convenient light which stimulates this transition. So a large number of new effects can be obtained from such experiments.

In conclusion, this book brings a solid brick to the high potential of magnetic nanoparticles by means of a good balance between experimental observations and theoretical thoughts. First of all, the full richness of the nanoparticle world is well described and the new world of nanoparticle networks introduced. For the theoretical parts the study of dipolar interactions versus exchange both in their static properties and in their dynamic properties is brought to a high level, which includes a deep understanding of magnetic domains, walls, lines, even for macroscopic samples as well as the conception and production of high-frequency metamaterials. For the experimental part the link between real-space elastic properties and spin properties of nanoparticles is well evidenced in several different cases, and so too the link with the unavoidable dipolar interactions competing with exchange and anisotropy. It must be underlined that this book results from many collaborations which extend far from the registered places and which have lasted during long times beyond standard frontiers.

Jean-Claude Levy
December 2015

Chapter 1

Magnetic Structures of 2D and 3D Nanoparticles

Jean-Claude Serge Levy

Laboratoire Matériaux et Phénomènes Quantiques, UMR 7162,
CNRS Université Paris Diderot, 10 r. A. Domon et L. Duquet, 75013 Paris, France
jean-claude.levy@univ-paris-diderot.fr

This chapter is intended for students as well as for researchers. The topic of magnetic nanoparticles is at least as rich in new applications as the general topic of tailored nanoparticles and nanosystems is since the magnetic specificity of long-ranged dipolar interactions makes the magnetic properties of such samples very dependent on both sample size and sample shape and thus induces a large amount of different properties. So this specificity produces a large diversified panel of tentative dynamic applications.

1.1 Introduction

Interest in magnetic particles started a very long time ago with the prehistoric observation of magnetic materials and of their attractive properties. These materials were found near a small city of Turkey called Manisa, from which the words *magnet* and *magnetism* come

---

Magnetic Structures of 2D and 3D Nanoparticles: Properties and Applications
Edited by Jean-Claude Levy
Copyright © 2016 Pan Stanford Publishing Pte. Ltd.
ISBN 978-981-4613-67-5 (Hardcover), 978-981-4613-68-2 (eBook)
www.panstanford.com

from. The early use of attractive magnets, of magnetic compasses, and later of magnetic tapes evidenced the general and vivid interest in magnetic particles along centuries.

Nowadays a large variety of sizes, shapes, chemical compositions, and structures of magnetic nanoparticles is available in the laboratory, and new applications such as externally driven biochemical actors [179], spintronic components [44], and high-frequency components [86] have already started gaining attraction. So the interest in magnetic nanoparticles is really very active, and the recent Nobel Prize attributed for giant magnetoresistance (GMR) in layered materials with a nanometer thickness [7], i.e., a 1D version of magnetic nanoparticles, validates the hopes in further practical applications in the near future, since the technology of nanomaterials is progressing fast. Among the most hopeful applications is cancer therapy by means of magnetically driven nanoparticles acting either by selective biochemical coating [49] or by thermal heating [26], and in that case the active external field can be only a radiofrequency field [33]. The useful driving process of magnetic nanoparticles at the nanometer scale extends the more traditional interest in magnetic printing on the basis of magnetic ink controlled by an extra magnetic field [10]. And this driving process also works in the case of ferrofluids in a more complex manner [11]. This still active process remains in progress and requires attention since both the magnetic ground state of such nanoparticles and their excited states are not clearly known. Magnetic memories at the nanometer scale [124] can be written and read and are the other active goal of applied research. For instance, applications of magnetic nanoparticles to spintronics with a minimal functional energy use as well as a high space concentration [76] are also exciting tools in our era of practical devices with both low energy cost and low size. In analogy with the work done recently in photonics with new material systems with specially designed optical index [197], systems made of tailored magnetic nanoparticles are also good candidates for deriving such new metamaterials for magnetic index [129], under the name of "magnonic" materials as they were recently introduced [131]. According to this new scope, numerous high-frequency magnonic devices [88] have been already proposed and among them even emitters were produced.

So understanding both the magnetic structures of nanoparticles and their dynamics is a real challenge for finding new applications. And the number of such applications is very large.

On the theoretical ground, in Heisenberg systems, i.e., approximately insulating magnetic systems, magnetic structures result from the competition between local interactions such as magnetocrystalline anisotropy [128], an exchange [51] which is usually restricted to nearest neighbors and extends sometimes up to next-nearest neighbors, and long-ranged dipole–dipole or dipolar interactions [93], which cannot be effectively shrunk via a cutoff process because of their rather slow decrease with distance. And the dipolar field created by one spin not only extends far away from its source in and out of the sample but is oriented in all directions with a null total average as evidenced by both attractive and repulsive properties of magnets. As a consequence, dipolar interactions introduce a highly complex behavior of magnetization in large samples. This very specific case of frustration [181] is known even in quite simpler cases of magnetic interaction to lead to complex magnetic arrangements known as spin glasses [153], where the ground state is highly degenerate and strong energy barriers occur between these approximately degenerate ground states which can remain frozen during cooling. Even in metallic components where more or less free conduction electrons interact with fixed spin-polarized ions, this long-ranged dipolar contribution cannot be neglected. And in the case of metallic magnets where free electrons can partially screen the dipolar effect, other long-ranged magnetic interactions occur like Ruderman–Kittel–Kasuya–Yosida (RKKY) oscillating interactions, which are due to the finiteness of the Fermi volume in reciprocal space [144]. So RKKY interactions also contribute to similar magnetic behaviors in metallic systems.

A major theoretical point lies in the fact that dipolar interactions within a sample depend on both sample size and sample shape. The dependence in size is obvious since the resulting dipolar field which is active on a site results from the sum of all the other site contributions. On the other hand, the dependence on sample shape is well known experimentally and theoretically from the evidence of the demagnetizing field which is induced from the dipolar contribution [159] and strongly depends on sample shape. This shape effect results from the orientation-dependent dipolar field, as already noticed. From that point of view it is quite obvious that a general theory of magnetic structures in 3D nanoparticles cannot be found. For true macroscopic 3D samples the situation is a little bit simpler since the sample size is quite larger than the exchange length which scales dipolar effects, i.e., the size of magnetization variations. Then nearly uniform Weiss domains occupy most of the macroscopic sample space, as initially observed by Weiss [192], a century ago, while walls and other topological defects occur between Weiss domains within the rather restricted rest of the sample in order to fulfill dipolar constraints. Earlier, Landau and Lifshitz pointed out that this space partition is due to dipolar interactions [92]. Thus the search for an optimal periodical structure for magnets as followed by Luttinger and Tisza [111] cannot deal with finite samples as they are and more especially as nanoparticles obviously are.

In the very special case of basically 2D samples such as small dots, recent numerical simulations of micromagnetism [45], with Monte–Carlo slow cooling [37] or Langevin's simulations [31] evidenced the stability of topological defects such as magnetic vortices, antivortices, and so on. These predictions helped experimentalists in observing such structures from Kerr effect and from other microscopic techniques with magnetic sensitivity [34]. So the actual problems of magnetic structures of nanoparticles in their ground state are the following:

1. What happens in large 2D samples? The previously mentioned observations already suggest that a combination of several topological defects occurs in their magnetic structures within some topological association rules.
2. What happens in true 3D samples of different shapes and different sizes? And here there is a special interest in basic units since larger samples can be analyzed as a combination of basic units, as proposed before about 2D samples. Of course, more complex structures can also be interesting for applications.

These two kinds of structures define the main goals of this chapter. Of course. an obvious secondary goal is the study of the elementary excitations within these magnetic structures: spin waves, also called magnons [16]. Magnons can appear as spin motions from stable or metastable states [16]. Thus a basic study of magnons is required for stability considerations [115], and magnons can be observed from several resonance techniques [80]. It must be added that the excitation spectra of such samples reveal a strong sensitivity to their underlying structural complexity even when dipolar effects are neglected [126]. And this structural sensitivity which is added to the strong dipolar shape effect suggests the need for a systematic study of geometric complexity in order to derive applications. So, dynamic considerations are necessary in order to understand the very complex situation of magnetic structures of nanoparticles. As already noticed, low-energy magnetic structures (LEMS) are usually quite numerous for a nanoparticle, so according to the value of the observation time and temperature, a "superparamagnetism" effect where its magnetic structure flips from one LEMS to another LEMS and so on is often observed in magnetic nanoparticles [134]. Such a study of superparamagnetism is also involved in the study of dynamics.

Thus the next point to consider is the required choice of sample shape and structure in order to obtain an overview of this hard geometric problem. As just noticed before, several ways deserve attention in the design of new samples and are listed below.

A first way consists in looking at different shapes from 2D mono-layer to 3D cube or sphere or column made from a homogeneous sample. This is already a rich topic [187].

Then practically it is often useful to introduce on these nanoparticles a convenient coating by surfactants in order to protect the basic sample from aging effects or from other physical or chemical effects. And this coating can be magnetically active by itself. This is also the case of nanoparticles with shell and core where shells are issued from segregation effects or from extra deposits [109]. This structure generalizes surface effects in thin films. In such a case several steps of building this coating or segregation within the nanoparticle can be considered in a more or less artificial process. Exactly as in the celebrated case of magnetoresistance multilayers [7], multi-coating steps can be considered, and there is already a large amount of literature devoted to this topic [110].

With a similar concept of breaking the infinite symmetry of a continuous medium and thus inducing a new internal complexity, new material arrangements such as quasicrystals [95], which are composed of several chemical elements and have no translational symmetry, can be considered as new magnetic nanoparticle [85]. Since there are several classes of quasicrystals [121], according to their different unconventional rotational symmetry, different applications must be investigated. And unconventional symmetry can compete with dipolar interaction leading to new magnetic structures [184].

A rather similar research also playing on the level of sample complexity introduces fractals [47], i.e., samples without translational symmetry and with internal self similarity. These fractals can be either deterministic or can include a more or less random contribution at quite different levels. The resulting disorder may help in order to take optimal advantage of this new geometry. There are also several ways to play with this complexity either with dense matter where complementary matter to the fractal structure is magnetic or not, with the interplay of different exchange values. This introduces a very extensive work of which already one-dimensional Fibonacci (1D) arrays of nanoparticles [175] has been proposed. Special cases of fractals with a high level of randomness are percolation clusters [62] which can be reached by diffusion limited aggregation [195, 196], a rather general process for aggregating samples. This also defines a large class of possible samples with fractal properties.

Since dipolar effects are still active even if there is no contact between magnetic materials, various lattices of magnetic nanoparticles have also been studied in order to reinforce these dipolar effects, such as 1D arrays of nanoparticles [82], 2D arrays of nanoparticles [90] and 3D arrays of nanoparticles [140] are already available, while quasicrystalline arrays of nanoparticles [63] were already introduced at least theoretically.

So the amount of work already devoted to magnetic nanoparticles is considerable. In this book numerous ways of using nanoparticles are considered [186]. In the present chapter we shall focus our theoretical interest on dipolar effects and more especially on dense nanoparticles even if the formalism of energy minimization is suited for all samples.

The first part (1.2) of this chapter is devoted to a short historical view of experimental techniques and numerical simulations which are now available for detecting magnetic structures in nanoparticles. It includes the ways used to produce magnetic nanoparticles, to observe their structures and magnetic properties as well as the principle of numerical simulations for nanoparticle structure and nanoparticle magnetism. The theoretical part (1.3) firstly introduces spin, angular momentum, and magnetic interactions. Then the structure of magnetic materials and basic magnetic states are considered. The variational treatment of energy minimization when dealing with 2D and 3D magnetic nanoparticles in presence of dipolar interactions follows with a special interest devoted to basic units, i.e., small enough samples. Part 1.4 deals with dynamic effects such as magnetic excitations and critical effects in nanoparticles. And before deriving the conclusions, part 1.5 mentions some of the numerous parallel topics where nanoparticles are submitted to similar long-ranged interactions, from nuclear spins up to liquid crystals and multiferroic materials as well as neural networks with long-ranged synaptic couplings.

1.2 Experimental Status

This part includes reviews of experiments as well as of numerical simulations, and reviews of sample preparations as well as of sample structural and magnetic observations. All these reviews are given with some historical order.

1.2.1 Nanoparticle Preparation

There are several ways to prepare magnetic nanoparticles. The first one to be used historically is a physical method starting from very macroscopic blocks: grinding or ball milling [39]. It consists in making a powder from the initial constituents, as done currently with pestle and mortar, up to a nanometer size. Several other physical methods start from gaseous independent elements sent together through a vacuum chamber in order to aggregate them either in flight or on convenient traps. Instead of vacuum, a low-pressure inert gas can occupy the chamber. This physical preparation is the case of jet beam clustering [160], sputtering on a convenient surface [25], and molecular beam epitaxy [29], which are based on different stimulations of gas condensation. Numerous chemical methods of nanoparticle preparation consist in developing chemical reactions within a liquid bath at a rather low temperature and then selecting the first steps of aggregation of the precipitated products as nanoparticles [169]. There are also electrochemical methods in which the chemical reduction of cations is activated and controlled by the occurrence of electric potentials and fields [136] during electrolysis.

1.2.1.1 Magnetic nanoparticles from grinding

This old method, also called ball milling, derives from the traditional pestle and mortar way of making powder. Modern mills [12, 137] can turn at high speed and moderate temperature. Extremely hard particles are also introduced within the initial mixture in order to break the macroscopic magnetic matter into very small nanometer units during the activated motion. Because of violent contacts between nanoparticles themselves and with balls and mill walls, the so produced nanoparticles generally include a lot of mechanical and chemical defects. So these samples are far from being uniform and from one to the other different compositions and textures appear. So there is a large defect distribution according to samples. Another question about resulting nanoparticles is size selection, which can be achieved by a further analysis such as mass spectrometry of such powders [200]. However, in spite of this very simple process, very small nanometer sizes have been obtained by grinding for metallic materials as well as for insulators. And such basic grinding methods can be completed by further chemical treatments such as oxidation or reduction and physical or chemical rejection of inconvenient products. An advantage of these grinding methods is the production of a large number of nanoparticles at once. So this method remains a basic way for preparing magnetic nanoparticles which can be either metallic or insulators such as oxides.

1.2.1.2 Ionic or atomic jet beam clustering

This method consists in emitting the matter for nanoparticle as a gas, i.e., at a high temperature enough produced by a local furnace, and sending this gas through a convenient nozzle within an empty space or a low pressure space filled with inert gas in order to produce collisions and finally clustering of the atoms or ions contained in this gas beam [202]. The resulting clusters can be ionized and further analyzed in size from mass spectrometry and then cooled down. With a slow cooling as it occurs within an inert gas chamber, clusters can be nearly defect-free, both mechanically and chemically, a strong advantage of this method. The size distributions of these atomic beam clusters usually show very small nanometer sizes up to very atomic sizes. This size effect strongly depends upon the nozzle shapes and sizes. So the nozzle choice is quite useful in order to select a cluster size range.

The size distributions of very small clusters also show "magical" numbers of cluster elements which are observed as peaks in the size distribution [46]. These magical numbers occur when clusters are made of nearly full shells of atoms or ions according to atomic coordination, exactly as it occurs for nuclei made of full shells of nucleons with the example of silicon and iron for instance [59]. Quite similarly to the well-known case of nuclei made of nucleons, these magical numbers have also a part on cluster stability as observed from in flight laser excitation which can reach cluster splitting [53]. "Magical" clusters, i.e., clusters containing a magical number of atoms, are more stable than average clusters. These magical numbers are characteristic of three-dimensional (3D) geometry and of atom–atom interactions as discussed in the part devoted to structural simulations.

For nuclei, on the reverse hand, low stability is associated with rare earths and actinides, something like an anti-magical effect occurring when the external shell is unfilled. For atomic clusters a similar low stability is also observed for these anti-magical numbers [145]. Of course, these magical or anti-magical effects slowly disappear when the cluster size is largely increased since the nanoparticle structure becomes then more and more crystalline-like [146] and then cluster growth becomes a nearly continuous process.

A similar smearing out effect of this short range limit of aggregation occurs for neutrons in neutron stars [6].

Atomic cluster properties such as stability or electric and magnetic susceptibility can be observed from laser light absorption either during flight [20, 53] or from deposited clusters over an inert substrate. Alloyed nanoparticles can also be obtained when using a convenient, initially mixed gas [185]. Practically these clusters are often ionized, which is more convenient for accelerating them and for mass spectrometry selection. So there has been already an extended analysis of the atomic structures of these ionized or neutral clusters with comparison to simulated models [22].

While very small clusters with an atom number $(n) < 300$ develop quasicrystalline symmetry with fivefold axis, for instance, larger clusters develop more usual crystalline symmetry. This is well in agreement with the observation of magical effects and with the results of numerical clustering simulations as shown in what follows.

This atomic or ionic beam method is very interesting for producing very small, nearly perfect clusters in their ground state and analyzing them. However, the amount of produced clusters from this method is not large. So this method is not very convenient for large-scale applications.

1.2.1.3 Sputtering on convenient surfaces or within voids close to the surface

Sputtering is an already conventional method for obtaining thin films deposited on surfaces as it has been widely used for producing first thin films [133] and more recently multilayers [204]. Atomic beams propagate in a vacuum chamber with a moderate vacuum before reaching the deposition surface target. These surfaces are seriously cooled down so gas condensation occurs quite near the surface with a final deposition and diffusion of the matter at the surface. This is now a traditional process to decorate surfaces and thus to prepare two-dimensional (2D) nanoparticles, i.e., ultrathin films on a substrate.

Surface rearrangements induced by the existence of a free surface or induced by some amount of deposits have been extensively studied, especially for semiconductors but also quite generally for surface science [2]. Several experimental techniques revealed the very nature of surface rearrangements. First low-energy electronic diffraction (LEED) revealed the new symmetry of surface rearrangements from 2D diffraction patterns [15], since low-energy electrons have a short penetration depth and see mainly surface. High-resolution electron microscopy (HREM) [158] with improved electronic computing and scanning electron microscopy (SEM) [113], where the sample is moved with atomic steps, also reached the atomic resolution at the surface by considering the reflected beam. More recently surface tunneling microscopy (STM), which uses the displacements of an atomically tiny electrode to derive an atom-by-atom map, revealed the atom-by-atom structure at the surface even with heights [14]. Other channeling techniques were also used to reveal successfully these surface rearrangements [50]. So the sputtering method was early developed on such a well-informed basis. It leads to the production of 2D dots where the basic lateral extension of these dots is usually restricted by the use of lithographic masks [84]. Smaller lateral dot sizes can be also reached from a careful analysis of surface rearrangements, as discussed later. The purity of these 2D dots is usually high when there is no substrate diffusion back towards the deposit. And there are also some elastic stresses induced by the lattice mismatch between deposit and substrate as well as by the specific interaction between deposit and substrate. These are the two limits of this production of 2D dots. Since the amount of such effective 2D nanoparticles produced at once is large, this sputtering process is well used practically.

Elastic deformations of the surface deposit have been largely observed by the powerful LEED, HREM, SEM, and STM techniques and can be also used for practical applications. As a matter of fact it has been noticed that on some surfaces, because of both lattice mismatch and weak interaction between deposited matter and substrate, the deposited matter appears on the substrate as islands with a rather porous shape in a Stranski–Krastanov way [186]. Between these islands, more or less large voids and pores are observed. And these 3D voids and pores can be later used to trap an extra gas in order to form 3D clusters of this extra matter. The existence of this distribution of 3D defects near the solid surface enables to fill them with some extra sputtered gas and thus to produce 3D clusters within these traps.

The purity of these 3D trapped clusters is limited by substrate diffusion within cluster matter, which can be selected to be as poor as possible. Mechanically the contact between substrate and cluster can be selected to be more or less efficient. So rather homogeneous 3D clusters can be obtained in such a way. One basic point for this 2D production of 3D clusters is that cluster matter must scatter through substrate in order to reach these trapping sites. Practically this means that the porous film is thin enough in order to avoid a strong diffusion limitation.

When comparing these trapped clusters to clusters issued from collisions in the atomic beam, it must be noticed that filling traps avoids most of the magical effects. Here for trapped clusters, two special steps occur, first a filled monolayer within the trap and later at higher flux a nearly filled trap. So magical effects can appear only if traps are large enough in order that approximately free clusters can appear within large voids.

So by sputtering methods both 2D dots, i.e., basically monolayer islands, and real 3D clusters trapped within convenient voids can be produced. It must be added that various trap shapes can be produced and thus selected. So different nanoparticle geometries can be produced such as elongated ellipsoid or even rods, curved rods. By selecting some parts of the beam of magnetic matter with convenient masks, different patterns of 2D dots can be obtained on the substrate [142].

For instance, alumina deposited on typical metals is known to be porous because of the occurrence of several mismatches between substrate and deposit [130]. This effect is then used to produce 3D clusters of selected sizes by sputtering techniques over these so deposited porous layers. Filling these apparent voids with freshly deposited matter makes then supported clusters, like grains surrounded by an inactive matrix. There are also quite numerous ways for preparing such clusters since many porous materials can be used for that purpose. As noticed before, very thin films of these porous systems must be used in order to avoid inhomogeneity in the third dimension. Natural porous systems such as rocks can be used and among them natural and artificial zeolites exhibit well-defined void sizes [178].

Porous systems can be also produced in a first experimental step over a regular surface. Among these porous materials carbon tubes share also the advantage of well-defined void sizes [91]. For defining this trapping process an ultrathin film is enough so graphene has been also used as a porous media for its sites [30].

So produced magnetic clusters are both well localized and fixed. The substrate temperature is also well easily controlled at a low level, which ensures that nanoparticles have only low energy defects from a mechanical point of view. Chemical contamination can also be controlled in order to stay at a low level.

In such experimental conditions the resulting sample is a 2D array of 3D clusters surrounded by an inert matrix which could be removed by chemical ways. One advantage of this production localized on traps is that large amounts of rather uniform nanoparticles can be produced at once. Experimental observations on these nanoparticles can be done either on a single cluster with a local analysis tool such as STM and HREM, or with a large number of clusters together as in the case of optical observation by Kerr effect [193]. Already electric conductivity experiments on Coulomb nanoparticles produced within porous alumina layers were achieved successfully with evidence for quantified electric charge effects in very thin nanoparticles [52].

The locations of trapped clusters are due to the very nature of the porous system. So there is a distribution of distances between clusters which can include some randomness according to the porous system. As a consequence magnetic interactions between these clusters can be estimated. Finally the choice of a convenient porous thin film as substrate enables the experimentalist to select the amplitude of these interactions with a large flexibility.

Another advantage of the production of such 2D arrays of 3D magnetic nanoparticles within an inert matrix film is the possibility for applied shear and stress on nanoparticles by mechanical action on this matrix. Even acoustic modulation can be conceived. Of course, similar effects are available for 2D dots deposited on surfaces and were already used [23].

1.2.1.4 Molecular beam epitaxy on convenient surfaces

Molecular beam epitaxy is just an improved way of sputtering method with the use of both a high level of vacuum and of well-defined crystallographic substrates [29]. So both 2D dots and 3D clusters can be produced according to this method. The main advantage of this method lies in its high level of selectivity since surfaces and layers below are very well defined. So it is possible to take advantage of the known detailed structure of the substrate surface for tailoring a 2D deposit on it.

For instance, in the case of gold, Au(111) exhibits a surface rearrangement which is well known from STM [8]. This rearrangement involves a long-ranged herringbone-like structure with intrinsic defects, i.e., very local parts with a different atomic spacing than other parts. A medium-ranged part of this long-ranged rearrangement is shown in Fig. 1.1, with a chevron structure where kinks are alternatively stretched or compressed according to successive lines. Of course, the extended arrangement does not favor any special direction and so results from the imbrication of such parts. During further deposition the intrinsic defects shown in Fig. 1.1 are primarily occupied as it occurs in the case of cobalt

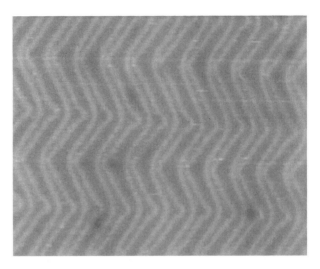

Figure 1.1 Au(111) 22× reconstruction, 80 nm × 70 nm deposited at room temperature [8].

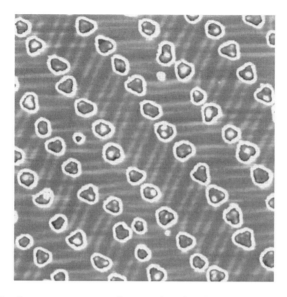

Figure 1.2 Room temperature deposited Co/Au (111). The lower herringbone gold structure is also seen 80 nm × 80 nm [139].

deposition on reconstructed Au(111) at a low flux level [139] with the production of basically triangular dots.

Such a use of a surface rearrangement defines a general process for obtaining a covering with a network of distant 2D nanodots. This quite regular network of cobalt is shown in Fig. 1.2 with just a few extra deposits. It must be noticed in Fig. 1.2 that the so deposited islands are different according to the level of stress or compression they bear. A few more small islands appear out of this network.

Since there are many available surface rearrangements for pure surfaces or for covered surfaces [143], such a low covering process defines many ways of preparing networks of distant 2D nanoparticles. For instance, on conveniently cut gold surfaces such as Au (11,10,10) distant steps with Au (111) parts occur, and these steps enable to produce cobalt wires just at the boundary between them [139], i.e., at 10 atomic distances. So in that case a regular 1D array of nanowires is produced. Of course, as noticed earlier, combinations of these singular arrangements can be introduced. So the amount of accessible networks is quite large.

Among the currently deposited nanoparticles, cobalt is the more frequent basis for 2D nanodots. This is due to both its large magnetization and its strong anisotropy [191]. Iron and nickel as well as iron oxides are also good candidates for magnetic nanoparticles.

1.2.1.5 Chemical preparation of magnetic nanoparticles

The initial steps of chemical precipitates in a reaction within a fluid are obviously ultrasmall nanoparticles which act as seeds for further deposits since the chemical potentials for reduction of a single ion differ from that for a singly ionized cluster and this further chemical potential depends on cluster size. So the basic points to form magnetic nanoparticles are first to initiate cluster formation and then to deposit on these initial clusters magnetic materials with a low precipitation velocity in order to be able to select convenient sizes of the finally grown clusters. For that chemical action, solvents with both a high electrical permittivity and a rather high temperature stability such as polyols are required [176]. There are several control parameters which enable us both to initiate numerous initial clusters within this solvent and to slow down the growth reaction: ion concentrations and redox level. And these parameters can be easily varied quickly during the reaction time. Platinum ions are often used to initiate cluster formation [176]. Reactions are easily stopped by a further change in ion concentration, in temperature. In another paper [156] the reader will see how to obtain anisotropic nanoparticles by controlling their directional growth, and this can be obtained by introducing a convenient external magnetic field.

One advantage of these chemical methods is the easy possibility of making alloyed clusters by adding convenient ions simultaneously in the liquid. And since temperature is high enough while being not so large, rearrangements occur within the nanoparticles. On the other hand there is always a possibility of occurrence of chemical defects if the liquid matter is not very pure. About the size distribution of such nanoparticles, there is also a rather large distribution since many clusters grow simultaneously with a large distribution of "external" conditions when looked at a very local scale. The advantage of this method is the production of a large amount of 3D magnetic nanoparticles at once.

1.2.1.6 Electrochemical preparation of magnetic nanoparticles

Compared with the chemical methods used for preparing nanoparticles, electrochemical methods add one control parameter: electric field. For that, electrodes are required and finally most of the electric field is located close to electrodes because of ion screening effect. Thus the deposit occurs just at the electrodes and so it is basically a 2D deposit of nanoparticles on an electrode. Of course the use of a tricky porous electrode would enable to obtain 3D nanoparticles. So chemical and electrochemical methods can be compared to atomic beam methods and sputtering methods respectively.

As noticed earlier about chemical methods, the advantages of electrochemical methods consist in a rather low temperature process, with a large amount of nanoparticles produced at once. The limits of these methods are due to the chemical purity within their liquid mixture and so the possibility for chemical defects and for a size distribution since many nanoparticles are produced in parallel ways on various initial seeds. For instance, in the case of cobalt deposited on gold, Au (111), electrochemical methods produce quite nice results [151] which are comparable to those obtained by physical methods.

1.2.2 *Simulated Nanoparticles: Structures*

The main advantage of numerical simulations over analytical theories based on a continuous approach consists in that the discrete character of the formation and growth of nanoparticles, step by step, atom by atom, is well taken into account in numerical computations. Three simulation methods are commonly used: Monte Carlo methods (MC) [13], molecular dynamics (MD) [69], and density functional theory (DFT) [83].

In Monte Carlo methods, which were introduced early [123], the interacting potentials between units are first defined. Then an initial configuration of the system of N atoms is built and introduced. It can be a random configuration fulfilling some rules of convenient density while avoiding overlaps, or a test configuration of which relaxation is observed. Basic individual motions are defined and applied or rejected according to the Metropolis rule [123], which consists in taking for sure motions which lower the total energy and admitting according to some probability defined from temperature level motions which increase this total energy. This probability decreases exponentially with the ratio of excess energy divided by temperature. So the convergence toward a final configuration and convergence time depend on the choice of a realistic initial configuration and of fitted elementary motions. Of course, the ground state is obtained from a series of relaxations which are done for slowly decreasing temperatures. And realistic configurations at finite temperature, up to melting point, can be observed even if kinetic energy is neglected [18].

In molecular dynamics, interactions and initial ground state are also defined as well as a unit time, and atomic motions are defined from solving equations of motions with some Langevin random part which takes into account thermal motion. So this method considers a regular variation with time and is well fitted for following the time evolution of thermal excitations at the expense of computer time. This method takes into account kinetic energy and can be used up to melting point with realistic results [166].

Density functional theory (DFT) take into account quantum mechanics of electrons. This leads to more careful calculations, always at the expense of computer time. So DFT results are rather restricted to small clusters or to bulk constructions with admitted symmetry. They are a precious tool and can even deal with magnetic interactions from a very basic approach. And since computers are more and more efficient, such calculations are more and more extended.

As expected, the number of simulations done on nanoparticles is already very large. From the previous considerations on experimental data, there are three kinds of basic nanoparticle structures:

- supported 2D structures which are realistic but strongly dependent on surface rearrangements which can be due either to the substrate itself or to the substrate–deposit interaction. In that case interface rearrangement depends on deposit density.

- free 2D structures which are an extrapolation of the previous case on a perfectly "soft" substrate without any atomic roughness.
- free 3D structures which are the case of clusters in vacuum and an extrapolation of the cases of clusters within a gas or a liquid or in a trap, i.e., with completely "soft" interactions with their environments.

1.2.2.1 Simulated supported nanoparticles

First the substrate effect must be accounted for. For instance, pure gold and platinum surfaces are known to be rearranged. So deposits on these surfaces must also deal with these conditions. A deposit on a substrate induces several mismatches: first there is a lattice mismatch due to the difference in atomic volumes and lattice parameters. Then there is an energy mismatch since binding energies are not the same. Next there is an elasticity mismatch since bulk moduli are not the same as due to different curvatures of the interaction potential. Moreover, elasticity is a real 3D problem, so there are several elasticity mismatches such as shear mismatches, which must be solved within the sample. So this problem is quite complex and the present introduction of a hierarchy of such mismatches induces a hierarchy of rearrangement scales which is even well observed for pure gold Au (111) with the so-called herringbone structure [8]. One basic idea in this problem is that after a large enough number of layers the deposited matter forgets the very nature of the substrate and starts to heal the defects which were induced by the existence of an interface. This is demonstrated in several papers [168].

1.2.2.2 Simulated free 2D nanoparticles

This case corresponds to a perfectly flat surface and so to a simple general problem. Here a rather basic theoretical approach with pair potentials which generalize Lennard–Jones potentials [166] is considered. This approach enables us to consider simultaneously the three basic characteristics of atom–atom interaction: the interatomic distance and so the atomic volume, the pair interaction energy well, and the basic curvature of the interaction potential, i.e., the bulk modulus. Of course, other elasticity parameters can be derived for a crystalline structure from this effective pair potential by means of lattice sums [58, 166]. So in this basic picture the nature of the nanoparticle chemical element is characterized by an *atomic stiffness parameter*, S, linked with the ratio of bulk modulus over binding energy as compared to lattice atomic volume [100]. This dimensionless number $S = 9vB/\varepsilon$, where B is the bulk modulus, v the atomic volume in the dense phase, and ε the binding energy in this phase that strongly varies from one chemical element to the other. The atomic stiffness parameter S, which was also called anharmonicity parameter [141], characterizes the extension of the effective atom–atom interaction and its numerical values vary from 50 for Hg to 10 for Ce. A large value of S means a sharp interface between adjacent atoms while a low value of S means a smooth interface between adjacent atoms. This strong variation enables defining several classes of elements according to their atomic stiffness: very stiff ones ($S > 48$) Hg and Cd; noble metals ($43 > S > 36$) Au, Os, Ir, Pt, Re, Pd, Zn, and Rh; stiff metals ($36 > S > 31$) Pb, Mo, Ru, Tc, Ag, and W; moderately stiff metals ($31 > S > 23$) Ti, Tl, Co, Ni, Fe, Cu, Ta, and In; and finally soft metals ($21 > S > 10$) U, Al, Zr, Yr, Mn, Be, and Ce. Of course, this basic classification also appears in many structural as well as thermal properties [100].

The Lennard–Jones potential V_{LJ} and the generalized LJ potentials V_{nm} are

$$V_{LJ}(r) = V_{n=2, m=6}(r) \tag{1.1}$$

$$V_{n,m}(r) = \varepsilon \frac{\left(\frac{\sigma}{r}\right)^{nm} - n\left(\frac{\sigma}{r}\right)^{m}}{n - 1}$$

where σ is the interatomic distance. So considering the 2D aggregation with pair potentials of various atomic stiffness characterized here by numbers m and n, where for copper $m = 5$ and $n = 1,1$ and $m = 15, n = 2$ or $m = 30, n = 2$ for a coverage, i.e., a surface density: 0,23, Ghazali and Levy obtained the 2D cluster binding energy versus temperature diagram shown in Fig. 1.3 from extensive Monte Carlo simulations [57]. The total energy of a nanoparticle per atom corresponds to the effective coordination number at the considered

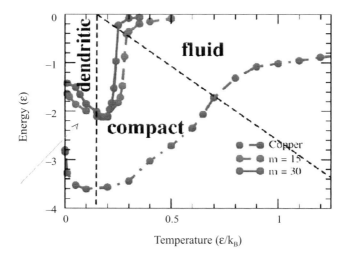

Figure 1.3 Energy versus temperature for several 2D clusters interacting via more or less smooth pair potentials after long MC relaxation [57].

temperature. So this figure shows the evolution of coordination with temperature for realistic 2D configurations since the chosen density is not too low for aggregation and not large enough to induce explosive overlap.

The interest in large values of m, i.e., of S, which is proportional to m^2, lies in the description of very hard units, which is the case of large units with interactions restricted at the surface such as colloids. Colloidal aggregation is an obvious limit case.

At low temperature ($T < 0, 1 = T_c$), where the temperature unit is the depth ε of the pair potential well as in the following, whatever the atomic stiffness S be, a dendrite fractal behavior is observed after a long diffusion time. So the 2D aggregation is mainly controlled by diffusion at such low temperatures. Consistently an increase in temperature increases the coordination level and so the total energy decreases while the binding energy increases. So the fractal dimension of such nanoparticles increases with temperature. This anomalous thermodynamic behavior is pointed out by a light line on Fig. 1.3. Equilibrium is not reached, never reached, since diffusion is a quite slow process, while simultaneously aggregated atoms can escape from the nanoparticles. The comparison of the three curves evidenced the easy occurrence of fractal structures for colloids as observed [147].

At higher temperatures $(0,1 < T < T_m)$ a compact behavior of clusters is observed in dense islands with sixfold symmetry. Then the binding energy decreases when the temperature is further increased since thermal agitation decreases the binding level and the coordination. Thus the effective coordination level is lowered at this step of practical equilibrium. The melting temperature T_m depends on the "atomic" stiffness S, while the aggregation temperature T_c for compact structure, 0,1, does not depend on the atomic interaction. As expected, the melting temperature T_m increases when the atomic stiffness S decreases, i.e., when the interaction is more extended. And so in 3D, the stiffest metal in this classification, mercury is liquid at room temperature.

At higher temperatures, moving smooth islands are broken into smaller and smaller islands resulting in a final low coordination. Because of the increase of border effects the coordination level decreases with temperature. At different coverage levels, similar results are obtained. So this picture does not depend strongly on coverage level.

This general view over 2D condensation with pair potentials explains that fractal structures of various coordination levels occur at low temperature deposition even during long observations, and even with strong interaction between deposited atoms. This can be useful for producing devices with random fractals of any fractal dimension.

Dealing here with pair potentials which just act on the distance between particles, the only dense 2D structure is the triangular one with sixfold symmetry since its filling factor is 0.91. This leaves a very small place for holes. Other possible crystalline 2D structures have a quite lower filling factor. So there is no competition between crystalline orders or even local order, and the triangular structure is the only dense one.

Of course, other interactions than pair potentials have been considered in the literature. This is the case of embedded atom model (EAM) [36] and also of numerous models involving three-body interactions [170]. These models use several parameters for their definition and thus are more accurate than the simple model introduced here. However, this model well introduces the evidence for fractals and the transition between a solid 2D state and a fluid 2D state, as shown in Fig. 1.3 [57].

1.2.2.3 Simulated three-dimensional nanoparticles

There have been a lot of simulations of free 3D nanoparticles and clusters since it is the basic step for 3D aggregation. A lot have been done with MC computations [188] using Lennard–Jones pair potentials as well as with more sophisticated ones. MD computations have been done with interest for dynamics and even melting [117]. And numerous DFT calculations dealt with very small clusters which exhibit original symmetry for their ground states [119].

The first point to notice is that in 3D the filling factor for crystalline structures, i.e., the ratio between the atomic occupied volume and the total volume per atom, is 0.73 for both face-centered cubic (fcc) and hexagonal compact packing (hcp), the highest value for 3D crystalline structures. So there is a basic degeneracy and this filling factor leaves the room for defects and rearrangements. Moreover, when looking at a very local scale and defining the solid angle on which an atom is seen by its neighbor gives an optimal number of neighbors, which is 14, quite more than the coordination number 12 for fcc and hcp. So in 3D, several structures compete as possible ground states, and for small clusters noncrystalline structures can occur. This defines a special motivation for such studies and especially about nanoparticles.

Before going further in this analysis, it must be said that the basic idea of the previous paragraph on 2D structures, i.e., the occurrence at low temperatures of fractal arrangements with average coordination and fractal dimensions increasing with temperature, remains true in 3D. The basis for this is that the competition between diffusion and aggregation as controlled by pair potentials is quite limited at low temperature. And this picture with interactions is more realistic than the sticking process considered in diffusion-limited aggregation (DLA), which has been largely considered in the literature [195, 196]. So fractals of arbitrary dimensions can be obtained at low temperature and must be frozen in order to avoid any further evolution. But, since there is also large competition between several possible equilibrium configurations, this occurrence of fractals within a 3D space was not deeply investigated. The attention was focused on the equilibrium configurations of small nanoparticles as a function of their size.

Here we want to focus first on cluster magical properties. These structural properties do not depend strongly on the nature of the interaction when leading to dense structures since they mainly depend on geometry. Starting with very small clusters, the most stable structures are the more symmetric ones. That is a good introduction to magical clusters and so magical numbers. So for four atoms the regular tetrahedron is stable [119]. For seven atoms the most stable configuration is a regular planar pentagon surrounded by two atoms on the top and the bottom [119]. This unusual fivefold symmetry is found also in the icosahedron of 13 atoms [35] where the icosahedral symmetry group Y_h contains 120 symmetry operations, more than the octahedral group O_h, the largest crystalline symmetry group which contains 48 symmetry operations. Then from that, cluster sizes 7 or 13 up to the largest sizes which favor crystalline symmetry, icosahedral symmetry is expected to occur as a rich local symmetry. According to this remark, a large cluster structure with this icosahedral symmetry was built from a variational minimization of energy [105]. An interesting feature of this cluster shown is in Fig. 1.4: this icosahedral symmetry is present everywhere in the cluster at the expense of defects such as holes or voids of different sizes [105].

A cut normal to a fivefold symmetry axis of such a cluster with 1681 atoms described as spheres is shown in Fig. 1.4. An obvious approximate self-similarity can be noticed in this figure since the central figure of one atom surrounded by 10 neighbors is reproduced 10 times around it, and the external layer shows 10 centered pentagons. Another feature observed in Fig. 1.4 is the occurrence of long rows of atoms with evidence for holes.

In Fig. 1.4 the central symmetry of the cluster shows the occurrence of a very dense central part followed by successive less or more filled parts. This is the basis for the occurrence of shells and magical numbers as observed by looking at the local elasticity coefficients [158] of such a structure submitted to Lennard–Jones

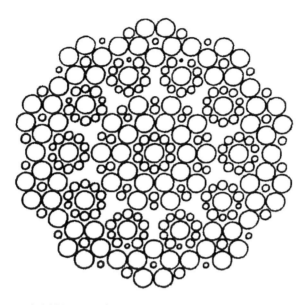

Figure 1.4 A 1681 atom cluster with icosahedral symmetry. Central cut with fivefold symmetry [105].

pair potentials. These out-of-equilibrium elastic coefficients [106] are derived from the potentials derivatives taken at cluster positions. For instance the first coefficient, p, measures an internal hydrostatic pressure and is issued from first-order potential derivatives, i.e., from unbalanced forces. This coefficient p takes just 4 or 5 different values alternating between themselves when looking at the 525 central atoms of the structure shown in Fig. 1.4 [106]. This occurrence of just a few repeated levels confirms the already pointed out approximate self-similarity of this structure made of successive shells. There is a first shell of 12 atoms surrounding a central one, in an icosahedron of 13 atoms. Then there is a second shell of 30 atoms, of which 16 are less stable. So it defines a next magic number: 27. The next shell contains 20 atoms. It would define a magic number: 47. In this series of shells a high level of instability is reached when there are 270 atoms, a number which recalls the radioactive properties of actinides.

A few years after these studies of icosahedral clusters [105, 106], similar macroscopic structures with icosahedral or pentagonal symmetry were observed in alloyed samples and called quasicrystalline structures [152]. Since that time a lot of quenched alloys have been studied and among them numerous noncrystalline symmetries were observed. So nowadays there is a large class of quasicrystalline structures [161]. In these so-called quasicrystalline samples, soft elements in the meaning defined previously about atomic stiffness, such as Mn, Ce, and Al, are included, so these "soft" elements which can be located in the holes of quasi-crystalline structures can easily bear local shear and stress in order to stabilize these structures.

The next point to consider when looking at larger cluster sizes is the structural transition from icosahedral clusters to clusters with crystalline symmetry. Since in 3D there are several crystalline symmetries with high coordination competing together—12 nearest neighbors for fcc symmetry as well as for hcp and 8 nearest neighbors + 6 close next-nearest neighbors for body-centered cubic (bcc)—there are also later expected transitions between these symmetries, i.e., transitions from a crystalline symmetry to another crystalline symmetry.

So first the transition from icosahedral symmetry to dense hcp/fcc structure with packing faults is shown to occur between an icosahedral cluster of 270 atoms and a dense packed cluster of 459 atoms in Fig. 1.5 [55] which is obtained after long MC relaxation with classical Lennard–Jones interaction when starting from a initial random configuration. In Fig. 1.5 the trick used to obtain rather organized views, consisting in defining the cluster framework from the very MC computational result by taking three nearest neighbors; among them, two define an axis, the third one defines an axis normal to the atom plane, and the third axis is normal to the two previous ones. This choice allows a careful examination of these structures.

These two structures were obtained when starting from random configurations with a rather low density which avoids expulsions in the MC process. Rows of atoms are observed for both samples. For the sample with 270 atoms, the fivefold symmetry is obvious. So the transition from quasi-crystalline structures toward crystalline structures occurs in the range of several hundred atoms. This has been observed for other potentials.

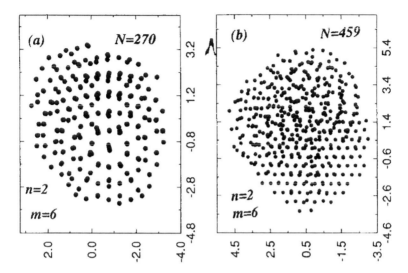

Figure 1.5 *xy* projections of two relaxed configurations with Lennard–Jones interactions for (a) 270 atoms and (b) 459 atoms. The fivefold symmetry of a cluster of 270 atoms (a) is clear, while the 459 atoms cluster is composed of 2 twinned parts with a mixing of hcp and fcc structures [55].

The advantage of choosing a local framework is also obvious for experimental techniques, but here this local framework is defined from inside points.

The next point concerns the transitions between different crystalline symmetries. An example is shown in Fig. 1.6 when comparing the energy of samples issued after a long MC relaxation process of 527 hcp atoms, 531 fcc atoms, and 531 bcc atoms in the case of soft atoms: $n = 1, 1$ and $m = 6$. These long relaxations at moderate temperatures start from parts of crystalline structures which are so optimized without modifying their initial symmetry.

Direct computations starting from random configurations lead to configurations of higher energy. The example shown in Fig. 1.6 well shows the interplay between different configurations. Such an interplay between different crystalline configurations often occurs among metals, with a basic high temperature stability of bcc phases as noticed in the literature [4].

As a conclusion of this short review of the nanoparticle structures as deduced from numerical computations and compared

28 | Magnetic Structures of 2D and 3D Nanoparticles

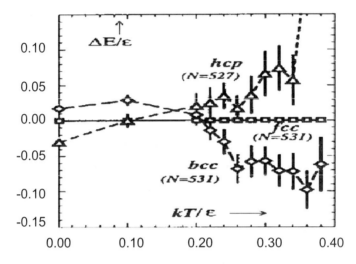

Figure 1.6 The comparison of the difference in reduced energy per atom versus reduced temperature between 527 atoms in an initial hcp configuration (triangles), 531 atoms in an initial bcc configuration (diamonds) as compared to 531 atoms in an initial fcc configuration (squares). Error bars are deduced from MC fluctuations. At low temperature hcp configuration is stable, while at high temperature bcc configuration is stable [55].

to direct observations, it must be noticed that nanoparticles are well organized even at very low sizes, at least in their low temperature configuration, while defects are mainly local. This organization, or self-organization, occurs with optimal symmetry, which is not obvious and depends on the interaction and was evidenced rather recently. Diffraction patterns well confirm this intense symmetry which appears in simulations.

Before concluding this section devoted to structural simulations and going to magnetic observations, it seems useful to introduce magnetic simulation of nanoparticle aggregation under dipolar forces since it is linked both with aggregation problems and with magnetism. It well concludes this part on aggregation and growth and well introduces the specific problems of dipolar forces in magnetism.

1.2.2.4 Simulated aggregation of magnetic nanoparticles

Most of the work done on the simulation of aggregation of magnetic nanoparticles was done on 2D spaces because of the already obtained complexity of the results and of the fact that some general rules already appear on these 2D results [56]. It must be added that 2D aggregation of magnetic particles has a real meaning as independent magnetic flat particles or electric flat dipoles can be put over a liquid surface and observed directly or under stress or mechanic waves. This has been extensively studied [154].

As in the previous part, the basic tool for understanding such aggregation process is MC computations where magnetic disks with arbitrary spin orientation are introduced in a random initial configuration when avoiding overlap. The interaction is given by the dipole–dipole interaction (2) between disk i and j, which depends both on the distance between disks i and j and on their mutual spin orientations:

$$H_d = \sum_{i \neq j} \frac{\vec{S}_i * \vec{S}_j}{r_{ij}^3} - 3 \sum_{i \neq j} \frac{(\vec{S}_i * \vec{r}_{ij})(\vec{S}_j * \vec{r}_{ij})}{r_{ij}^5} \qquad (1.2)$$

Each MC step involves both a restricted motion and a restricted change in 3D spin orientation. The independence of disks is ensured by a repulsive interaction as a function of distance with a power law with exponent -8. The results of relaxation at moderate temperature strongly depend on initial disk concentration within a central closed disk. Of course, a slow growth process occurs. So no equilibrium is reached within even large computer times while natural concentration ever increases. It must be added that experimentally many such 3D observations were done from electron microscopy [176] where similar slowly changing structures appear while 2D studies can be achieved optically [154].

At low concentration the first steps of aggregation give lines of magnetic disks where spins are collinear with particle lines with a north pole following a south pole and so on. And since in loops there is no end effect, large loops are stable in spite of curvature effects.

It can be noticed that during this process the length of lines increases at the expense of the number of lines. The number of closed lines also slowly increases, as well as the size of closed loops. However, the interaction between lines seem rather weak. As

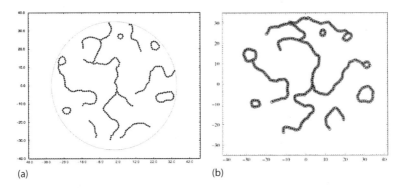

Figure 1.7 Snapshots of 331 magnetic nanoparticles (a) after 3×10^6 MC steps and (b) 6×10^6 MC. Note that the number of independent lines slowly decreases during relaxation. In part b, the arrows give the magnetization direction [56].

expected from dipolar field, straight lines are favored and curvature needs to be paid.

At higher density with magnetic particles always confined within a disk there is a forced interaction between lines, as seen in Fig. 1.8 [56].

Lines of magnetic particles are often parallel and so define magnetic domains with uniform magnetization. These domains are oriented in many directions with a resulting lack of total magnetization. At a medium scale these domains define vortices and other topological defects. The occurrence of vortex and other topological defects is characteristic of these 2D patterns. Here several hyperbolic antivortices can be observed. The defect at the vortex core, i.e., the lack of a central particle, as well as at the antivortex core is associated with the linear rigidity, which avoids curvature and especially strong curvature.

At intermediate 2D concentrations, magnetic nanoparticles aggregate with lines and vortices and antivortices [56]. The transition is easily understood. These patterns are quite similar to the magnetic patterns observed for 2D samples, as shown later. So it suggests the idea that 3D patterns of magnetic particles interacting through dipolar forces are quite similar to magnetic patterns of 3D samples with the same confined geometry. This justifies discussing

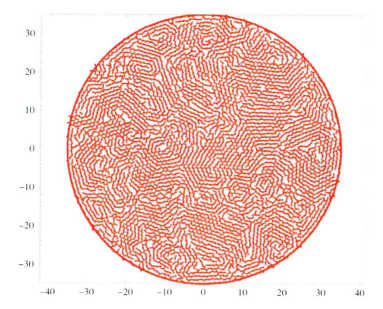

Figure 1.8 Snapshot of 4000 magnetic particles after MC relaxation at low temperature. Magnetic moments are shown by red arrows on the on-line version. Notice the occurrence of vortices made of magnetic domains consisting of parallel lines while vortex centers are unoccupied [56].

these structures of nanoparticle networks later when looking at 3D magnetic structures.

1.2.3 Experimental Evidence of the Magnetic Structure of Nanoparticles

Several techniques are used for obtaining direct magnetic structure images of nanoparticles. Several techniques also exist for observing the magnetic excitations of these nanoparticles. As a matter of fact there is a long history of magnetic imaging and this history can be more or less classified according to its resolution level, starting from physical tracks nearly followed by optical means with a wavelength resolution up to the more recent atomic resolution of surface techniques or the nearly atomic resolution of spin holography and also of polarized scattering with photon, electron, or neutron beam improved by polarization detection, nanometric displacements, and numerical help. As noticed earlier about atomic structures, there is much more imaging information about 2D structures, i.e., surface problems, than about 3D realistic samples because 2D images are easy to produce and understand. A more or less historical order is used for simultaneously reporting basic experimental techniques and typical results on these magnetic structures. Magnetic excitations in such small particles can be detected by spin resonance techniques and Brillouin light scattering, which are reported later with their basic results after looking at numerical simulations of magnetic structures. It must be added that our physical interest is not restricted here just to nanoparticles since larger-sized samples can also reveal interesting features of magnetism in spite of more complex arrangements. This remark reintroduces the interest in classical results on large samples.

1.2.3.1 Bitter powder and other tracking techniques

The basis of these surface techniques consists in depositing on the sample surface small magnetic particles which are sensitive to the local magnetization and its stray field. So these particles are aggregated where a strong gradient of magnetization attracts them. Classical examples of such moving particles are compass or iron filings, and now Bitter colloids [71], which are both microscopic and able to detect the in-plane component of magnetization. Observations are done under optical microscope and thus the resolution is not better than 1 μm. These already classical results reveal the complexity of magnetic structures [65]. In Fig. 1.9 several samples exhibit typical results on large samples [65]. Figure 1.9a,b shows two pictures of the surface of a Fe sample doped with 4% Si where the surface makes an angle of $3°$ with [001] plane. Figure 1.9a and 1.9b are at different scales. The tree-like structure of magnetic domains is made of alternated magnetic stripes decorated with branches which are oriented at $90°$ with these stripes.

Figures 1.9c and 1.9d show domains in a Co sample. In Fig. 1.9c the hexagonal axis is perpendicular to the surface while it is within the surface in Fig. 1.9d. The firework structure of magnetic domains in Fig. 1.9c involves a specific localization, while Fig. 1.9d shows how branches are divided into internal branches and so on. Domain walls appear as straight lines, and these lines make a small angle between

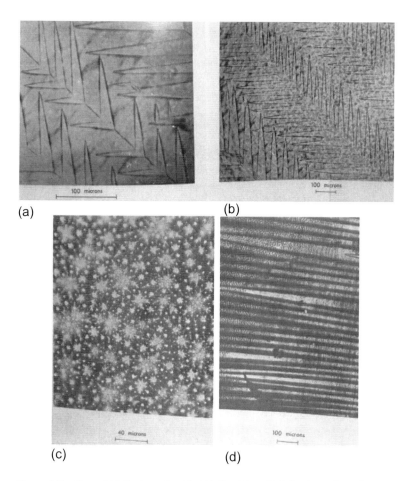

Figure 1.9 (a and b) Fe doped with 4% Si with a slightly slanted surface at 3° from [001] surface. The two different scales show parallel branches and stripes. (c and d) Hexagonal Co. In c the surface is normal to the hexagonal axis which is in the plane of d [65].

themselves, thus leading to strong local singularities. There is a lot of branching, especially in Fig. 1.9c, where domains are not widely extended except the main black one.

Figure 1.10 shows the magnetic structure of two samples of Ni ferrite doped with 0.1% Co by using the technique of colloidal Bitter powder [65]. Figure 1.10a is taken in plane [001] with a field orthogonal to this plane, while Fig. 1.10b shows different sections with a [001] plane at the top in presence of a weak [001] magnetic

34 | *Magnetic Structures of 2D and 3D Nanoparticles*

a

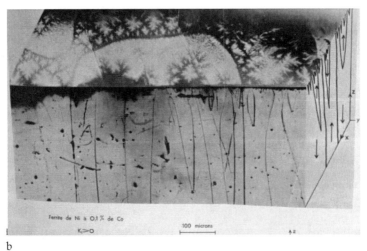

b

Figure 1.10 Ni ferrite doped with 0.1% Co. (a) View of [001] plane with a labyrinthine structure decorated with dendrites. (b) Three simultaneous views showing the interplay of magnetic domains with a snowflake-like top view [65].

field. These nice, tapestry-like figures and especially Fig. 1.10a show a lot of white branches, sub-branches, and sub-sub-branches, a tree structure, immersed within the black sea of the main magnetic domain. A new feature brought by Fig. 1.10a is the appearance of wall curvature within a nice maze, as also seen in the top view of Fig. 1.10b.

The general occurrence of resulting labyrinthine structures and of vortex, antivortex, and other topological defects as a consequence of such correlated domain and wall curvatures is well confirmed by the study of smaller samples, as shown later from the results obtained by recent techniques. The appearance of labyrinthine structures in thin films is well known in the literature [24]. The appearance of vortex and other topological defects in magnetic structures is more recently known but already well confirmed [87]. Of course, labyrinths and topological defects are complementary observations when samples exhibit different crystalline anisotropy [99].

1.2.3.2 Faraday and Kerr effect: Magneto-optical Kerr effect (MOKE)

The electromagnetic field of light interacts with the spin magnetic moments, giving rise to a magnetic susceptibility. This leads to the absorption and polarization of light by transmission, the so-called Faraday effect [74], and by reflection, the Kerr effect, also called magneto-optical Kerr effect (MOKE) [77]. From the previous observations it sounds that global Faraday effect is hard to understand even when samples do not absorb light too strongly since it results from the interaction with many different layers, while Kerr effect is just sensitive to a thin surface layer. So Kerr effect reveals the state of magnetization at the surface with a wavelength resolution. And so MOKE was largely used for thin films and ultrathin films with perpendicular anisotropy which show bubble-like magnetic domains [68], considered for magnetic memories [17]. At this time MOKE revealed the existence of magnetic stripes, chevrons, and labyrinths [125], as well as of magnetic bubbles. Now, when considering small samples, the advantage of MOKE remains its capacity of fast integration of the magnetization component over the

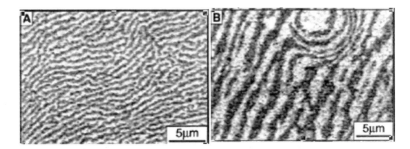

Figure 1.11 SEMPA view of two parts of Fe film (2–3 layers) on Cu (001) [182].

sample. So now MOKE is mainly used to observe hysteresis cycles of nanoparticles [107] in the presence of variable external fields, usually in parallel with numerical micromagnetic simulations with the basic code OOMMF [45]. Such a combination of experiment and simulation enables the experimentalist to explain the singularities of hysteresis loops by changes in magnetic configuration.

1.2.3.3 Scanning electron microscopy with polarized analysis (SEMPA)

This scattering method with polarized electrons takes advantage of the short wavelength of kinetic electrons in order to obtain high resolution for magnetic structures. As a scanning method it also has the advantage of high resolution determined from the accuracy of sample motion and thus belongs to the new class of surface experiments where resolution is determined by the mechanical accuracy of displacement.

Figure 1.11 well shows both the stripe structure and the transition from stripe structure toward labyrinthine structure with a vortex-like structure within an ultrathin film [182].

The first result in Fig. 1.11 shows the stripe structure of a Fe film of 2 or 3 layers deposited over Cu (001) [182]. Figure 1.12 shows SEMPA results on a V film of a soft magnetic material [67]. This specific V-shape well induces the localization of a vortex close to the knee. This expected result is confirmed by micromagnetic simulations with OOMMF [67]. During the hysteresis loop the vortex

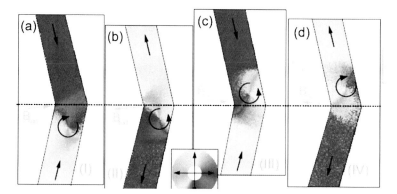

Figure 1.12 SEMPA analysis of a V-shaped, 18 nm thick, 400 nm wide, soft magnetic Co39Fe54Si7 (atomic percentages), with different in-plane external fields shown in gray in (a), (b), (c), and (d), and in color online [67].

chirality changes, and so too its accurate localization near the knee. Extrapolating from Figs. 1.11 and 1.12 the SEMPA resolution can be as low as 10 nm and flat samples are easily observed from that technique.

1.2.3.4 Photoemission electron microscopy (PEEM)

PEEM takes advantage of X-ray scattering with X-ray magnetic circular dichroism (XMCD) to induce polarized secondary electrons which are analyzed by means of electron microscopy according to spectroscopic photoemission at low-energy electron microscopy (SPELEEM) [148]. Using electron microscopy with low-energy electrons of moderate wavelength restricts the spatial resolution to about 30 nm. Figure 1.13 shows three rings of cobalt, 15 nm thick, observed by PEEM [81]. The inner diameter is 900 nm and the outer diameter 1200 nm. The bottom ring and the right ring are in a vortex configuration with a positive chirality, while the left ring is in "onion" configuration—i.e., it is made of two halves, each one a half-vortex of opposite chirality [27]. This is an experimental proof that onion configuration and vortex configuration are of very similar energy and thus are both stable or metastable, as also confirmed by micromagnetic computations.

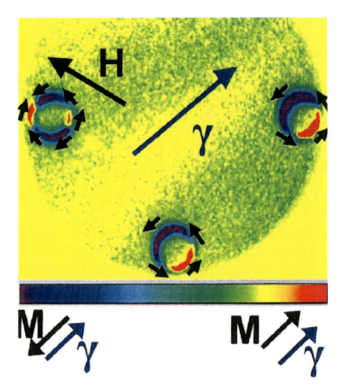

Figure 1.13 PEEM image of three polycrystalline Co rings. The bottom and right rings are in the counterclockwise vortex state, and the left ring is in the onion state pointing along the direction of the applied field **H**. The field of view is 10 μm across and the blue arrow points in the direction of the photon beam. The color scale indicates the direction of the magnetization with reference to the direction of the incoming photon beam magnetization from parallel to the incoming photon beam in red to antiparallel purple (color online) [81].

The advantage of using rings as observed in Fig. 1.13 consists in avoiding the strong change in magnetization direction which occurs at the vortex core in a full disk. This strong variation costs exchange energy in a full disk, and as a matter of fact, magnetization direction of the core escapes in the third dimension, giving rise to a polarized vortex core with a non-zero magnetization in the third dimension. In a ring, no such escape occurs, except for the vortex or antivortex located within the ring.

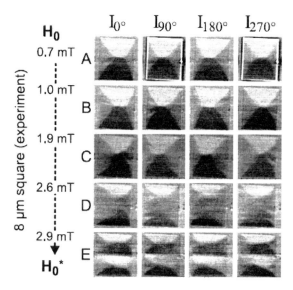

Figure 1.14 Temporal XMCD images of the magnetization (at 0°, 90°, 180°, and 270° of excitation phase) for increasing excitation field amplitude H_0 on permalloy structures of 8 µm [162].

1.2.3.5 X-ray magnetic circular dichroism (XMCD)

As already noticed in the previous part about photoemission, the scattering of convenient X-rays is sensitive to the magnetization orientation of the target. This selective scattering occurs when X-rays have a frequency close to a resonant frequency of the sample. Then, an analysis of the polarization of the transmitted X-ray beam can be used to reveal the state of magnetization of the sample. Using a narrow X-ray synchrotron beam and accurate motional techniques enables one to obtain both high spatial resolution and fast time resolution, as shown in Fig. 1.14 [162].

This fast technique enables each magnetic state during hysteresis loops to be observed. As already noticed, when comparing previous figures, there is a strong similarity between the images coming from different experimental techniques even if the magnetic constituents are different. The main difference comes from local anisotropy, but general shapes are quite similar and are a proof for an effective universality.

40 | *Magnetic Structures of 2D and 3D Nanoparticles*

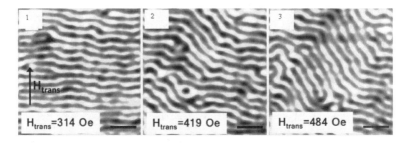

Figure 1.15 MFM on Fe$_{0.8}$Ga$_{0.2}$ film 65 nm thick. Scale bar: 300 nm. Three kinds of stripe structure according to the field level during a hysteresis loop, from parallel stripes (1) to chevrons (2) and labyrinths (3) [16] in color online.

1.2.3.6 Magnetic force microscopy (MFM)

MFM consists in using a magnetic tip when recording surface properties with now classical surface techniques such as measuring the cantilever displacement, which depends upon the local magnetic field at the tip. It is by now a classical technique [70]. Two examples of such studies are shown here. First the transition from parallel stripes toward a labyrinthine structure under increasing fields is shown on a thin film of Fe$_{0.8}$Ga$_{0.2}$ [165] in Fig. 1.15.

This now standard MFM method can be improved in several ways according to the chosen tip. A quite recent idea consists in using a tip with a color centre, i.e., an extended fictitious hydrogen atom located within a matrix, and taking advantage of the Zeeman level splitting of this color centre in presence of a magnetic field. The energy difference is detected by resonance techniques. This method was used to obtain Fig. 1.16 about the vortex core of a permalloy disk of about 1000 nm diameter. Here the color centre is a single nitrogen-vacancy defect (NV centre) in diamond and the tip also contains a radiofrequency antenna in order to observe the electron spin resonance (ESR) in this centre and so to measure accurately the stray field on the tip, as reported in Fig. 1.16 [172]. The result confirms the small size of the vortex core with a radius of about 20 nm, while the vortex structure fills the entire disk.

These recent results show that MFM techniques can take advantage of numerous new detectors in order to reveal the local

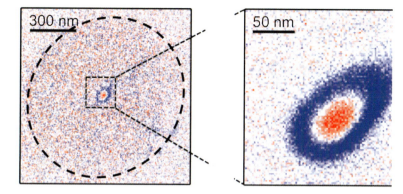

Figure 1.16 A permalloy disk of 500 nm radius with a border shown on the left part is analyzed by means of MFM. The AFM tip is functionalized with a 40 nm nanodiamond hosting a single NV centre, and a radiofrequency (rf) antenna is used to perform optical detection of the NV defect ESR transition. Two different levels of magnetic field are pointed out in blue and red (in color online). These colored rings show how narrow is the vortex core within this permalloy disk [172].

magnetic structure of samples, at least the stray fields quite close from the samples.

1.2.3.7 X-ray holography

With the new possibility of X-ray free-electron (XFEL) synchrotron, which can take advantage of magnetic dichroism, atomic resolution by means of holography as well as femtosecond resolution can be reached [190]. The practical point consists in avoiding damage which occurs at a rather low fluence threshold since X-rays are energetic and strongly coupled to matter and so lead to irreversible damage. Some authors succeeded in avoiding such a radiation damage, as can be observed by further inspection, later. The result shown in Fig. 1.17 gives a classical picture with labyrinths made of stripes, a classical result for thin films.

As a conclusion of this long list of sophisticated experiments on basically 2D magnetic nanoparticles with evidence for labyrinthine structures or their dual representation, i.e., vortex and other topological defects, the lack of experiment on 3D magnetic nanoparticles must be noticed. As a matter of fact, a rather recent experiment

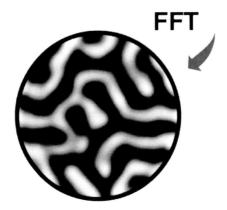

Figure 1.17 X-ray holography of a 1000 nm wide disk composed of Ta1:5nmPd3nm[Co0:5nm,Pd0:7nm]$_{40}$Pd2nm multilayers. The result shown is obtained from fast Fourier transform (FFT). Note the labyrinthine structure made of magnetic stripes [190].

revealed that a 3D magnetic nanometric nanoparticle is made of several magnetic domains [118]. And in this experiment, only the external, surface skin of these domains is viewed. So the real problem of the magnetic structure of 3D nanoparticles is still open. Thus the interest in numerical simulations is quite obvious. Moreover, there are numerous numerical approaches of magnetic stable configurations which also lead to dynamics. That is why we are studying them before analyzing the experimental data on dynamics within a magnetic nanoparticle.

1.2.4 *Numerical Evidence of the Magnetic Structure of Nanoparticles*

1.2.4.1 Numerical methods for determining the magnetic ground state of a nanoparticle

The numerical methods used for this determination are the now standard Monte Carlo simulation with moderate individual spin rotation as the basic local change [38], the method derived from molecular dynamics, i.e., the Langevin equation for magnetic variation [31], and a standard numerical approach of micromagnetism based on the resolution of the Landau–Lifshitz–Gilbert equation with damping: OOMMF [45]. The Monte Carlo method uses an explicit form of energy as a function of spins and geometry, and with convenient annealing and slow cooling gives a rather fast result for a large number of spins. Langevin's methods well follow the configuration evolution and so inform about the system dynamics and its spin excitations, magnons in the general meaning. The numerical resolution of micromagnetic equations introduces an artificial mesh, which is useful for studying large samples but without real nature. Most of these calculations were performed for 2D samples or effective 2D samples such as thin films and very thin films. Real 3 D treatments are less numerous.

1.2.4.2 Monte Carlo studies of 2D magnetic nanoparticles

These studies started with the progress in surface science and were devoted first to Ising spins with a spin component normal to the sample plane [112] with the reference of yttrium garnet thin films and their numerous applications such as magnetic bubbles [17]. Stripes, chevrons, and labyrinthine structures were observed. A typical result is shown in Fig. 1.18, where only dipolar forces are taken into account [183]. This case corresponds to an effective large size, since dipolar effects increase with size. So such a figure corresponds to a size of several micrometers instead of several nanometers for a nanoparticle. With this reading, the final low-temperature result of slow-cooled Monte Carlo simulations without external field shown in Fig. 1.18 well demonstrates the simultaneous occurrence of stripes, chevrons, and labyrinthine structures at a large scale, as observed on Figs. 1.10a, 1.11, 1.15, and 1.17.

The next figure is devoted to Heisenberg spins with uniaxial out-of-plane anisotropy and exchange as well as dipolar forces [183] for the same square of spins. In this figure the occurrence of large domains must be noticed as well as the general labyrinthine shape, which is similar to that in Fig. 1.18 but at a larger scale. Now, spins evolve in the full 3D space. As already noticed, when looking at experimental observations, where there is a large variation of in-plane magnetization, spins escape to the third dimension, i.e., out of the plane, as is obvious from the inset. Moreover, the general arrangement of domains suggests the occurrence of a vortex. In

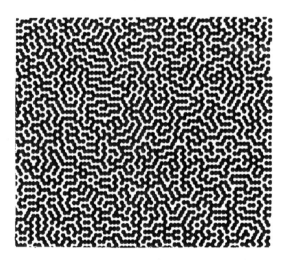

Figure 1.18 The low-temperature result of a Monte Carlo slow cooling of a square of 10192 Ising spins. Up spins are shown as black circles and down spins as white circles. Notice the various geometrical shapes of the atomic stripes [183].

other words, there is continuity between Ising spin labyrinths and Heisenberg spins vortex and other topological defects. This observation is confirmed when considering different values of uniaxial anisotropy. Here we shall observe similar results from Langevin's simulations without any anisotropy, but with different values of the ratio d between dipolar interaction and exchange. As a matter of fact, an increase in d corresponds to an increase in sample size for a given sample or to a decrease in exchange between different samples. Thus, the correspondence between Ising labyrinths and the Heisenberg vortex and topological defects is a relevant one.

1.2.4.3 Langevin's simulations of 2D magnetic nanoparticles

As already noticed, the advantage of Langevin's simulations is their ability to follow the evolution of a state with thermal fluctuations possibly mixed to an irreversible evolution. This is quite useful to detect relatively slow evolutions such as gyrotropic vortex motion [41], or even the full spectrum of magnon excitations in

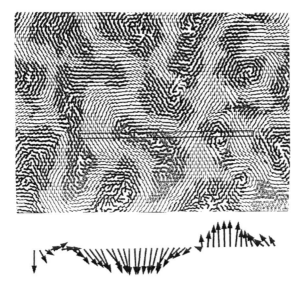

Figure 1.19 The in-plane spin components are shown as black arrows when the out-of-plane component is up and gray arrows when the out-of-plane component is down. The rectangular cut shown within the main figure leads to the down inset with spin projections on a vertical plane [183].

presence of dipolar interactions [42]. This sensitivity has for price computer time. So Langevin's simulations show often nearly out-of-equilibrium snapshots which also reveal further dynamics. Here two snapshots are reported for two large samples with a square of 16,384 Heisenberg spins interacting with a ratio d between dipolar interaction and exchange. In the original paper [40] attention was given to a phase diagram at low temperature as a function of d and L, L being the square size. As a matter of fact, metastable states also occur as onion states already quoted both in simulations and in observations. Thus this existence of nearly degenerate states must be taken into account, so such phase diagrams are just guidelines since several phases can coexist in the same conditions. For very low values of d which are realistic for samples, a nearly uniform domain appears with magnetization parallel to a main diagonal, with closure effects at the borders in order to obtain a local magnetization parallel to the border. Slowly increasing the d value leads to the appearance of a central vortex with a polar core.

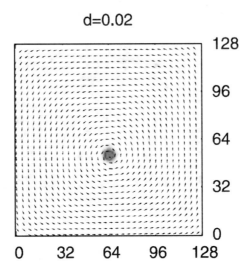

Figure 1.20 A Langevin's snapshot at low temperature with $d = 0.02$ (in color online). The arrow gives the in-plane spin projection averaged over a 4×4 local square, while color gives the local out-of-plane component. Notice the four main domains. In this snapshot the vortex is close to the center and some noise appears in the out-of-plane contribution as a result of thermal excitation. When averaging several close snapshots, these out-of-plane fluctuations disappear. A longer observation proves a gyrotropic vortex motion [40].

The nearly central vortex of Fig. 1.20 is made of four domains with 90° walls on diagonals. A further increase of the d value leads to simultaneously nearly in-plane vortex and other topological defects as shown in Fig. 1.21. The transition from one single polar vortex state or polar antivortex state which are nearly degenerate toward a set of in-plane topological defects as seen in Fig. 1.21 is of course a little bit noisy. A careful study of this transition requires the consideration of many initial conditions. This observation also means that out-of-equilibrium metastable states have a long life and can be used for applications. With an increased d value, the demagnetizing field increases so all spins become in-plane ones. This increase also induces the existence of several topological defects: chiral vortex, antivortex, and higher-order defects [170]. The organization of these topological defects

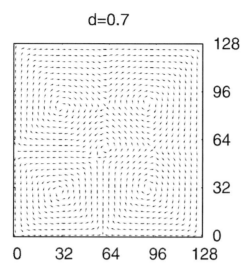

Figure 1.21 A Langevin's snapshot at low temperature with $d = 0.7$ (in color online). The arrow gives the in-plane spin projection averaged over a 4×4 local square, while color gives the out-of-plane component. Notice how all in-plane domains and walls with several topological defects are rather well ordered [40].

follows simple topological rules: between two vortices with the same chirality, there is an antivortex, while a double circle defect occurs between two vortices with opposite chirality. This is clearly seen in Fig. 1.21, where the starting point of a lattice of topological defects, as well as the occurrence of extra defects, can be observed.

In Fig. 1.21, 5 vortices appear, three of which are clockwise while two are counterclockwise, and three antivortices appear between vortices of the same chirality. There is an approximate balance between chiralities. The leading point of this structure is the weakness of all multipoles deduced from the global structure. The deduced structure anneals all the lowest multipoles and thus realizes a discrete screening of the dipolar field on the spin network. So the magnetic structure of 2D magnetic nanoparticles sounds clear, and this is confirmed by many numerical micromagnetic experiments, with OOMMF and other finite elements methods [170]. About real 3D magnetic nanoparticles, there is not the same convergence of experiments and simulations on their magnetic

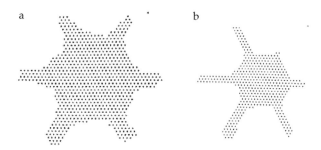

Figure 1.22 Equatorial cuts of a sphere of 47,443 Ising spins: black up-spins, and white down-spins, after a long MC low-temperature relaxation in presence of a vertical external field, (a) close to the coercive field $H/J = -3$, and (b) at a lower field $H/J = -4$. Notice the shrinking of domains according to dense crystalline lines and the expansion of white domains [102].

structures, but just some data. Similar observational difficulties occur about hexatic order [127], which is clear only for two-dimensional samples and for very thin films.

1.2.4.4 Monte Carlo studies of 3D magnetic nanoparticles

Basically, these MC calculations are very similar to the ones done for 2D magnetic nanoparticles with convenient changes in spatial environment. The question of representation depends on size. Looking at small nanoparticles, the sight of a few cuts enables one to understand the interplay of different parts, i.e., magnetic domains, while looking at rather large nanoparticles, only singularities must be focused. As for 2D nanoparticles, we start the study with Ising spins in presence of an external field parallel to Ising spins before studying Heisenberg spins. The results of an equatorial cut shown in Fig. 1.22 after a long, low-temperature relaxation with two different values of the external field, the first one close to the coercive field (a) and the other one at a higher field (b) well indicate the formation of domains [102]. These domains keep their basic structures under high fields, just with a shrinking of domains with magnetization opposite to the external field and an expansion of favored domains. It can be noticed that dipolar interaction generates a magnetocrystalline anisotropy, as observed here on wall shapes and theoretically proved [99].

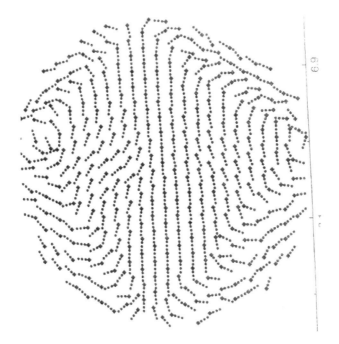

Figure 1.23 Equatorial cut of the relaxed MC low-temperature configuration of a sphere of 5947 Heisenberg spins $D/(Ja^3) = 0.1$. The in-plane spin components are shown as arrows. Note the occurrence of domains, walls, and vortices [102].

As noticed for 2D magnetic nanoparticles, going from Ising spins to Heisenberg spins leads to a change from a labyrinthine structure of domains, as seen in Fig. 1.22 toward a network of topological defects. Such a network is also observed for 3D nanoparticles at the expense of longer relaxation times since the resulting structure is more complex [102]. This appearance of topological defects is clearly seen in Fig. 1.23, where the planar spin components of an equatorial cut of 5947 spins after a long MC relaxation at low temperature without external field are reported. The observation of extended central domains as well as of topological defects close to the border under closure effects is quite obvious [102].

Already in this rather small sample a 3D map of singularities would help understand the spatial organization of these singularities. Of course there is an interest in larger samples and in the response to an external field, as well as in dynamics, but there is no known accurate work on that. It must be added that several numerical methods are also used to solve that kind of micromagnetic problems. They are based upon finite elements resolution of Landau Lifshitz Ginzburg equations of spin motion and are reported below.

1.2.4.5 Micromagnetic and OOMMF studies of magnetic nanoparticles

Since the Mermin–Wagner remark on the instability of magnetism with pure exchange in 2D samples [122], there has been a lot of interest in dipolar interactions in 2D samples since dipolar interactions introduce a natural anisotropy and thus a gap in the spin wave spectrum which stabilizes 2D magnetism [198]. This strong effort leads to numerical computations based on the solution of the Landau Lifshitz Gilbert equation by finite elements methods, also called micromagnetism [3], in order to understand the resulting 2D magnetic structure. Among these methods, Donahue and colleagues proposed an object-oriented micromagnetic framework, also called OOMMF [45], largely used in the literature for 2D and 3D magnetic nanoparticles as well. Such a calculation is reported in Fig. 1.24 about the magnetization reversal of a magnetic pillar [21].

This figure aims to prove the ability to deal with 3D features in spite of difficulties. However, details of topological defects are not obtained, and there is no symmetry in this magnetization reversal. Several papers of this team were devoted to comparisons between different results of numerical approaches [174].

So the next points to consider are the experimental observations on excited magnetic states in nanoparticles and the results of numerical simulations on excited states in magnetic nanoparticles.

1.2.5 Experimental Evidence of the Magnetic Excitations of Nanoparticles

There are two basic experiments for detecting magnetic excitations in nanoparticles: magnetic resonance and Brillouin light scattering.

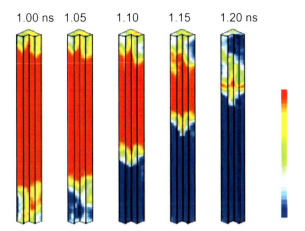

Figure 1.24 Magnetization reversal in a magnetic pillar after a rapid reversal of the field. The magnets have a square cross-section, but are shown in a one-quarter cutaway view. Here the z component of the magnetization is shown, with red representing the metastable orientation and blue representing the equilibrium orientation. Note the nucleation of both end caps, but at different times [21].

In magnetic resonance a radiofrequency electromagnetic field interacts with the magnetic sample. The absorption of this field of a given frequency reveals a resonance, i.e., a peak in the magnetic susceptibility at this frequency. This already classical experiment [79] also works with nanoparticles when using nanoantennas for emission and reception. Now standard techniques of modulation and lock in detection give access to high resolution. According to the waveguide mode or antenna, magnetic components can be selected. When looking at a set of magnetic nanoparticles, a direct study of magnetic susceptibility or magnetic permittivity is used [89].

The other technique, Brillouin light scattering, is a scattering technique using an incident laser beam and analyzing the scattered beam. This technique, which has the advantage of lasers–well-defined frequency and possible power–also fulfills selection rules since the magnetic excitation has excitation energy and also one wavevector or a few wavevectors in the case of a wave packet [73]. Such selection rules are quite useful, especially for 2D samples, where laser beam direction and polarization can be selected as well as for the observed light.

The newness of excitations in magnetic nanoparticles comes from the non-uniformity of magnetization within these samples when there is not a very large saturating external field. And the understanding of these specific new modes requires the use of numerical simulations in order to be checked. Several specific excitations can be observed.

First, at low frequency, a motion of the topological defect as a whole quasiparticle, vortex [32], or antivortex [189] can be observed. This motion of the topological vortex with its neighboring domains is observed for polar topological defects and is called a gyrotropic motion since the dipolar field created by the out-of-equilibrium instantaneous distribution of magnetization in the sample acts on the polar component with a resulting precession motion. This slow motion obviously depends on the vortex position as well as the corresponding dipolar field does. This frequency variation is well confirmed by experience [75].

Then there are magnetic excitations which are localized within the topological defect. So these excitations are quite localized within the topological defect and its core and thus as a consequence of localization they occur at higher frequency than gyrotropic motion, i.e., at frequencies comparable to that of usual magnons [9].

1.2.5.1 Gyrotropic vortex (antivortex) motion

These slow motions were early observed [32], just after the observations of polar vortex and antivortex. Micromagnetic calculations and OOMMF calculations helped in this early work. Following the interpretation of this gyrotropic motion as a precession motion of a single effective fictitious particle, the topological defect, a well potential model was used as defined by Thiele [173] a long time ago about magnetic bubble motion.

This gyrotropic vortex motion is shown on Fig. 1.25 without external field [32].

As already noticed, complementary simulations are quite necessary to identify such modes.

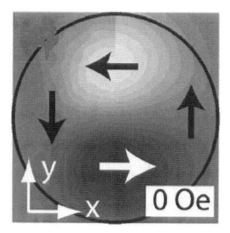

Figure 1.25 The In-plane magnetization configuration for a 2μ disk (arrows) and image of the z component of the magnetization (shading) is seen 80 ps after the arrival of the field pulse. The split frame shows both experimental (left) and simulated (right) images in a field of 0 Oe [32].

1.2.5.2 Spin wave motion within a topological defect

These spin wave motions for effective 2D topological defects are rather similar to usual spin waves when taking into account the fact that the vortex closure brings an angular constraint to stationary spin waves and thus several modes with an integer number of nodes around the defect can be seen. This angular quantification is clearly seen in Fig. 1.26, where the principle of the absorption saturation experiment during a long pulse is reported [135].

Quite similarly spin wave resonances were observed in nanometric nanoparticles with various shapes [120] by means of the analysis of magnetic permittivity.

1.2.6 Numerical Evidence of the Magnetic Excitations of Nanoparticles

As already mentioned, such numerical evidence is quite necessary to recognize the observations. There are two ways for simulating magnetic excitations in nanoparticles: the direct resolution of dynamic equations of spin motion either by Langevin's formulation

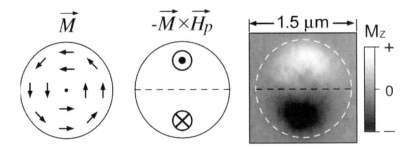

Figure 1.26 Schematic of the vortex magnetization (left), the torque exerted by the in-plane pulse (middle), and polar Kerr image immediately after the pulse for a 1.5 μ diameter disk (right) [135].

or by micromagnetic ways and on the other hand the use of the linearized equations of spin motion from an assumed ground state. These equations define a dynamical matrix of which magnons are eigenvectors. In the first case only a few well separate modes are easy to observe while in the other case a full spectrum of spin waves is derived. In the first case, gyrotropic motion is easily observed while in the second case very complex modes with angular and radial variation for 2D samples as well as full 3D variations for 3D samples can be distinguished.

1.2.6.1 Gyrotropic vortex and antivortex motion simulated

In order to simulate a vortex or antivortex motion, it is necessary to start from an out-of-equilibrium state where the topological defect is put out of its stable place and so to introduce a fictitious state with this topological nature. The first steps of relaxation will save this topological nature and the resulting defect motion during this initial time is quite small. So the effective gyrotropic motion can be observed as shown in Fig. 1.27 for three temperatures from Langevin's simulations with Gilbert damping [41].

The vortex gyrotropic motion is perfectly regular at low temperature and gives rise to an helix written on a cone because of effective damping due to Gilbert damping and interaction with the bath. This motion is perturbed by thermal magnon scattering at higher temperatures. This precession mode is due to the polar

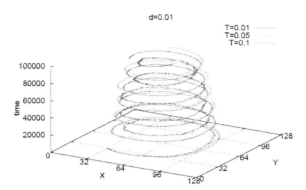

Figure 1.27 Vortex core motion for $d = 0.01$ at three temperatures, $T = 0.01$, 0.05, and 0.1. The X and Y axes give the position of the vortex core on the planar square lattice and the vertical axis shows time (color online) [41].

part of the vortex which is localized at the vortex core. So the concept of a macrospin localized at the vortex core centre has been quite fruitful to look at these motions [66]. Within these ideas the concept of a Thiele's well potential for this macrospin was developed according to previous models used for magnetic bubble memories [189]. As a matter of fact, Langevin's simulations are quite useful to derive such an effective potential, as shown in Fig. 1.28, where at step t of the iteration process the spin set has a total energy E and a vortex position (X, Y). Thus the plot $E(X, Y)$ directly gives the shape of this potential well when taking advantage of different starting points in order to fully describe this well [41], as shown in Fig. 1.28.

There is a vortex expulsion when the starting vortex point lies too close the border $x \leq 1/8$. This means a strongly non harmonic well potential close the border. This is explained since when the vortex is close to the border, one of the four magnetic domains of the vortex collapses. That leads to a strong variation of the energy potential and thus to anharmonic effects which are also observed [66].

An antivortex has also a polar core, as shown in Fig. 1.29a [189]. Thus the same reasoning as before on macrospin precession on vortex works, so there is also a gyrotropic motion of antivortex, and as in the case of vortices the chirality of this motion depends

56 | Magnetic Structures of 2D and 3D Nanoparticles

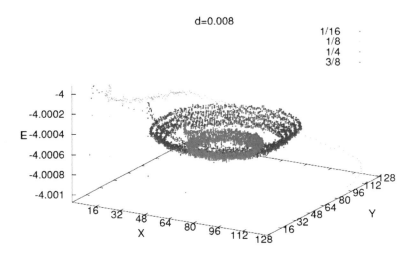

Figure 1.28 The vortex potential $E(X, Y)$, i.e., the total energy as a function of vortex position from several Langevin's trajectories at $d = 0.008$ for a 128×128 square (in color online) [41], where the starting points distance x from the border on the half diagonal is noted. Note the vortex expulsion for $x \leq 1/8$.

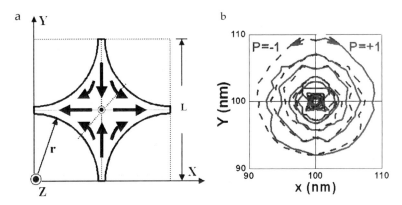

Figure 1.29 (a) Asteroid particles with a magnetic antivortex. Solid arrows represent the direction of equilibrium magnetization inside the asteroid sample. The core region is indicated by the small circle about the centre, and the dot at the centre signifies a core pointing upward out of the plane. (b) Trajectories of the antivortex core around its equilibrium position (in color online). The solid, clockwise helix is for core polarity $P = +1$, while the dashed, counterclockwise path is for core $P = -1$. $L = 200$ nm and $t = 20$ nm [189].

on antivortex polarity as shown on Fig. 1.29b [189], where Gilbert damping also leads to an helix written on a cone.

When sample is larger or when the material has a higher ratio d of dipolar versus exchange, several topological defects appear such as vortices, antivortices, and so on. As a consequence of the various gyrotropic effects, several such gyrotropic motions must coexist. That leads to several contradictory motions and finally to domain blocking. This analysis of the superparamagnetism of magnetic nanoparticles shows that blocking temperatures depend on sample size. This dipolar consequence is confirmed by a careful analysis of magnetic dynamics in such samples close to the ground state [155].

1.2.6.2 Spin wave simulations in presence of topological defects

As already noticed, these spin wave simulations are basically done when starting from linearized equations of spin motion derived from an approximate ground state which can be deduced from either basic relaxation from one of the previously quoted methods or by mere intuition. As a matter of fact, as before most of the known calculations deal with effective 2D samples, since 3D samples are quite more complex.

The approximate stability of the chosen spin configuration is confirmed by the analysis of the eigenfrequencies. If there is no zero frequency or only one, the configuration is at least metastable. This remark enables to deduce stability domains [116] for such configurations.

In Fig. 1.30 the 15 first modes of a stable in-plane vortex configuration taken at the threshold are shown, with the in-plane contribution as well as out of plane configuration. In Fig. 1.30 the progressive increase of the order of the azimuthal symmetry must be noticed as well as some split degeneracy for modes with high symmetry. There are modes with twofold symmetry ($i = 2, 3$) and only one nodal line, modes with twofold symmetry ($i = 4, 5$) and two nodal lines, modes with threefold symmetry ($i = 6, 7$), modes with fourfold symmetry ($i = 8, 9, 10$), modes with fivefold symmetry ($i = 11, 12$), and so on. This quite progressive evolution is characteristic of such materials. This progression is broken by the

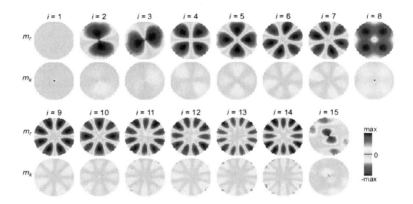

Figure 1.30 (Color online). Distribution of the in-plane (top profile) and out-of-plane (bottom profile) amplitudes of elementary magnetic moment precession for the first fifteen modes with in-plane vortex configuration [116].

emergence of a mode linked with the polar excitation of the vortex core.

A similar work was done in connection with the observation of thermally induced polarity flips in Langevin's simulations [42] when starting from a polar configuration.

These modes classified according to the usual frequency rule well confirm their activity in the core.

A similar research of active spin wave modes was produced about modes located within an asteroid particle where an antivortex is stabilized. The result of these spin wave computations for observable resonant spin waves is shown in Fig. 1.32 [189].

As in the previous figures about spin waves, a progression according to symmetry level must be focused.

Before leaving the field of experimental observation and numerical simulations, it must be added that there is a tremendous number of experimental work on non linear properties, in accordance with out-of equilibrium observations. For instance, thermal Seebeck effects are often observed experimentally [19]. Another class of experiments deals with magnetoresistance effects [48] and with the general coupling between resistivity and magnetic field with the example of current induced torque [187]. In these experiments,

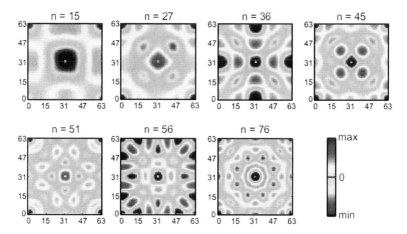

Figure 1.31 Profiles of the out-of-plane oscillation amplitude for the most intense modes in agreement with the frequencies of effective modes active on Langevin'simulations of polarity spontaneous reversal ($d = 0.04$) [42].

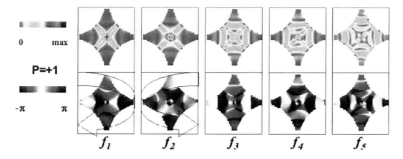

Figure 1.32 Spectral amplitude (top) and spectral phase (bottom) at resonance for the first five azimuthal spin wave modes in the $P = +1$ case at the resonant frequencies $f_1 = 3.8$ GHz, $f_2 = 4.7$ GHz, $f_3 = 6.3$ GHz, $f_4 = 6.7$ GHz, and $f_5 = 8.4$ GHz [32].

due to dipolar effects, there is a strong geometrical dependency. So there is a large amount of geometrical tricks. And once more the knowledge of the magnetic structure and of magnetic excitations would be quite useful.

1.3 Theory

As previously introduced our main goal is to deal with the competition between exchange and dipolar interactions about magnetic structures and magnetic excitations in nanoparticles, but before studying these problems it is necessary to recall the general framework of magnetism: magnetization carriers and magnetic interactions.

1.3.1 *Magnetic Carriers*

Quite obviously electrons carry a spin which gives rise to magnetization. But there are insulators where all electrons are localized, while for metals there are both nearly free electrons and also localized electrons of spin $\vec{S}_i$ localized on site i [203]. In that case magnetism comes from the interplay between localized electrons and nearly free conduction electrons in a Stoner–Wolfarth approach [164]. Later we will focus on a Heisenberg approach of magnetic insulators [51], but in these starting parts both Heisenberg spins and Stoner–Wolfarth spins are considered.

1.3.2 *Magnetic Interactions*

Several magnetic interactions are active upon magnetic carriers, i.e., electrons. First exchange and superexchange have a basic quantum origin from molecular bonding [5] with a local nature. Anisotropy comes from a quantum relativistic origin, the spin-orbit effect [61] and so is also very local. Dipolar effects are classical and long-ranged [94] as nearly as gravitational interaction and Coulomb forces are. If dipolar effects can be screened by free electron clouds in metals, the finiteness of Fermi sea has a consequence on electron and spin polarization, it induces a long ranged polarization effect which results in a long ranged spin-spin coupling by means of Ruderman–Kittel–Kasuya–Yosida interaction called RKKY interaction [199]. So even for metals there is a competition between short ranged and long ranged interactions. This competition leads to the so complex domain organization observed in all magnetic materials.

1.3.2.1 Exchange and superexchange

Exchange is involved in the chemical binding because of anticommutation Pauli's rule. So between neighboring spin sites there is tendency towards parallel spin coupling, called ferromagnetism or antiparallel spin coupling, called antiferromagnetism. Both couplings are schematized by the effective exchange Hamiltonian:

$$H_{\text{exc}} = -(1/2) \sum_{ij} J_{ij} \vec{S}_i * \vec{S}_j \qquad (1.3)$$

Here sites i and j are nearest neighbors and the factor one half is due to pair effect. In some cases and especially in the case of oxides where a strong coupling between metallic ions occurs by the way of oxygen ion, the exchange coupling is more extended and thus there is a superexchange with further neighbors [5]. And even in this rather special case, exchange is quite localized to a few neighbors.

1.3.2.2 Anisotropy

As introduced before anisotropy has a relativistic quantum origin in the atom because of an effective spin orbit coupling. So it is a quite local property. In the case of a uniaxial anisotropy of axis z it reads

$$H_{\text{an}} = \sum_{i} I_i S_{iz}^2 \qquad (1.4)$$

Because of coupling with exchange by means of binding anisotropy can be extended to neighbors, and it can take more complex writing such as in the case of Dzyaloshinskii–Moriya coupling [201]. However, anisotropy remains local interaction. Since anisotropy has an atomic origin, anisotropy can be different at the border or near the border from what it is in the bulk. In the case of surfaces of thin films the term surface anisotropy was introduced [60]. Rather similarly in 2D samples, corner anisotropy as well as edge anisotropy was introduced [104]. Since relativistic the quantum computation of electronic structures is still hard to be done by density functional theory (DFT) [72], such assumptions sound to be plausible but are not fully confirmed.

1.3.2.3 Dipolar interactions

Dipolar interactions are directly issued from Maxwell equations [94]. Between two spins $\vec{S}_i$ and $\vec{S}_{ji}$ located at sites i and j, with respective positions $\vec{r}_i$ and $\vec{r}_j$ they are written up to a numerical factor as

$$H_d = \sum_{i \neq j} \frac{\vec{S}_i * \vec{S}_j}{r_{ij}^3} - 3 \sum_{i \neq j} \frac{(\vec{S}_i * \vec{r}_{ij})(\vec{S}_j * \vec{r}_{ij})}{r_{ij}^5} \tag{1.5}$$

This long-ranged character leads to shape effects such as the demagnetizing field [159]. It also leads to an effective anisotropy since spin and space are coupled [99].

1.3.2.4 RKKY interactions

In RKKY interactions a local spin polarizes the electron Fermi sea and this polarized Fermi sea interacts with another local spin giving a long ranged interaction which has a r^{-2} power law and oscillates with a wavevector which is basically Fermi wavevector. Of course, this interaction depends on the details of the Fermi surface which is not necessarily simple and thus on the considered material.

1.3.2.5 Zeeman effect

In presence of an external field $\vec{H}$, there is a Zeeman effect on each spin of the sample which is written up to a numerical factor:

$$H_Z = \sum_i \vec{H} * \vec{S}_i \tag{1.6}$$

1.3.3 *Determination of the ground state*

Since each spin has a unit length, only two free variables, its Euler angles, describe this spin. So with N spins, there are $2N$ free variables. This simple remark allows us to evaluate the difficulty in finding the ground state for some typical cases.

Firstly in the basic case of a local interaction with z neighbors, the case with exchange anisotropy and external field, the total Hamiltonian contains zN terms. So if $z > 2$, all these terms cannot be optimized together. So the linear chain with only nearest neighbor interactions is a very specific case which can be solved unambiguously.

In the case with long ranged pair interactions as it occurs with dipolar interactions, there are $N(N-1)/2$ terms to obtain an optimized sum. So this is a quite more complex problem than the case with exchange only and numerical ways are quite useful and necessary to solve this problem. However, analytical approaches are interesting since they give an overview on such specific problems which are size and shape dependent.

There are two analytical ways for searching the ground state.

Firstly, a static way consists in considering an assumed ground state and adding to it selected variations in order to consider all possible configurations. If the assumed ground state is a real one, all these variations must lead to an increase of energy. And with infinitesimal variations a metastable state does not change its energy. The resulting equation for no energy change characterizes metastable states. Then from this metastable state, the dynamical matrix can be computed and its eigenvalues and eigenvectors. The eigenvalues must differ from zero.

In front of this static variational treatment a dynamic treatment consists in searching directly the excited states of an assumed ground state. A requirement for a ground state is that there is no more than one eigen mode for the dynamical matrix with zero frequency. If more frequency modes have zero frequency a condensation of such zero frequency modes will lead to a new ground state.

These methods are similar in their principle but different in their applications. The advantage of the variational method is the arbitrariness of the assumed ground state while the dynamical method requires an explicit writing of the assumed ground state and so is restricted in looking at the stability of this assumed ground state with some insight on its possible changes by mode condensation if a zero frequency mode, a soft mode as usually called, appears. In this chapter this variational treatment is used while in the chapter done by S. Mamica the dynamical approach is used.

The first point in this variational approach consists in defining the parameters of each configuration, here the spin angles and so later the parameters of configuration variations.

For configurations there is a spin field $\vec{S}_i$ defined on each point of the sample, i.e., on a lattice. In order to introduce the spin difference from one site to a neighboring site, i.e., the local properties, it is useful to introduce discrete derivatives $(\vec{S}_i - \vec{S}_j)$, $(-2\vec{S}_i + \vec{S}_j + \vec{S}_k)$ and so on. Quite obviously when dealing with long ranged interactions as dipolar interactions, all high-order field derivatives must be introduced in order to manage this long ranged neighboring. So the next point is to consider a continuous spin field with all its derivatives. This is the classical Landau tricky approach with the advantage of considering general functions or distributions even if they are extensions of effective suites on nanoparticle sites. In order to take advantage of this new spin field formulation, the energy must be written consistently with this local approach. This is achieved by using Taylor Mac Laurin expansion of the spin field when neglecting the rest:

$$\vec{S}_j = \sum_{p,q,r} \frac{x_{ij}^p y_{ij}^q z_{ij}^r}{p!q!r!} \frac{\partial^{p+q+r}}{\partial x^p \partial y^q \partial z^r} \vec{S}_i \tag{1.7}$$

This Taylor expression leads to write the total Hamiltonian as a sum of local contributions on each site.

$$H = \sum_i H_i \tag{1.8}$$

Since anisotropy is local, there is no change in the anisotropy Hamiltonian. Similarly there is no change in the already local Zeeman term. For exchange with nearest neighbors a second-order expansion is enough and the local exchange Hamiltonian reads when taking into account usual cubic symmetry laws, i.e., neglecting surface effects:

$$H_{exc,i} = -J_i a^2 \vec{S}_i * \Delta \vec{S}_i \tag{1.9}$$

This classical Landau expression is usually rewritten when using integration by parts, with up to a numerical factor:

$$H_{exc} = J_i a^2 \left(\nabla \vec{S}_i \right)^2 \tag{1.10}$$

Of course, this expression neglects surface contributions as they occur in integration by parts. It leads when neglecting dipolar contributions and other long-ranged contributions to uniform

Theory | 65

configurations, i.e., $\nabla \vec{S} = 0$ with possible surface rearrangements [97, 98].

If exchange is more extended, it is required to introduce higher derivatives in the local exchange Hamiltonian and so more complex bulk spin arrangements can occur such as helimagnetism for instance.

For dipolar interactions, an expansion up to high order is always required. So lattice sums as introduced by Lennard-Jones about the calculation of cluster energy [163] are useful since they resume the geometrical contribution from the nanoparticle. Here we need isotropic nanoparticle sums $I_{p,q,r,i}$

$$I_{p,q,r;i} = \sum_j \frac{x_{ij}^p y_{ij}^q z_{ij}^r}{p!\,q!\,r!\left(x_{ij}^2 + y_{ij}^2 + z_{ij}^2\right)^{3/2}} \tag{1.11}$$

And anisotropic nanoparticle sums $L_{p,q,r,\alpha,\beta,i}$

$$L_{p,q,r,\alpha,\beta;i} = \sum_j \frac{r_{\alpha,ij} r_{\beta,ij} x_{ij}^p y_{ij}^q z_{ij}^r}{p!\,q!\,r!\left(x_{ij}^2 + y_{ij}^2 + z_{ij}^2\right)^{5/2}} \tag{1.11bis}$$

These two terms are site-dependent and summations are extended over sites j different from i. In order to appreciate their variation as a function of parameters it is useful to evaluate them in the case of a continuous large sample characterized by size L. If $p + q + r > 0$, these two nanoparticle sums grow like L^{p+q+r}, thus high-order terms cannot be neglected. Even if $p + q + r = 0$, i.e., $p = q = r = 0$ 3D nanoparticle sums grow like $\ln L$.

With this writing the dipolar Hamiltonian becomes a local one as expected:

$$H_d = \sum_i H_{d,i}$$

$$H_{d,i} = I_{p,q,r,i} \left(\vec{S}_i * \frac{\partial^{p+q+r}}{\partial x^p \partial y^q \partial z^r} \vec{S}_i \right)$$

$$-3 \sum_{\alpha,\beta} L_{p,q,r,\alpha,\beta,i} \left(S_{\alpha,i} \frac{\partial^{p+q+r}}{\partial x^p \partial y^q \partial z^r} S_{\beta i} \right) \tag{1.12}$$

For spin variations it is interesting to introduce also local spin variations by means of Dirac delta functions on site k, when satisfying the rule of unitary spins:

$$\vec{S}_i = \vec{S}_{i,0} + C\delta\,(i - k)\,\vec{S}_{k,0} \wedge \vec{n} \tag{1.13}$$

Here the vector $\vec{n}$ is a unit vector. So $3N$ variations are introduced instead of $2N$ independent free variations. This redundancy is quite useful for calculations. Of course, infinitesimal variations are deduced from an infinitesimal value of C. Field derivatives are easily derived as distributions with the usual rules for distributions [150]:

$$\partial \vec{S}_i = \partial \vec{S}_{i,0} + C\delta'\,(i-k)\,\vec{S}_{k,0} \wedge \vec{n} + C\delta\,(i-k)\,\partial \vec{S}_{k,0} \wedge \vec{n} \quad \text{(1.13bis)}$$

And so on.

In the previous parts on experimental and numerical results, it was shown that there is a large difference between the cases of magnetic structures of 2D and 3D nanoparticles. Moreover the amount of results about really 3D nanoparticles is rather small while there is a very large amount of data about effective 2D nanoparticles. So we will start to study 2D magnetic nanoparticles in detail before looking at realistic 3D nanoparticles.

With the presence of topological defects in 2D nanoparticles, it is interesting to look at the arrangement of such topological defects together as due to topological rules. Another consequence of this remark consists at looking firstly at very small samples in 2D and 3D before considering larger samples. This avoids the complication of topological arrangements to be considered later.

1.3.3.1 Ground state of 2D nanoparticles

For 2D nanoparticles, nanoparticles sums must be rewritten since there is no more z axis [159].

$$I_{p,q;i} = \sum_j \frac{x_{ij}^p y_{ij}^q}{p!q!\left(x_{ij}^2 + y_{ij}^2\right)^{3/2}}$$

$$L_{p,q,\alpha,\beta;i} = \sum_j \frac{r_{\alpha,ij} r_{\beta,ij} x_{ij}^p y_{ij}^q}{p!q!\left(x_{ij}^2 + y_{ij}^2\right)^{5/2}} \quad \text{(1.14)}$$

The first point is to consider the lowest-order terms $I_{0,0}$ and $L_{0,0,\alpha,\alpha}$ which give rise to anisotropy as due to dipolar contribution, in other words to the demagnetizing field [159]. These terms are calculated exactly for an infinite 2D lattice and they depend on the lattice symmetry [159], a result due to the discreteness. So dipolar interaction introduces in-plane anisotropy.

2D demagnetizing field and dipolar induced anisotropy on a 2D lattice

For a simple square lattice, one obtains for this dipolar induced anisotropy:

$$H_{s;i} = I_{s;0,0} S_i^2 - 3 L_{s;0,0,1,1} \left(S_{i,x}^2 + S_{i,y}^2 \right) \tag{1.15}$$

With

$$I_{s;0,0} = \frac{4}{a^3} \left[\varsigma(3) + \frac{1}{2^{3/2}} + 2 \sum_{m=2}^{\infty} \frac{1}{\left(m^2 + 1\right)^{3/2}} \right. \\ \left. + \sum_{m=2;n=2}^{\infty,\infty} \frac{1}{\left(m^2 + n^2\right)^{3/2}} \right]$$

Where $\varsigma(3)$ is the Riemann zeta function of 3 [1] and

$$L_{s;0,0} = \frac{4}{a^3} \left[\varsigma(3) + \sum_{m=1;n=1}^{\infty,\infty} \frac{m^2}{\left(m^2 + n^2\right)^{5/2}} \right] \tag{1.15bis}$$

From that point x and y axes are strictly equivalent as obvious from the square symmetry.

For a hexagonal lattice with the same atomic density, this dipolar induced anisotropy reads:

$$H_{h;i} = I_{h;0,0} S_i^2 - 3 \left(L_{h;0,0,1,1} S_{i,x}^2 + L_{h;0,0,2,2} S_{i,y}^2 \right) \tag{1.16}$$

Now axes x and y are no more equivalent, there is a demagnetizing field which leads spins to be not only in the plane but also in a dense direction of atoms. After calculations this in-plane dipolar induced anisotropy reads

$$H_{h';i} = \frac{1.723}{a^3} \left(S_{i,x}^2 - S_{i,y}^2 \right) \tag{1.17}$$

Continuous treatment of dipolar induced anisotropy

As already suggested, nanoparticles sums can be calculated approximately within a continuum approach of a ring of size L without center, i.e., without an internal disk of radius a, with a radial integral K_{pq} and an azimuthal integral N_{pq} [99]. This calculation is approached both because of the continuous approach and because of the non uniformity of nanoparticle sums in finite samples. It reads up to numerical factors:

$$I_{p,q} = K_{p,q} N_{p,q} = K_{p+q} N_{p,q} \text{ with } K_n = \frac{L^{n-1} - a^{n-1}}{a^2(n-1)} \text{ for } n > 1 \quad (1.18)$$

and

$$N_{p,q} = (p!q!)^{-1} \int_0^{2\pi} \cos^p \theta \sin^q \theta \, d\theta$$

With this approximation of uniform lattice sums the first term of the local dipolar Hamiltonian reads

$$H_{d:0} = \pi K_0 \left(-S_{ix}^2 - S_{iy}^2 + 2S_{iz}^2\right) \quad (1.19)$$

This is the classical demagnetizing field effect for an ultrathin film.

The next non zero local Hamiltonian is the second-order term because of inversion symmetry in this quite extended sample:

$$\frac{4H_{d,2}}{\pi K_2} = 4\vec{S} * \left(\frac{\partial^2 \vec{S}}{\partial x^2} + \frac{\partial^2 \vec{S}}{\partial y^2}\right) - 6S_x \frac{\partial^2 S_y}{\partial x \partial y} - 6S_y \frac{\partial^2 S_x}{\partial x \partial y}$$

$$-9S_x \frac{\partial^2 S_x}{\partial x^2} - S_x \frac{\partial^2 S_x}{\partial y^2} - S_y \frac{\partial^2 S_y}{\partial x^2} - 9S_y \frac{\partial^2 S_y}{\partial y^2} \quad (1.20)$$

Since from Eq. (1.18) the term K_2 is very large for large samples, the energy minimization just gives the partial derivative equation:

$$4\vec{S} * \left(\frac{\partial^2 \vec{S}}{\partial x^2} + \frac{\partial^2 \vec{S}}{\partial y^2}\right) - 6S_x \frac{\partial^2 S_y}{\partial x \partial y} - 6S_y \frac{\partial^2 S_x}{\partial x \partial y}$$

$$-9S_x \frac{\partial^2 S_x}{\partial x^2} - S_x \frac{\partial^2 S_x}{\partial y^2} - S_y \frac{\partial^2 S_y}{\partial x^2} - 9S_y \frac{\partial^2 S_y}{\partial y^2} = 0 \quad (1.21)$$

There are also higher-order terms of this local dipolar Hamiltonian, they give rise to non linear effects. The first one always with these approximations is the term of order 4:

$$\frac{64H_{d,4}}{\pi K_4} = 2S_z \frac{\partial^4 S_z}{\partial x^4} - 3S_x \frac{\partial^4 S_x}{\partial x^4} + S_y \frac{\partial^4 S_y}{\partial x^4}$$

$$+ 2S_z \frac{\partial^4 S_z}{\partial y^4} + S_x \frac{\partial^4 S_x}{\partial y^4} - 3S_y \frac{\partial^4 S_y}{\partial y^4}$$

$$+ 4S_z \frac{\partial^4 S_z}{\partial x^2 \partial y^2} - 2S_x \frac{\partial^4 S_x}{\partial x^2 \partial y^2} - 2S_y \frac{\partial^4 S_y}{\partial x^2 \partial y^2}$$

$$- 4S_x \frac{\partial^4 S_y}{\partial x \partial y^3} - 4S_x \frac{\partial^4 S_y}{\partial x^3 \partial y} - 4S_y \frac{\partial^4 S_x}{\partial x \partial y^3} - 4S_y \frac{\partial^4 S_x}{\partial x^3 \partial y}$$

$$(1.22)$$

Equation (1.22) shows that non linear effects are restricted at the place where there are large variations of the spin directions, i.e., where there are large high-order spin derivatives. This is a quite restrictive condition which is well observed experimentally and there are several experiments with such non linear effects such as magnon creation or annihilation [149]. Moreover, since K_4 is a term of order L^3, it is very large and the right term of Eq. (1.22) must be strictly equal to zero. Of course, the same remark is valid for higher-order Hamiltonian terms. So a large number of partial derivative equations like Eq. (1.21), right order of Eq. (1.22) equal to zero and so on must be fulfilled. This is the discrete screening effect which implies a full series of partial effects with a multipolar expansion. This is well observed in numerical experiments where for instance there is the same number of both local chiralities, i.e., where all multipoles are close to zero.

Intrinsic topological defects of the 2D ground state

Always with this continuum approach of the ground state for large samples, the out-of-plane contribution also called polar contribution can be neglected since it occurs within a quite restricted space, the topological defect core. And this polar core is due to the competition between exchange and dipolar interactions. So if exchange is neglected at this scale or if one can neglect the short scale effects, this approximation is perfectly consistent. So each spin is characterized by an azimuthal angle θ with

$$S_x = \cos\theta \quad S_y = \sin\theta \quad S_z = 0 \tag{1.23}$$

So partial derivative Eq. (1.22) on spin components becomes a partial derivative equation on a scalar, the azimuthal angle θ [99]. Using now polar coordinates in the plane: ($x = r\cos\phi$, $y = r\sin\phi$) leads to a partial derivative equation reported in [99]. From the numerical and experimental evidence, at a low scale the spin organization does not depend on the radial distance r. So the first ground state Eq. (1.22) now reads

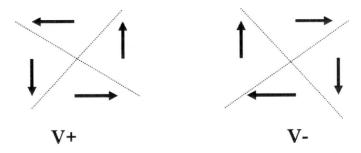

Figure 1.33 Vortex made of four domains. On the left positive chirality, i.e., counterclockwise vortex, on the right negative chirality, clockwise vortex.

$$\{7 \sin [2(\theta - \phi)] + \sin [2(\theta + \phi)]\} \frac{d^2\theta}{d\phi^2}$$

$$- \{2 + 2 \cos [2(\theta + \phi)] + 7 \cos [2(\theta - \phi)]\} \left[\frac{d\theta}{d\phi}\right]^2 = 0 \quad (1.24)$$

Thus there is a strong modulation of the azimuthal angle. Then neglecting this modulation there is the classical solution $\theta =$ const. the ferromagnetic arrangement within a domain and other solutions. When looking at a small sample the energy must be optimized and then there are approximate solutions like these ones:

$$\theta = \phi \pm \pi/2 + n\pi \quad (1.25)$$

and

$$\theta = -\phi \pm \pi/2 + n\pi \quad (1.26)$$

Equation (1.25) means a vortex with two possible chiralities as shown in Fig. 1.33. Figures 1.33 and 1.34 are written when considering only a square of four points surrounding a (0, 0) center: (1, 0), (0, 1), (−1, 0), and (0, −1) for the spin location.

With the usual partition in four domains also indicated it corresponds to the domains within a magnetic square dot. Equation (1.26) means an antivortex also called a hyperbolic defect with two possible chiralities, as shown in Fig. 1.34. Here these states of chirality mean states of symmetry.

Looking at higher-order terms introduces more complex configurations involving the arrangement of several topological defects [99].

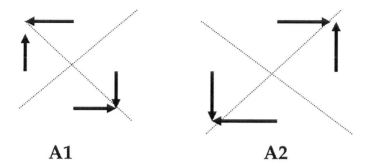

Figure 1.34 Antivortex, also called hyperbolic defect, made of four domains. Two symmetric configurations A1 and A2 are shown.

Instead of following this complex analytical way with numerous complex partial derivative equations it sounds better to look at possible arrangements within a square which satisfy closure relations, i.e., with magnetization parallel to the border.

Natural 2D closure structures

Always with this four domain structures, the only freedom comes from the spin sense since orientation is blocked. So there are two vortices where spins are rotating together. Then there is the case where three spins are rotating together while the next one has opposite sense, this is the 1–3 case. The next case consists of two spins rotating in the same direction followed by two spins rotating in the opposite direction. This is the so-called "onion" or "leaf" case. Finally, there is the antivortex case where two following spins have opposite senses. All these configurations were observed experimentally and numerically. So they can be an extended basis for typical structures, as shown in Fig. 1.35 where the top spin is chosen to be in the Ox direction.

All these typical structures fulfill the border condition and contain four domains and four 90° walls, thus there are quite close in total energy. However, the 1–3 structure has a resulting magnetization parallel to the Ox axis with the amplitude of 2 domains while onion states have a resulting magnetization parallel to one of

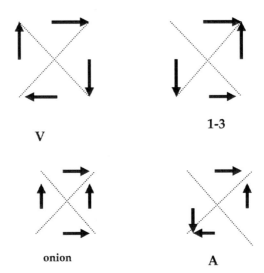

Figure 1.35 The typical four-domain spin textures with top spin parallel to Ox axis. On the left clockwise vortex V, the 1–3 structure with one spin is oriented opposite to the rotating sense of the others. The onion or leaf structure is characterized by two half vortex, and finally in the antivortex A each spin sense is opposite to the next one.

the bisectors with the amplitude of $2\sqrt{2}$ domains. It means that in presence of strong dipolar effect, the most stable configurations are equally vortex and antivortex while 1–3 structure is a little bit higher in energy and onion structure is the higher in energy of these four structures. This is in agreement with experimental observation since onion structure is only observed in small 2D units, so with effective low dipolar interaction. This classification of spin configurations according to resulting magnetization is thus important.

1.3.3.2 Ground state of 3D nanoparticles

At the level zero of the Taylor expansion of the dipolar interaction, the effect of demagnetizing field only occurs as induced by the anisotropic character of dipolar interaction. The set of optimization equations reads, within the assumption of uniform nanoparticle sums:

$$(B - C)S_z S_y = 0$$
$$(C - A)S_x S_z = 0$$
$$(A - B)S_y S_x = 0 \qquad (1.27)$$

With

$$A = L_{000,11} \approx \sum \frac{x^2}{r^5}$$

$$B = L_{000,22} \approx \sum \frac{y^2}{r^5}$$

$$C = L_{000,33} \approx \sum \frac{z^2}{r^5} \qquad (1.27\text{bis})$$

Here only orders of magnitude are considered. For a flat sample in the xy plane, magnetization is in-plane within this continuous model well in agreement with the previous calculations.

At the level two of the Taylor expansion of the dipolar Hamiltonian, always considering nanoparticle sums as uniform over the sample, a set of three equations over the spin field components and their partial derivatives is obtained. The first equation of this set reads as a matrix product:

$$A - 3D \ B - 3G \ C - 3J \ -6E \ -6F \ \lambda \begin{pmatrix} \frac{\partial^2 S_x}{\partial x^2} \\ \frac{\partial^2 S_x}{\partial y^2} \\ \frac{\partial^2 S_x}{\partial z^2} \\ \frac{\partial^2 S_y}{\partial x \partial y} \\ \frac{\partial^2 S_z}{\partial x \partial z} \\ S_x \end{pmatrix} = 0 \qquad (1.28)$$

Here coefficients A, B, and C are issued from isotropic nanoparticle sums I while coefficients D, E, ... L come from anisotropic nanoparticle sums L, and λ is a free parameter. Two other equations with similar coefficients are derived according to circular permutation, they correspond to the other orientations of the unit vector $\vec{n}$ introduced in spin variations. These equations complete the set of second-order equations. It must be noticed that while lattice sums depend on lattice geometry and lattice parameter, here nanoparticle sums and related coefficients depend also on sample

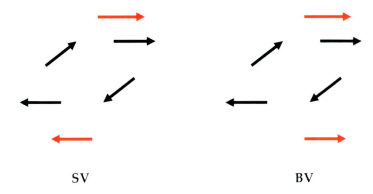

Figure 1.36 (Color online) On the left the 6 domains of S-shaped vortex SV and on the right the 6 domains of bent vortex BV. Top and bottom magnetizations are shown in red.

shape as shown in Eq. (1.27bis). Thus Eqs. (1.28) defines a very large class of spin fields submitted to second-order partial derivative equations.

This set of properties of 3D magnetic structures, Eqs. (1.28) are quite more complex than the set obtained for a 2D sample [99] which can be deduced from the more general 3D case from Eqs. (1.27) and (1.28). The Fourier transforms of Eqs. (1.28) leads to a characteristic equation of order 6, which means the occurrence of static wavy like deformations in the ground state. Thus to go beyond these generalities it is required to look at very small samples and to consider natural 3D closure structures as basic units.

Natural 3D closure structures

Following the previous scheme we consider the octagon of the centers of six faces of a cube $(1, 0, 0)$, $(0, 1, 0)$, $(-1, 0, 0)$, $(0, -1, 0)$, $(0, 0, 1)$, and $(0, 0, -1)$ surrounding the origin. These six points define six elementary closure domains with spin parallel to the faces.

So in each case only two domains must be added to the four cases of 2D closure structures. For the 2D vortex of given chirality, it means as seen in Fig. 1.36 two 3D structures: S-shaped vortex and bent vortex.

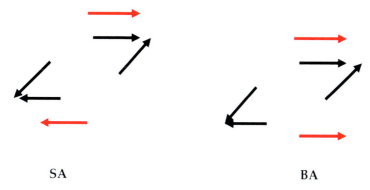

Figure 1.37 (Color online) The 6 domains of a S shaped antivortex SA on the left and of a bent antivortex BA on the right. The in-plane antivortex is shown with black magnetization while the magnetizations of top and bottom domains are shown in red.

Since the bent vortex has a resulting magnetization it is less stable in the case of strong dipolar contribution than the S-shaped vortex as discussed previously about spin configurations of 2D structures.

The other 2D stable structure with strong dipolar effect is the antivortex. Quite similarly to the 2D vortex case it gives rise to a S-shaped antivortex SA without resulting magnetization and to a bent antivortex BA with resulting magnetization. And as before the S-shaped antivortex is more stable than the bent antivortex in the case of a strong dipolar effect. SA and BA are shown in Fig. 1.37.

The same representation of the octagon of 6 domains is also used for the 1–3 structure and the onion state by addition of 2 extra top and bottom domains. It must be added that such arrangements are not symmetric in x, y, z axes. Thus other arrangements with in-plane structures yz, zx must also occur as observed in recent numerical simulations [43].

The next 2D cases: onions and 1–3 are less stable with strong dipolar effect since they have non-zero magnetization than vortex and antivortex. It also means a non-equivalency of in-plane x and y axes. Thus each such 2D structure leads to four 3D structures where S-shaped 3D structures are more stable than bent-shaped ones with strong dipolar interaction since they bear no extra magnetization.

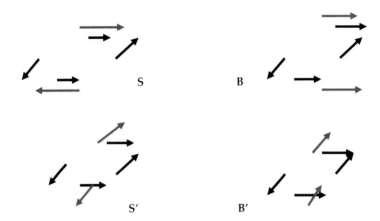

Figure 1.38 (Color online) The four 3D structures issued from 1–3 2D structures. On the left are the S-shaped structures S and S′, and on the right the bent structures B and B′. At the top, for structures S and B magnetization of top and bottom domain is parallel or antiparallel to the 1–3 magnetization, while at the bottom for S′ and B′ structures magnetization of top and bottom domain is perpendicular to the 1–3 magnetization.

These eight possible structures are shown in Figs. 1.38 and 1.39 according to the same representation as in Figs. 1.36 and 1.37.

The main difference between onions and 1–3 structures comes from the magnetization direction and intensity. For 1–3 structures the resulting magnetization is parallel to one axis and corresponds to two domains, while for onions the resulting magnetization is according to the bisector and corresponds to that of $2\sqrt{2}$ domains. This difference in intensity shows that onions are less stable than 1–3 structures in the case of strong dipolar effects. So we report first the structures issued from 1–3 structures in Fig. 1.38. Figure 1.39 is devoted to structures issued from onions 2D structures.

As shown in Fig. 1.38, there is a further classification of structures issued from 1–3 2D structures according to the resulting magnetization value. S and S′ structures share the resulting magnetization of two domains. B′ structures have the magnetization of $2\sqrt{2}$ domains. Finally, B structures have the magnetization of four domains. This is useful to derive the stability for strong dipolar effect. Then S and S′ structures are the more stable, while B′ is a little bit less stable. B is the less stable of these four structures.

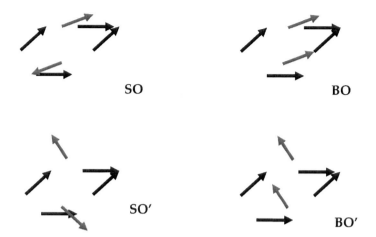

Figure 1.39 (Color online) The four basic 3D structures issued from onion 2D structures. On the left-hand side, the S-shaped structures SO and SO' are shown, and on the right the bent structures BO and BO'. At the top, for structures SO and BO magnetization of top and at the bottom domain is parallel or antiparallel to the first bisector, while on bottom for SO' and BO' structures magnetization of top and bottom domain is perpendicular to the first bisector.

About the structures issued from 2D onions there is a similar hierarchy deduced from the magnetization resulting value. According to the notation shown in Fig. 1.39, the two S-shaped structures SO and SO' share the same magnetization, that of $2\sqrt{2}$ domains. So SO and SO' have the same stability as B'. Then the bent structure BO' has a magnetization of $2\sqrt{3}$ domains. Finally, BO structure has a resulting magnetization corresponding to that of $2 + 2\sqrt{2}$ domains. This BO structure is less stable in the presence of strong dipolar effect.

As a global conclusion on the elementary structures of a cubic nanoparticle, S-shaped vortex S and S-shaped antivortex SA are the more stable ones in the presence of strong dipolar effect. It must be added that since the three axes are equivalent, similar structures deduced from cubic symmetry are equivalent. This remark defines three sets of equivalent structures with central axis z, x, and y, respectively. Since also both chiralities must be considered as equivalent, this symmetry multiplies the number of relevant sets by 2. So there are a large number of degenerate individual low-energy units. In the case of strong dipolar effects these low-energy basic units are S-shaped vortex S and S-shaped antivortex SA. Their arrangement must satisfy symmetry rules. Another constraint on the arrangement of these 6-domain units comes from magnetocrystalline anisotropy induced by dipolar interaction. This anisotropy was clearly shown in 2D samples [99] in Eq. 1.17 and is easily generalized. Thus, dense axes are preferred. All these features, including S-shaped vortex and antivortex with several orientations and dense axis alignments as well as isotropy for cubic samples well appear in recent simulations reported in this book [43] as building rules for realistic singularity lines.

The last point to consider in this theoretical part is excited states which can be also called dynamics of these spin configurations.

1.3.4 Excited states

As already seen from the research about ground states, an analytic treatment of excited states is very hard to be obtained for different reasons. First, there are several degenerate ground states or nearly degenerate approximate ground states which make a complex arrangement of excited states. Second, the experimental and numerical studies about effective 2D samples evidenced the occurrence of global low-frequency modes as gyrotropic vortex motion. With the previous theoretical evidence of lines of singularities in 3D nanoparticles shown in Figs. 1.36–1.39, gyrotropic motion of singularity lines must also occur in 3D samples where long-ranged interactions enable coordinated spin motions. So an analytic study of a more simple system is of interest. This is the case of a one dimensional model suggested by layered samples which can take an arbitrary 3D shape. Such systems were already investigated [138]. The advantage of such systems consists in avoiding rotational spin motion such as gyroscopic ones since in 1D there is no room for rotation and in avoiding degeneracy effects which are also due to symmetry. However, 1D models of layered samples well take into account the long ranged coupling. This long range coupling makes all 1D sites different since their long-ranged environment must be

Theory | 79

considered. This leads to specific effects of localization and of nature of excited states.

1.3.4.1 Effective "1D" excited states

As a matter of fact, in models of layered materials, useful approximations lead to the generic spin deviation equation of motion for an internal layer n:

$$(a - E) u_n - b (u_{n+1} + u_{n-1}) - c (u_{n+2} + u_{n-2})$$
$$- d (u_{n+3} + u_{n-3}) + \ldots = 0 \qquad (1.29)$$

Here the coefficients a, b, c, d, must be calculated according to the model [138] and they can depend on the layer number n. Assuming no such layer number dependency in a first step leads to write for an assumed spin wave $u_n = C \exp [iknа_0 - i\varpi t]$, the dispersion relation:

$$E = a - 2b \cos (ka_0) - 2c \cos (2ka_0) - 2d \cos (3ka_0)$$
$$- 2e \cos (4ka_0) - \ldots \qquad (1.30)$$

According to the values of the coefficients $a, b, c, d, \ldots$ this dispersion relation is not monotonous in the wavenumber k. So one frequency, i.e., one mode does not correspond to a single wavenumber and wavelength but to several wavenumbers and wavelengths. In other words, the eigen modes of this dynamical matrix are no more sine-like and result from the superposition of several sine-like modes. So localization can occur.

The analysis of the dynamical matrix can be done directly.

From the system of Eq. 1.29 the set of successive equations on spin motions is written:

$$(a - E) u_n - b (u_{n+1} + u_{n-1}) - c (u_{n+2} + u_{n-2})$$
$$- d (u_{n+3} + u_{n-3}) + \ldots = 0$$
$$(a - E) u_{n+1} - b (u_{n+2} + u_n) - c (u_{n+3} + u_{n-1})$$
$$- d (u_{n+4} + u_{n-2}) + \ldots = 0 \qquad (1.31)$$

80 | *Magnetic Structures of 2D and 3D Nanoparticles*

And the writing of the dynamical matrix follows:

$$
\begin{pmatrix}
a-E & -b & -c & -d & -e & . & & . & . \\
-b & a-E & -b & -c & -d & -e & & . & . \\
-c & -b & a-E & -b & -c & -d & & -e & . \\
-d & -c & -b & a-E & -b & -c & & -d & -e \\
-e & -d & -c & -b & a-E & -b & & -c & -d \\
. & -e & -d & -c & -b & a-E & & -b & -c \\
& . & . & -e & -d & -c & -b & a-E & -b \\
& . & . & & . & -e & -d & -c & -b & a-E
\end{pmatrix} \times
$$

$$
\begin{pmatrix} u_1 \\ u_2 \\ u_3 \\ u_4 \\ u_5 \\ u_6 \\ u_7 \\ u_8 \end{pmatrix} = \mathbf{A} * \mathbf{u} = \begin{pmatrix} 0 \\ 0 \\ 0 \\ 0 \\ 0 \\ 0 \\ 0 \\ 0 \end{pmatrix} \tag{1.32}
$$

Since a matrix consisting of two symmetric diagonals has powers which cover more diagonals, the matrix $\mathbf{A}$ can be deduced from the basis of the identity matrix $\mathbf{I}$ and the powers of a simple low diagonal ($n \times n$) matrix $\mathbf{d}$, up to corrective "surface" terms located at borders here shown for a (4×4) matrix :

$$
d = \begin{pmatrix} 0 & 1 & 0 & 0 \\ 1 & 0 & 1 & 0 \\ 0 & 1 & 0 & 1 \\ 0 & 0 & 1 & 0 \end{pmatrix} \tag{1.33}
$$

$\mathbf{A}$ is a polynomial form of $\mathbf{d}$ and $\mathbf{I}$ with real coefficients. As done usually for polynomial forms, it can be factorized with real or complex roots and this also leads to derive the corresponding Green function matrix A^{-1} as a sum of Green function matrices for simple three diagonal matrices $\mathbf{d}'$. Since such Green function matrices were previously calculated with sine-like behaviors [101], the resulting Green function matrix A^{-1} results from the superposition of several sine-like behaviors with different wavenumbers in perfect agreement with the considerations derived from the dispersion relation (30). So the complexity of excited modes is well established

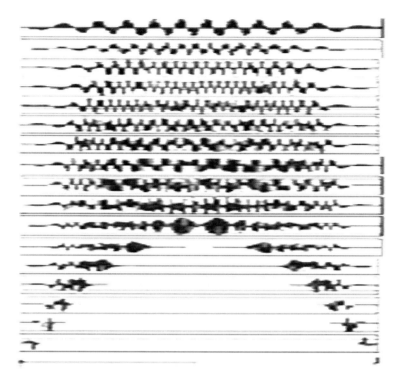

Figure 1.40 Numerically calculated mode profiles are shown with increasing frequency from the bottom to the top in a rod of length 200. The lowest mode is the mode number 1. The following mode numbers are, respectively, 15, 30, 40, 50, 60, 70, 80, and so on, up to 180. All these modes are localized either close to the surfaces ($n < 80$) or near the center of the rod ($80 < n < 180$) [138].

even in these effective 1D models. It must be added that in real 2D samples, for instance, the transition from a ferromagnetic structure toward a vortex structure and toward a polar vortex structure with a core well also exhibits strong localization effects for such a soft mode. And in 3D real spin structures are more complex as seen before, so they require quite complex soft modes when starting from ferromagnetic configurations.

Practically, these dynamical matrices for different effective 1D configurations were solved numerically [138]. They exhibit quite numerous modes with singular spatial extensions. This is reported in Fig. 1.40, where 18 typical modes in an effective 1D picture of a magnetic rod of length 200 atomic layers are shown. These modes largely differ from classical sine-like modes and exhibit localization properties not only at borders.

In the simple case when neglecting dipolar interactions, exchange magnons depend on the nanoparticle geometry because of spin wave quantization in finite particles. This gives classical selection rules for thin films [103]. For spherical nanoparticles these quantization properties define selection rules on radial wavefunction as well as on azimuthal wavefunction [120]. For cylindrical nanoparticles there are also selection rules [120]. For 2D magnetic disks also, selection rules are observed [114].

As a conclusion about excited states in 3D magnetic nanoparticles, this short part well introduces the richness of the properties of excited states such as localization, without defect as due to dipolar interactions. These localization properties can have applications in magnonic.

1.4 Conclusions

There are numerous applications of 2D and 3D magnetic nanoparticles, from printing to medical therapy as well as high-frequency devices or memory units. This richness in applications is well balanced by the richness in observations and in simulations, especially for 2D or effective 2D magnetic nanoparticles. One part of this special interest is due to the well-known appearance of magnetic vortices and magnetic antivortices as stable features of nanoparticles. The arrangement of vortex, antivortex, and other singularities follows topological rules. Real 3D magnetic nanoparticles also have many applications that are already developed, but the understanding of their magnetic structure and of their magnetic excitation is inadequate because of its complexity. So it is quite useful to consider very small units with just a few intrinsic topological defects.

Here, focusing the interest on dipolar interactions, the transition from 2D magnetic nanoparticles toward 3D magnetic nanoparticles always restricted to a small size in order to avoid entering a forest of singularities enabling us to derive in a theoretical approach the basic units of 3D magnetic nanoparticles: S-shaped vortices and S-shaped antivortices as well as several related low-energy metastable states. This chapter introduces further numerical and experimental work on such restricted nanoparticles.

Of course, there are not only dipolar interactions and exchange ones in magnetism, but also anisotropy and other interactions issuing from spin–orbit coupling such as Dzyaloshinskii–Moriya interactions, which play a strong part in magnetic nanoparticles since sites are far from being all equivalent there. On the other hand, in metals or even doped semiconductors, magnetic interactions are well screened by electronic motions. Such different interactions introduce a further complexity in the magnetic structures of nanoparticles, as observed under the generic name of skyrmions [78]. The competition between these numerous interactions leads to a very large program of structures and excitations. However, the work on dipolar interactions is already rich enough and promising to be studied completely.

A further interest of this study of the competition between long-ranged dipolar interactions and short-ranged exchange interactions within nanoparticles comes from the numerous analogous situations it suggests. The richness of configurations such as vortex and antivortex and of excitations such as gyrotropic vortex motion, for instance, is quite stimulating.

First, there is the case of nuclear spins which interact between themselves by dipolar interactions only [96], even if they interact also with other magnetic carriers.

Then there is the case of Bose–Einstein condensates (BECs), where nuclear spins are present [194]. Then dipolar interactions must be taken into account and lead to magnetic structures and magnetic excitations. It must be added that in the BEC case the concept of bent vortex was introduced [54].

For a long time, the word magnetic vortex had the unique meaning of a superconductivity effect. And as a matter of fact, the Abrikosov lattice of vortex involves dipolar interaction between flux tubes [108]. So now it seems that considerations on superconductivity in extended 3D samples must take into account similar occurrences of S-shaped vortex, for instance. Several people have already proposed the existence of twisted vortex as well as of a glass of entangled vortex [132]. From the present analysis it would be interesting to look at nanometric samples.

Dipolar interactions are not restricted to magnetism. Electric dipolar interactions are quite effective. So the previous considerations could be applied to ferroelectric domains [64] as well as to multiferroic materials [28], where dipolar magnetic interaction occurs as well as dipolar electric interaction.

Dipolar interactions are also basic intermolecular interactions. Dipole-induced dipole interactions are the origin of Van der Waals interactions [180]. So dipolar interactions must be accounted for in many molecular problems. This is the case for liquid crystals, where many observed structures evidenced vortices and other topological defects [177]. Thus, it would be of interest to consider also the consequence of dipolar interactions in liquid crystals.

The specificity of dipolar interactions in a crystalline nanomagnet is the occurrence of a few dense lines which obviously pin the singularities and so lead to specific S-shaped lines which interplay together. Thus the crystalline structure induces a vortex stiffness which stabilizes linear segments. So noncrystalline structures such as quasicrystals, glasses, or fractal sets can avoid this stiffening and favor more complex structures. This is also of interest for low-frequency vortex motions.

Acknowledgments

The present work results from a rather long study involving numerous collaborations in many fields. An initial step about magnetic structures was taken with J.-L. Porteseil, P. Molho, and Y. Souche from Grenoble, and J. Gouzerh from Paris and Bellevue.

Later, with D. Mercier (Paris and Saint-Quentin), the late A. Ghazali (Paris), E. Vedmedenko (Paris, Halle, Hamburg), F. Mertens (Bayreuth), and Ph. Depondt (Paris), the modeling of these structures was improved in several steps.

Experiments from G. Suran and coworkers, and then from the late J. Jakubowitz (Oxford), and coworkers were quite stimulating.

As regards magnetic excitations, the work started a long time ago with E. Ilisca, J.-L. Motchane, E. Gallais, T. H. Diep, and

D. Mercier (Paris). With H. Puszkarski, M. Krawczyk, and S. Mamica (Poznan), this work was extended from spin waves toward magnonic structures. With P. J.-M. Monceau, the work on magnetic excitations was extended to fractal structures. The connection with the experimental work on 3D magnetic nanoparticles of F. Fievet and coworkers was also very fruitful.

References

1. Abramowitz, M., Stegun, I.A. Eds (1964). *Handbook of mathematical functions*, NBS Washington.
2. Aguilar, M., Oliva, A.I., Quintana, P., and Pena, J.L. (1997). Dynamic phenomena in the surface of gold thin films: macroscopic surface rearrangements, *Surf. Sci.*, 380:91–99.
3. Aharoni, A. (1996). *Introduction to the theory of ferromagnetism*, Clarendon Press.
4. Alexander, S., and McTague, J. (1978). Should all crystals be bcc? Landau theory of solidification and crystal nucleation, *Phys. Rev. Lett.*, 41:702–705.
5. Anderson, P.W. (1959). New approach to the theory of superexchange interaction, *Phys. Rev.*, 115:2–13.
6. Baade, W., and Zwicky, F. (1934). On super-novae, *Proc. Nat. Ac. Sc.*, 20:254–259.
7. Baibich, M.N., Broto, J.M., Fert, A , Nguyen Van Dau, F., Petroff, F., Etienne, P., Creuzet, G., Friederich, A., and Chazelas, J. (1988). Giant Magnetoresistance of (001)Fe/(001)Cr Magnetic Superlattices, *Phys. Rev. Lett.*, 61:2472.
8. Barth, J.V., Brune, H., Ertl, G., and Behm, R.J. (1990). Scanning tunneling microscopy on the reconstructed Au (111) surface: Atomic structure, long-range superstructure, rotational domains and surface defects, *Phys. Rev. B*, 42:9307–93018.
9. Bauer, H.G., Sproll, M., Back, C.H., Woltersdorf, G. (2014). Vortex core reversal due to spin wave interference, *Phys. Rev. Lett.*, 112:077201–077204.
10. Berkowitz, A.E., Lahut, J.A., Meiklejohn, W.H., Skoda, R.E., and Wang, J.M. (1982). A High Speed Magnetic Printer, *IEEE Transactions on Magnetism*, MAG-18, 1796–1799.

11. Berkowitz, A.E., and Walter, J.M. (1983). Ferrofluids Prepared by Spark Erosion, *J. Magn. Magn. Mater.*, 39:75–79.

12. Bettge, M. Chatterjee, J., and Haik, J. (2004). Physically synthetized Ni-Cu nanoparticles for magnetic hyperthermia, *Biomagn. Res. Technol.*, 2:1–17

13. Binder, K. (1995). *The Monte-Carlo Method in Condensed Matter*, New York, Springer.

14. Binnig, G., Rohrer, H., Gerber, Ch., and Weibel, E. (1982). Surface studies by scanning tunneling microscopy, *Phys. Rev. Lett.*, 49:57–61.

15. Binnig, G., Rohrer, H., Gerber, Ch., and Weibel, E. (1983). 7×7 Reconstruction on Si(111) Resolved in Real Space, *Phys. Rev. Lett.*, 50:120–126.

16. Bloch, F. (1930). Magnons, *Z. Phys.*, 61:206–211.

17. Bobeck, A.H., and Scovil, H.E.D. (1971). Magnetic bubbles, *Sci. American* 224:78-91.

18. Bocchetti, V., and Diep, T.H. (2013). Monte Carlo simulation of melting and lattice relaxation of the (111) surface of silver, *Surf. Sci.* 614:46–62.

19. Brechet, S.D., Vetro, F.A., Papa, E., Barnes, S.E., and Ansermet, J.P. (2013). Evidence for a magnetic Seebeck effect, *Phys. Rev. Lett.*, 111:087205–087208.

20. Brechignac, C., Cahuzac, Ph., Carlier, F., and Leygnier, J. (1989). Collective excitation in closed-shell potassium cluster ions, *Chem. Phys. Lett.*, 164:433–437.

21. Brown, G., Novotny, M.A., and Rikvold, P.A. (2001). Langevin simulation of thermally activated magnetization reversal in nanoscale pillars, *Phys. Rev. B*, 64:134422, 1–14.

22. Campargue, R. (2001). *Atomic and molecular beams, the state of the art 2000*, Springer.

23. Campiglio, P., Repain, V., Chacon, Fruchart, O., Lagoute, J., Girard, Y., and Rousset, S. (2011). Quasi unidimensional growth of Co nanostructures on a strained Au (111) surface, *Surf. Sci.*, 605:1165–1170.

24. Cape, J.A., and Lehman, G.W. (1971). Magnetic domain structures in thin uniaxial plates with perpendicular easy axis, *J. Appl. Phys.*, 32:5732–5756; Seul, M., and Andelman, D. (1995). Domain shapes and patterns: the phenomenology of modulated phases, *Science*, 267:476–483.

25. Carcia, P.F. (1988). Perpendicular magnetic anisotropy in Pd/Co and Pt/Co thin film layered structures, *J. Appl. Phys.*, 63:5066–5071.

26. Carrey, J., Mehdaoui, B., and Respaud, M. (2011). Simple models for dynamic hysteresis loop calculations of magnetic single-domain nanoparticles: Application to magnetic hyperthermia optimization, *J. Appl. Phys.*, 109, 083921–17.

27. Castano, F.J., Ross, C.A., Frandsen, C., Eilez, A., Gil, D., Smith, H.I., Redjdal, M., and Humphrey, F.B. (2003). Metastable states in magnetic nanorings, *Phys. Rev. B*, 67:184425.

28. Chae, S.C., Horibe, Y., Jeong, D.Y., Rodan, S., Lee, N., Cheong, S.-W. (2010). Self-organization, condensation and annihilation of topological vortices and antivortices in a multiferroic, *Proc. Natl. Acad. Sci. USA*, 107:21366–21370.

29. Cho, A.Y., and Arthur, J.R. (1975). Molecular beam epitaxy, *Prog. Solid-State Chem.*, 10:157–191.

30. Choi, B.G., Chang, S.-J., Lee, Y.B., Bae, J.S., Kim, H.J., and Huh, Y.S. (2012). 3D heterostructured architectures of Co_3O_4 nanoparticles deposited on porous graphene surfaces for high performance of lithium ion batteries, *Nanoscale*, 4:5924–5930.

31. Chubykalo, O., Hannay, J.D., Wongsam, M., Chantrell, R.W., and Gonzalez, J.M. (2002). Langevin dynamic simulation of spin waves in a micromagnetic model, *Phys. Rev. B*, 65:184428–184439.

32. Compton, R.L., and Crowell, P.A. (2006). Dynamics of a pinned magnetic vortex, *Phys. Rev. Lett.*, 97:137202–137205.

33. Corr, S.J., Raoof, M., Wilson, L. J., and Curley, S.A. (2012). *Functional Nanoparticles for Bioanalysis, Nanomedicine and Bioelectronic Devices 2*, Chapter 6 Nanoparticles for noninvasive radiofrequency-induced cancer hyperthermia, ACS Symposium Series 1113.

34. Cowburn, R.P., and Welland, M.E. (1998). Micromagnetics of the single domain state of magnetic square structures, *Phys. Rev. B*, 58:9217–9221.

35. Coxeter, H.S.M. (1973). *Regular Polytopes*, 3rd ed., New York Dover.

36. Daw, M. S., and Baskes, M. (1984). Embedded-atom method: Derivation and application to impurities, surfaces, and other defects in metals, *Phys. Rev. B*, 29:6443–6453.

37. De'Bell, K., MacIsaac, A.B., Booth, I.N., and Whitehead, J.P. (1997). Dipolar-induced planar anisotropy in ultrathin magnetic films, *Phys. Rev. B*, 55:15108–15112.

38. De' Bell, K., MacIsaac, A.B., and Whitehead, J.P. (2000). Dipolar effects in magnetic thin films and quasi-two-dimensional systems, *Rev. Mod. Phys.*, 72:225–257.

88 | *Magnetic Structures of 2D and 3D Nanoparticles*

39. Del Bianco, L., Boscherini, F., Fiorini, A.L., Tamisari, M., Spizzo, F., Vitorri Antisari, V., and Piscopiello, E. (2008). Exchange bias and structural disorder in the nanogranular Ni/NiO system produced by milling and hydrogen reduction, *Phys. Rev. B*, 77:094408–094412.

40. Depondt, Ph., and Levy, J.-C.S. (2011). Spin-dynamics simulations of vortex precession in 2-D magnetic dots, *Phys. Lett. A*, 375:4085–4090.

41. Depondt, Ph., and Levy, J.-C.S. (2012). Temperature and vortex gyrotropic motion in 2-D magnetic nanodots by Langevin simulations, *Phys. Lett. A*, 376:3411–3416.

42. Depondt, Ph., and Levy, J.-C.S., Mamica, S. (2013). Vortex polarization dynamics in a square magnetic nanorod, *J. Phys. Condens. Matter*, 25:4660001–4660007.

43. Depondt, Ph. (2015). This volume.

44. Dieny, B., Sousa, R.C., Hérault, J., Papusoi, C., Prenat, G., Ebels, U., Houssameddine, D., Rodmacq, B., Auffret, S., Buda-Prejbeanu, L.D., Cyrille, M.C., Delaët, B., Redon, O., Ducruet, C., Nozières, J-P., and Prejbeanu, I.L. (2010). Spin-transfer effect and its use in spintronic components, *Int. J. Nanotechnol*, 7:591–614

45. Donahue, M.J., and Porter, D.G. (1999). OOMMF Object Oriented Micro Magnetic Framework, *User's guide 1.0*, Interagency Report Nistir 6376.

46. Douglass, D.C., Bucher, J.P., and Bloomfield, L.A. (1992). Magic numbers in the magnetic properties of gadolinium clusters, *Phys. Rev. Lett.*, 68:1774–1777.

47. Epstein, A.J. (2003). Organic-based magnets: Opportunities in photoinduced magnetism, spintronics, fractal magnetism, and beyond, *MRS Bull.*, 28:492–499.

48. Estevez, V., and Bascones, E. (2011). Robustness of the magnetoresistance of nanoparticle arrays, *Phys. Rev. B*, 84:075441–075448.

49. Fang, R.H., Hu, C.-M. J., Luk, B.T., Gao, W., Copp, J.A., Tai Y., O'Connor, D.E., and Zhang L. (2014). Cancer cell membrane-coated nanoparticles for anticancer vaccination and drug delivery, *Nano Lett.*, 14:2181–2188.

50. Feldman, L.C., Mayer, J.W., and Picraux, S.T. (1982). *Materials analysis by ion channeling*, Academic Press, New York.

51. Fetter, A.L., and Walecka, J.D. (2003). *Quantum Theory of Many Particles Systems*, Dover Publications.

52. Fettar, F. (1998). Thesis Université Paris Diderot.

53. Gangrskii, Y.P., Zemlyanoi, S.G., Markov, B.N., Myshinskii, G.V., Nesterenko, V.O., Vorykhalov, I.V., Izosimov, I.N., and Rimskii-Korsakov,

A.A. (1997). Production of heavy atomic clusters upon interaction of laser radiation with matter, *J.E.T.P.*, 85:42–47.

54. Garcia-Ripoll, J.J., and Perez-Garcia, V.M. (2001). Vortex bending and tightly packed vortex lattices in Bose-Einstein condensates, *Phys. Rev. A*, 64:053611, 1–7.

55. Ghazali, A., and Levy, J.-C.S. (1997). Structural transitions in clusters, *Phys. Lett. A*, 228:291–296.

56. Ghazali, A., and Levy, J.-C.S. (2003). Two-dimensional arrangements of magnetic particles, *Phys. Rev. B,* 67:064409-1-5.

57. Ghazali, A., and Levy, J.-C.S. (2004). A synthetic view of 2D aggregation, *Phys. Rev. E*, 69:061405-1-7.

58. Ghazali, A., and Levy, J.-C.S. (2006). Structure of 2D deposits by Monte-Carlo simulations: fractals, dendrites islands and fluids, *Surface Magnetism and Nanostructures*, A. Ghazali and J.-C.S. Levy (eds), Research Signpost Trivandrum, 19–48.

59. Goeppert-Mayer, M. (1950). Nuclear configurations in the spin-orbit coupling model. I. Empirical evidence, *Phys. Rev.*, 78:16–21.

60. Gradmann, U. (1986). Magnetic surface anisotropies, *J. Magn. Magn. Mat.*, 54–57: 733–736.

61. Griffiths, D.J. (2004). *Introduction to Quantum mechanics*, 2nd ed. Prentice Hall.

62. Grimmett, G. (1999). *Percolation*, Springer Berlin.

63. Grishin, S.V., Beginin, E.N., Morozova, M.A., Sharaevskii, Yu.P., and Nikitov, S.A. (2014). Self-generation of dissipative solitons in magnonic quasicrystal active ring resonator, *J. Appl. Phys.*, 115:053908-053908-8.

64. Gruverman, A., Wu, D., Fan, H.-J., Vrejoiu, I., Alexe, M., Harrison, R.J., and Scott, J.F. (2008). Vortex ferroelectric domains, *J. Phys.: Condens. Matter*, 20:342201–342205.

65. Guillaud, C., and Vautier, R. (1958). Les domaines élémentaires in *Colloque National de Magnetisme*, CNRS 24–74.

66. Guslienko, K.Y., Heredero, and R.H., Chubykalo-Fesenko, O. (2010). Nonlinear gyrotropic vortex dynamics in ferromagnetic dots, *Phys. Rev. B*, 82:014402, 1–7.

67. Hankemeier, S., Kobs, A., Frömter, and Oepen, H.P. (2010). Controlling the properties of vortex domain walls via magnetic seeding fields, *Phys. Rev. B*, 82:064414–064416.

68. Hasegawa, R. (1974). Static bubble domain properties of amorphous GdCo films, *J. Appl. Phys.*, 45:3109–3112.

69. Honeycutt, J.D., and Andersen, H.C. (1987). Molecular dynamics of melting and freezing of small Lennard-Jones Clusters, *J. Phys. Chem.*, 91:4950–4963.

70. Hopster, H., and Oepen, H.P. (2005). *Magnetic Microscopy of Nanostructures*, Springer.

71. Hubert, A., and Schäfer, R. (1998). *Magnetic Domains the analysis of magnetic microstructures*, Springer Berlin.

72. Jansen, H. (1988). Magnetic anisotropy in density functional theory, *Phys. Rev. B*, 38:8022–8029.

73. Jorzick, J., Demokritov, S.O., Mathieu, C., Hillebrands, B., Bartenlian, B., Chappert, C., Rousseaux, F., and Slavin, A.N. (1999). Brillouin light scattering from quantized spin waves in micron-size magnetic wires, *Phys. Rev. B*, 60:15194–200.

74. Kales, M.L. (1933). Modes in wave guides containing ferrites, *J. Appl. Phys.*, 24:601–604.

75. Kamionka, T., Martens, M., Drews, A., Krüger, B., Albrecht, O., and Meier, G. (2011). Influence of temperature on the gyrotropic eigenmode of magnetic vortices, *Phys. Rev. B*, 83:224424, 1–4.

76. Karmakar, S., Kumar, S., Rinaldi, R., and Maruccio, G. (2011). Nanoelectronics and spintronics with nanoparticles, *J. Phys. Conf. Ser.*, 292, 01 2002.

77. Kerr, J. (1877). On the rotation of the plane of polarization by reflection from the pole of a magnet, *Phil. Mag.*, 3:321–327.

78. Kiselev, N.S., Bogdanov, A.N., Schäfer, R., and Rossler, U.K. (2011). Chiral skyrmions in thin magnetic films: New objects for magnetic storage technologies, *J. Phys. D Appl. Phys.* 44:392001–292004; Romming, N., Hanneken, C. Menzel, M., Bickel, J.E., Wolter, B., Von Bergmann, K., Kubetzka, A., and Wiesendanger, R. (2013). Writing and deleting single magnetic skyrmions, *Science*, 341:636–639.

79. Kittel, C. (1948). On the theory of ferromagnetic resonance absorption, *Phys. Rev.* 73:155–156.

80. Kittel, C. (1958). Excitation of spin waves in a ferromagnet by a uniform rf field, *Phys. Rev.* 110:1295–1299.

81. Kläui, M., Vaz, C.A.F., Bland, J.A.C., Monchevsky, T.L., Unguris, J., Bauer, E., Cherifi, S., Heun, S., Locatelli, A., Heyderman, L.J., and Cui, Z. (2003). Direct observation of spin configurations and classification of switching processes in mesoscopic ferromagnetic rings, *Phys. Rev. B*, 68:134426–5.

82. Klos, J.W., Kumar, D., Krawczyk, M., and Barman, A. (2014). The influence of structural changes in a periodic antidot waveguide on the spin wave spectra, *Phys. Rev. B*, 89:014406–13.

83. Kohn, W., and Sham, L.J. (1965). Self-consistent equations including exchange and correlation effects, *Phys. Rev. A*, 140:1133–1138.

84. Kosiorek, A. Kandulski, W., Glaczynska, H., and Giersig, M. (2005). Fabrication of nanoscale rings, dots, and rods by combining shadow nanosphere lithography and annealed polystyrene nanosphere masks, *Small*, 1:439–444.

85. Kozelj, P., Jazbec, S., Vrtnik, S., Jelen, A., Dolinsek, J., Jagodic, M., Jaglicic, Z., Boulet, P., De Weerd, M.C., Ledieu, J., Dubois, J.M., and Fournee, V. (2013). Geometrically frustrated magnetism of spins on icosahedrak clusters: The Gd3Au3Sn4 quasicrystalline approximant, *Phys. Rev. B*, 88:214202–214207.

86. Krawczyk, M., Mamica, S., Mruczkiewicz, M., Klos, J.W., Tacchi, S., Madami, M., Gubbiotti, G., Duerr, G., and Grundler, D. (2013). Magnonic band structures in two-dimensional bi-component magnonic crystals with in-plane magnetization, *J. Phys. D: Appl. Phys.*, 46 495003:1–14.

87. Kronmüller, H., and Parkin, S.S.P. (2007). *Handbook of magnetism and advanced magnetic materials*, John Wiley.

88. Kruglyak, V.V., Demokritov, S.O., and Grundler, D. (2011). Magnonics, *J. Phys. D Appl. Phys.*, 43:264001.

89. Kruis, F.E., Fissan, H., and Peled, A. (1998). Synthesis of nanoparticles in the gas phase for electronic, optical and magnetic applications – a review, *J. Aerosol Sci.*, 29:511–535.

90. Kumar, D., Klos, J.W., Krawczyk, M., and Barman, A. (2014). Magnonic band structure, complete bandgap and collective spin wave excitation in nanoscale two-dimensional magnonic crystals, *J. Appl. Phys.*, 115:043917–043924.

91. Kyotani, T., Tsai, L.-F., and Tomita, A. (1996). Preparation of ultrafine carbon tubes in nanochannels of an anodic aluminum oxide film, *Chem. Mater.*, 8:2109–2113.

92. Landau, L.D., and Lifshitz, E.M. (1935). Dipolar origin of magnetic domains, *Phys Zeit Sow.*, 8:153.

93. Landau, L.D., and Lifshitz, E.M. (1960). *Electrodynamics of Continuous Media*, volume 6 of *Courses of Theoretical Physics*, Pergamon Press Oxford.

94. Landau, L., and Lifshitz, E.M. (1962). *The classical theory of fields*, 2nd ed., Pergamon Press.

95. Levine, D., and Steinhardt, P.J. (1984). Quasicrystals: A new class of ordered structures, *Phys. Rev. Lett.*, 63:1951–1953.

96. Levitt, M.H. (2001). *Spin dynamics: basis of nuclear magnetic resonance*, Wiley UK.

97. Levy, J.-C.S. (1979). Metastable rearrangements near the surface, *Surf. Sci.*, 86:760–764.

98. Levy, J.-C.S. (1981). Surface and interface magnons: magnetic structures near the surface, *Surf. Sci. Rep.*, 1:40–119.

99. Levy, J.-C.S. (2001). Dipolar induced magnetic anisotropy and magnetic topological defects in ultrathin films, *Phys. Rev. B*, 67:104409-1-7.

100. Levy, J.-C.S. (2009). Atomic stiffness parameter *Nanostructures and their magnetic properties*, ed. J.-C.S. Levy, Research Signpost Trivandrum, 1–19.

101. Levy, J.-C.S., Gallais, E., and Motchane, J.L. (1974). Thin ferromagnetic films: Spin waves and magnetization, *J. Phys. C*, 7:761–782.

102. Levy, J.-C.S., and Ghazali, A. (2001). Dipolar effects in magnetic nanostructures, *Nanostructured magnetic materials and their applications*, ed. Shi D., Springer, 183–202.

103. Levy, J.-C.S., Ilisca, E., and Motchane, J.L. (1972). Influence of surface anisotropy and n.n.n. coupling on spin wave spectrum, *Phys. Rev. B*, 5:1099–1108.

104. Levy, J.-C.S., Krawczyk, M., and Puzkarski, H. (2006). Magnons in Co dots, *J. Magn. Magn. Mat.*, 305:182–185.

105. Levy, J.-C.S., and Mercier, D. (1982). Amorphous structures: a local analysis, *J. Appl. Phys.* 53:7709–7711; Mercier, D., Levy, J.-C.S. (1983). Construction of amorphous structures, *Phys. Rev. B*, 27:1292–1302.

106. Levy, J.-C.S., and Mercier, D. (1984). Local elasticity properties of an amorphous structure: evidences for typical sites and shell structure. Dynamic stability, *J. Phys. (Paris)*, 45:291–301.

107. Li, S.P., Peyrade, D., Natali, M., Lebib, A., Chen, Y. Ebels, U., Buda, L.D., and Ounadjela, K. (2001). Flux closure structures in cobalt rings, *Phys. Rev. Lett.*, 86:1102–1105.

108. Lipavsky, P., Morawetz, K., Kolacek, J., and Brandt, E.H. (2008). Surface deformation caused by the Abrikosov vortex lattice, *Phys. Rev. B*, 77:184509-7.

109. Liu, W., Zhong, W., and Du, Y.V.V. (2008). Magnetic nanoparticles with core/shell structures, *J. Nanosci. Nanotechnol.*, 8:2781–2792.

110. Lu, H., Yi, G., Zhao, S., Chen, D., Guo, L.-H., and Cheng, J. (2004). Synthesis and characterization of multi-functional nanoparticles possessing magnetic, up-conversion fluorescence and bio-affinity properties, *J. Mater. Chem.*, 14:1336–1341.

111. Luttinger, J.M., and Tisza, L. (1946). Theory of dipole interaction in crystals, *Phys. Rev.*, 70, 1954–1963.

112. MacIsaac, A.B., Whitehead, J.P., Robinson, M.C., and De' Bell, K. (1995). Striped phases in two-dimensional dipolar ferromagnets, *Phys. Rev. B*, 51:16033–16045.

113. McMullan, D. (1988). Von Ardenne and the scanning electron microscope, *Proc Roy Microsc Soc.*, 23:283–288.

114. Mamica, S. (2013). Stabilization of the in-plane vortex state in two-dimensional circular nanorings, *J. Appl. Phys.*, 113:093901–4.

115. Mamica, S., Levy, J.C.S., Krawczyk, M., and Depondt, Ph. (2012). Stability of the Landau-state in square two-dimensional magnetic nanorings, *J. Appl, Phys.*, 112:043901–5.

116. Mamica, S., Levy, J.-C.S., and Krawczyk, M. (2014). Effects of the competition between the exchange and dipolar interactions in the spin wave spectrum of two-dimensional circularly magnetized nanodots, *J. Phys. D Applied Physics*, 47:5003–21.

117. Mandelshtam, V.A., Frantsuzov, P.A., and Calvo, P. (2006). Structural transitions and melting in LJ74-78 Lennard-Jones clusters from adaptative exchange Monte Carlo simulations, *J. Phys. Chem. A*, 110:5326–5332.

118. Manke, I., et al. (2010). Three-dimensional imaging of magnetic domains, *Nature Com.*, 1:125/1–6.

119. Martins, J.L., Buttet, J., and Carr, R. (1985). Electronic states and structural properties of sodium clusters, *Phys. Rev. B*, 31:1804–1811.

120. Mercier, D., Levy, J.-C.S., Viau, G., Fievet-Vincent, F., Fievet, F., Toneguzzo, P., and Acher, O. (2000). Magnetic resonance in spherical CoNi and FeCoNi nanoparticles, *Phys. Rev. B*, 62:532–45.

121. Mermin, N.D. (1992). The space groups of icosahedral quasicrystals and cubic, orthorhombic, monoclinic and triclinic crystals, *Rev. Mod. Phys.*, 64:3–48.

122. Mermin, N.D., and Wagner, H. (1966). Absence of ferromagnetism or antiferromagnetism in one- or two-dimensional isotropic Heisenberg models, *Phys. Rev. Lett.*, 17:1307–1340.

94 | *Magnetic Structures of 2D and 3D Nanoparticles*

123. Metropolis, N., Rosenbluth, A.W., Rosenbluth, M.N., Teller, A.H., and Teller, E. (1953). Equations of states: computation by fast machines, *J. Chem. Phys.*, 21:1087–1092.

124. Miyamachi, T., Gruber, M., Davesne, V., Bowen, M., Boukari, S., Joly, L., Scheurer, F., Rogez, G., Yamada, T.K., Ohressar, P., Beaurepaire, E., and Wulfheckel, W. (2012). Robust spin crossover and memristance across a single molecule, *Nat. Comm.*, 3:938–6.

125. Molho, P., Porteseil, J.L., Souche, Y., Gouzerh, J., and Levy, J.-C.S. (1987). Irreversible evolution of the topology of magnetic domains, *J. Appl. Phys.*, 61:4188–4192.

126. Monceau, P.J.-M., and Levy, J.-C.S. (2012). Effects of deterministic and random discrete scale invariance on spin wave spectra, *Physica E*, 44:1697–1702.

127. Murray, C.A., and Wenk, R.A. (1989). Microscopic particle motions and topological defects in two-dimensional hexatics and dense fluids, *Phys. Rev. Lett.* 62:1643–6; Park, J.-M., Lubensky, T.C. (1996). Topological defectsd on fluctuating surfaces: general properties and the Kosterlitz Thouless transition, *Phys. Rev E*, 53:2648–52; Reichhardt, C., Olson-Reichhardt, C.J. (2003). *Phys. Rev. Lett.*, 90:095504–7.

128. Néel, L. (1940). Le ferromagnétisme et l'état métallique, *J. Phys. Rad.*, 1:242–250.

129. Neusser, S., and Grundler, D. (2009). Magnonics: Spin waves on the Nanoscale, *Adv. Mater.*, 21:2927–2932.

130. Nielsch, K., Müller, F., Li, A.-P., and Gösele, U.(2000). Uniform nickel deposition into ordered alumina pores by pulsed electrodeposition, *Adv. Mater.*, 12:582–586.

131. Nikitov, S., and Tailhades, T. (2001). Spin waves in magnetic structures-magnonic crystals, *J. Magn. Magn. Mater.*, 236:320–330.

132. Obukhov, S.P., and Rubinstein, M. (1990). Topological glass transition in entangled flux state, *Phys. Rev. Lett.*, 65:1279–1282.

133. Ohring, M. (2002). *Materials science of thin films* (2ed) Academic Press.

134. Pankhurst, Q.A., Connolly, J., Jones, S.K., and Dobson, J. (2003). Applications of magnetic nanoparticles in biomedicine, *J. Phys. D: Appl. Phys.*, 36:R167–R181.

135. Park, J.P., and Crowell, P.A. (2005). Interactions of spin waves with a magnetic vortex, *Phys. Rev. Lett.*, 95:167201–4.

136. Pascal, C., Pascal, J.L., Favier, F., Elidrissi Moubtassim, M.L., and Payen, C. (1998). Electrochemical synthesis for the control of $\gamma-Fe_2O_3$

nanoparticle size, morphology, microstructure and magnetic behavior, *Chem. Mater.*, 11:141–147.

137. Principi, G. (2001). High-energy ball milling of some intermetallics, *Hyperfine Interactions*, 134:53–67.

138. Puszkarski, H., Levy, J.-C.S., and Krawczyk, M. (2005). Localization properties of pure magnetostatic modes in a cubic nanograin, *Phys. Rev B* 71:14421–14432; Puszkarski, H., Krawczyk, M., Levy, J.-C.S. (2007). Purely dipolar versus exchange dipolarmodes in cylindrical nanorods, *J. Appl. Phys.*, 101:02436–02446.

139. Repain, V., Baudot, G., Ellmer, H., and Rousset, S. (2002). Two-dimensional long-range ordered growth of uniform cobalt nanostructures on a Au(111) vicinal template, *Europhys. Lett.*, 58:73036–73040.

140. Romero Vivas, J., Mamica, S., Krawcyk, M., and Kruglyak, V.V. (2012). Investigation of spin wave damping in three-dimensional magnonic crystals using the plane wave method, *Phys. Rev. B*, 86:144417–144430.

141. Rose, J.H., Smith, J.R., Guinea, F., and Ferrante, J. (1984). Universal features of the equation of state of metals, *Phys. Rev. B*, 29:2963–2970.

142. Ross, C.A. (2001). Patterned magnetic recording media, *Annu. Rev. Mater. Res.*, 31:203–35.

143. Rousset, S., Repain, V., Baudot, G., Ellmer, H., Garreau, Y., Etgens, V., Berroir, J.-M., Croset, B., Sotto, M., Zeppenfeld, P., Ferré, J., Jamet, J.P., Chappert, C., and Lecoeur, J. (2002). Self-ordering on crystal surfaces: fundamentals and applications, *Proc. ACSIN-6 Materials Science & Engineering B*, 96:169–177.

144. Ruderman, M.A., and Kittel, C. (1954). Indirect exchange coupling of nuclear magnetic moments by conduction electrons, *Phys. Rev.*, 96:99–104.

145. Sattler, K. (1986). Clusters of atoms, *Physica Scripta*, 13:93–99.

146. Sattler, K. (1993). C60 and beyond: from magic numbers to new materials, *J. Appl. Phys.*, 32:1428–1432.

147. Schaefer, D.W., Martin, J.E., Wiltzius, P., and Cannell, D.S. (1984). Fractal geometry of colloidal aggregates, *Phys. Rev. Lett.*, 52:2371–2374.

148. Schmidt, T., Heun, S., Slezak, J., Diaz, J., Prince, K.C., Lilienkamp, G., and Bauer, E. (1998). SPELEEM: Combining LEEM and spectroscopic imaging, *Surf. Rev. Lett.*, 5:1287–1296.

149. Schultheiss, H., Vogt, K., and Hillebrands, B. (2012). Direct observation of non linear four-magnon scattering in spin-wave microconduits, *Phys. Rev. B*, 86:054414–7.

150. Schwartz, L. (1950). *Théorie des distributions*, Hermann.

151. Sellmyer, D.J., Zheng, M., and Skomski, R. (2001). Magnetism of Fe, Co, and Ni, nanowires in self assembled arrays, *J. Phys. Condens. Matter*, 13:R433–R460.

152. Shechtman, D., Blech, I., Gratias, D., and Cahn, J. (1984). Metallic phase with long-range orientational order and no translational symmetry, *Phys. Rev. Lett.*, 53:1951–1954.

153. Sherrington, David, and Kirkpatrick, Scott (1975), "Solvable model of a spin-glass," *Phys. Rev. Let.*, 35(26): 1792–1796.

154. Skjeltorp, A.T. (1983). One- and two-dimensional crystallization of magnetic holes, *Phys. Rev. Lett.*, 51:2306–2309.

155. Skumryev, V., Stoyanov, S. Zhang, Y., Hadjipanayis, Givord, D., and Nogues, J. (2003). Beating the superparamagnetic limit with exchange bias, *Nature*, 423:850–853.

156. Soumare, Y., Garcia, C., Maurer, T., Chaboussant, G., Ott, F., Fievet, F., Piquemal, J.-Y., and Viau, G. (2009). Kinetically controlled synthesis of cobalt nanorods with high magnetic coercivity, *Adv. Funct. Mater.*, 19:1971–1977.

157. Spence, J.C.H. (1988). *Experimental high-resolution electron microscopy*, Oxford University Press, New York.

158. Srolovitz, D., Maeda, K., Vitek, V., and Egami, T. (1981). Structural defects in amorphous solids statistical analysis of a computer model, *Phil. Mag. A*, 44:847–866.

159. Stancil, D.D. (1993). *Theory of Magnetostatic Waves*, Springer Berlin.

160. Stein, G.D. (1985). Cluster beam sources: predictions and limitations of the nucleation theory, *Surf Sci.*, 156:44–56.

161. Steurer, W. (2004). Twenty years of structure research on quasicrystals. Part I. Pentagonal, octagonal, decagonal and dodecagonal quasicrystals, *Z. Kristallogr.*, 219:391–446.

162. Stevenson, S.E., Moutafis, C., Heldt, G., Chopdekar, R.V., Quitmann, C., Heyderman, L.J., and Raabe, J. (2013). Dynamic stabilization of nonequilibrium domain configurations in magnetic squares with high amplitude excitations, *Phys. Rev. B*, 87:054423-7.

163. Stillinger, F.H. (2001). Lattice sums and their phase diagramimplications for the classical Lennard-Jones model, *J. Chem. Phys.*, 115:5208–5212.

164. Stoner, E.C., and Wolhlfhart, E.P. (1948). A mechanism of magnetic hysteresis in heterogeneous alloys, *Phil. Trans. Roy. Soc. A*, 240:599–642.

165. Tacchi, S., Fin, S., Carlotti, G., Gubbiotti, G., Madami, M., Barturen, M., Marangolo, M., Eddrief, M., Bisero, D., Rettori, A., and Pini, M.G. (2014). Rotatable magnetic anisotropy in a Fe 0,8 Ga0,2 thin film with stripe domains: dynamics versus statics, *Phys. Rev. B*, 89:024411–5.

166. Tan, S., Ghazali, A., and Levy, J.-C.S. (1997). Monte-Carlo simulation of Pb/Cu (100) surface superstructures, *Surf. Sci.* 377–379:15–17; Monte-Carlo simulation of epitaxial growth on a (111)-layer with mismatch, *Surf. Sci.*, 377–379:997–1000.

167. Tan, S., Ghazali, A., and Levy, J.-C.S. (1997). Pb/Cu (100) Surface superstructures: Monte-Carlo and molecular dynamics simulations, *Surf. Sci.*, 392:163–172.

168. Tan, S., Lam, P.-M., and Levy, J.-C.S. (2002). Monte-Carlo investigations of vertical correlations in self-organized multilayer growth of islands, *Physica A*, 303:105–112.

169. Tartaj, P. del Puerto Morales, M., Veintemillas-Verdaguer, S., Gonzalez-Carreno, T., and Serna C.J. (2003). The preparation of magnetic nanoparticles for applications in biomedicine, *J. Phys. D: Appl. Phys.*, 36:R182–197.

170. Tchernyshyov, O., and Chen, G.W. (2005). Fractional vortices and composite domain walls in flat nanomagnets, *Phys. Rev. Lett.*, 95:197204–207.

171. Tersoff, J. (1986). New empirical model for the structural properties of silicon, *Phys. Rev. Lett.*, 56:632–635.

172. Tetienne, J.-P., Hingant, T., Rondin, L. Rohart, S., Thiaville, A., Roch, J.-F., and Jacques, V. (2013). Quantitative stray field imaging of a magnetic vortex core, *Phys. Rev. B*, 88:214408–5.

173. Thiele, A.A. (1973). Steady-state motion of magnetic domains *Phys. Rev. Lett.*, 30:230–233.

174. Thompson, S.H., Brown, G., Kuhnle, A.D., Rikvold, P.A., and Novotny, M.A. (2009). Resolution-dependent mechanisms for bimodal switching-time distributions in simulated Fe nanopillars, *Phys. Rev. B*, 79:024429, 1–9.

175. Todd, J., Merlin, R., Clarke, R., Moharty, K.M., and Axe, J.D. (1986). Synchrotron X-Ray study of a Fibonacci lattice, *Phys. Rev. Lett.*, 57:1157–1160.

176. Toneguzzo, P., Viau, G., Acher, O., Fievet-Vincent, F., and Fievet, F. (1998). Monodisperse ferromagnetic particles for microwave applications, *Anv. Mater.*, 10:1032–1035.

177. Toth, G., Denniston, C., and Yeomans, J.M. (2002). Hydrodynamics of topological defects in nematic liquid crystals, *Phys. Rev. Lett.*, 88:105504–7.

178. Tschernich, R.W. (1992). *Zeolites of the world*, Geosciences Press.

179. Vallejo-Fernandez, G., Whear, O., Roca, A. G., Hussain, S., Timmis, J., Patel, V., and O'Grady, K. (2013). Mechanisms of hyperthermia in magnetic nanoparticles, *J. Phys. D: Appl. Phys.*, 46:312001–6.

180. Van Oss, C.J., and Chaudhury, M.K. (1988). Interfacial Lifshitz-van der Waals and polar interactions in macroscopic systems, *Chem. Rev.*, 88:927–941

181. Vannimenus, J., and Toulouse, G. (1977). Theory of the frustration effect. II Ising spins on the square lattice, *J. Phys. (Paris)*, 10:537.

182. Vaterlaus, A., Stamm, C., Maier, U., Pini, M.G., Politi, P., and Pescia, D. (2000). Two-step disordering of perpendicularly magnetized ultrathin films, *Phys. Rev. Lett.*, 84:2247–2250.

183. Vedmedenko, E. Y., Ghazali, A., and Levy, J.-C.S. (1998). Magnetic structures of Ising and vector spins monolayers by Monte-Carlo simulations, *Surf. Sci.*, 402–404:391–395.

184. Vedmedenko, E.Y. (2009). Modulated Multipolar Structures on Geometrically Frustrated lattices. In J.-C.S. Levy (ed), *Nanostructures and their magnetic properties*, Transworld Research, Kerala (2007). *Competing interactions and Patterns in nanoworld*, Wiley-VCH Berlin.

185. Veenstra, B.R., Jonkman, H.T., and Kommandeur, J. (1994). Formation of mixed binary clusters in a supersonic molecular beam: a nice case of RKK kinetics, *J. Phys. Chem.*, 98:3538–3543.

186. Venables, J. (2000). *Introduction to Surface and Thin Film Processes*. Cambridge: Cambridge University Press.

187. Waintal, X., and Parcollet, O. (2005). Current induced spin torque in a nanomagnet, *Phys. Rev. Lett.*, 111:024206–9.

188. Wales, D.J., and Doye, J.P.K. (1997). Global optimization by basin-hopping and the lowest energy structures of Lennard-Jones clusters containing up, to 110 atoms, *J. Phys. Chem. A*, 101:5111–5118.

189. Wang, H., and Campbell, C.E. (2007). Spin dynamics of a magnetic antivortex. Micromagnetic simulations, *Phys. Rev. B*, 76:220407–220410.

190. Wang, T. et al. (2012). Femtosecond single-shot imaging of nanoscale ferromagnetic order in Co/Pd multilayers using resonant X-ray holography, *Phys. Rev. Lett.*, 108:267403–6.

191. Weiss, N., Cren, T., Epple, M., Rusponi, S., Baudot, G., Rohart, S., Tejeda, A., Repain, V., Rousset, S., Ohresser, P., Scheurer, F., Bencock, P., and Brune, H. (2005). Uniform magnetic properties for an ultrahigh-density lattice of nonininteracting Co nanostructures, *Phys. Rev. Lett.*, 95:157204–4.

192. Weiss, P. (1906). *CR Ac. Sci.*, 143:1137.

193. Weller, D., Sun, S., Murray, C., Folks, L., and Moser, A. (2001). MOKE spectra and ultra high density data storage perspective if FePt nanomagnet array, *Trans. Magn.*, 37:2185–2187.

194. Wilson, R. M. Ronen, S., and Bohn, J.L. (2009). Stability and excitations of a dipolar Bose-Einstein condensate with a vortex, *Phys. Rev. A*, 79:013621–8

195. Witten, T.A., and Sander, L.M. (1981). Diffusion limited aggregation: a kinetic critical problem, *Phys. Rev. Lett.*, 47:1400–1403.

196. Witten, T.A., and Sander, L.M. (1983). Diffusion-limited aggregation, *Phys. Rev. B*, 27:5686–97.

197. Yablonovitch, E. (2001). Photonic Crystals: Semiconductors of light, *Scientific American*, 285:47–55.

198. Yafet, Y., Kwo, J., and Gyorgy, E.M. (1986). Dipole-dipole interactions and two-dimensional magnetism, *Phys. Rev B*, 33:6519–22.

199. Yafet, Y. (1987). Ruderman-Kittel-Kasuya-Yosida range function of a one dimensional free-electron gas, *Phys. Rev. B*, 36:3948–3949.

200. Yao, N., Chen, H., Lin, H., Deng, C., and Zhang, X. (2008). Enrichment of peptides in serum by C8- functionalized magnetic nanoparticles for direct matrix assisted laser desorption/ionization time of flight mass spectrometry analysis, *J. Chrom. A*, 1186:93–101.

201. Yi, L., Büttner, G., Usadel, K.D., and Yao, K.-L. (1993). Quantum Heisenberg spinglass with Dzyaloshinskii-Moriya interactions, *Phys. Rev. B*, 147:254–261.

202. Yokozeli, A., and Stein, G.D. (1978). A metal cluster generator for gas-phase electron diffraction and its application to bismuth, lead and indium: Variation in microcrystal structure with size, *J. Appl. Phys.*, 49:2224–2230.

203. Yosida, K. (1998). *Theory of magnetism*, 2nd ed., Springer.

204. Zhou, X.W., Wadley, H.N.G., Johnson, R.A., Larson, D.J., Tabat, N., Cerezo, A., Petford-Long, A.K., Smith, G.D., Clifton, P.H., Martens, R.L., and Kelly, T.F. (2001). Atomic scale structure of sputtered metal multilayers, *Acta mater.*, 49:4005–4015.

Chapter 2

Vortex Lines in Three-Dimensional Magnetic Nanodots by Langevin Simulation

Ph. Depondt[a,b] and J.-C. S. Lévy[c]

[a] *Sorbonne Universités, UPMC Univ. Paris 06, UMR 7588, INSP, F-75005, Paris, France*
[b] *CNRS, UMR 7588, INSP, F-75005, Paris, France*
[c] *MPQ, UMR CNRS 7162, Université Denis Diderot, Paris, France*
depondt@insp.jussieu.fr

The complex nature of magnetic interactions makes it difficult to obtain useful analytic results, especially when dipole–dipole interactions are present. These are quite interesting as they tend to produce structures such as vortices, which in turn have their own dynamics. Langevin simulations constitute a choice tool for such studies. Such simulations of three-dimensional magnetic nanodots with increasing thickness are shown here using the Landau–Lifshitz equation with exchange and dipolar interactions. The dynamics of parallelepipedic samples was thus studied at finite temperature. The vortex-cores set up in lines that tend to present patterns as thickness is increased.

Magnetic Structures of 2D and 3D Nanoparticles: Properties and Applications
Edited by Jean-Claude Levy
Copyright © 2016 Pan Stanford Publishing Pte. Ltd.
ISBN 978-981-4613-67-5 (Hardcover), 978-981-4613-68-2 (eBook)
www.panstanford.com

2.1 Introduction

Magnetic nanodots have a huge potential for applications such as data storage and manipulation and more generally an apparently unlimited number of nanodevices of various types and kinds. They are also quite fascinating for physicists because of the specific properties of the magnetic interaction, and moreover, when present, those of the dipole–dipole interaction. These systems tend to produce magnetic vortices which not only have the potential for data storage, but also are extremely interesting objects in their own respect, giving rise to, e.g., gyromagnetic precession, that is, a rotational motion of the vortex cores around the center of the sample.

A tremendous amount of work on vortices in such systems was therefore accomplished over the years: Monte Carlo [1–3] or micromagnetic simulations [4], vortex dynamics computations [5], theoretical studies of the dipole–dipole interaction [6] and experiments [7]. Gyromagnetic precession of a vortex was obtained by micromagnetic Landau–Lifshitz–Gilbert simulations [8]. The influence of a vortex and its polarity on spin waves in a magnetic disk [9] were observed. Reference [10] presents experiments and theory on magnon scattering by a vortex. Phase diagrams for a selection of two-dimensional lattices [11] were also obtained. Vortices and antivortices were observed experimentally [12] in two-dimensional microdots. Analytical calculations of vortex gyrotropic motion and micromagnetic simulations were done [13, 14], while gyrotropic motion was also observed experimentally [15–18]. The above selection is far from extensive.

Beyond the thriving interest in vortices and vortex dynamics, the question of the dimensionality of the observed system is interesting: most samples are thin layers, but how thin? Usually thin layers are not single atomic layers, so they are not strictly two-dimensional systems; however, it is often not only convenient but justified nevertheless to consider them as such. To what extent is that valid? This chapter, therefore, presents simulations of a system of increasing thickness in order to assess the transition, so to say, between a flat two-dimensional system and a three-dimensional nanoparticle.

Langevin-style simulations are well adapted to such interests as they allow to observe the full history of a sample, starting from some adequately chosen initial conditions, and both extensive structural and dynamical information is naturally obtained. If reasonable attention is devoted to computational speed and accuracy, this can be performed at relatively moderate cost in terms of human and computer resources.

In Section 2.2, the simulation model and methods are presented and Section 2.3 gives some significant results.

2.2 The Simulation

2.2.1 *Model*

We start with a $n_x \times n_y \times n_z$ simple cubic lattice, to each site of which is attached one spin. Each spin can rotate independently in any direction. The spin equations of motion are given by the Landau–Lifshitz equation:

$$\dot{\mathbf{s}}_\ell(t) = -\mathbf{s}_\ell(t) \times \mathbf{H}_\ell(t) \quad \forall \ell \tag{2.1}$$

where $\mathbf{s}_\ell$ is the vector associated with spin ℓ and $\mathbf{H}_\ell$ is the local field felt by spin ℓ. This is written in reduced units in which all constants are set to 1. If $\mathbf{H}_\ell$ does not vary in time, Eq. (2.1) ensures a precession motion of spin ℓ around the local field because of the cross-product; however, $\mathbf{H}_\ell$ is not a constant as it is the sum of the effects of all the other spins on spin ℓ, which, as they move, alter the field on ℓ. An external field can also be applied, but it will not be the case in this chapter. Equation (2.1) has no damping term at this stage, so it differs from the Landau–Lifshitz–Gilbert equation, which is also quite often used.

The spin–spin interactions are of two types. Firstly, the exchange interaction is a nearest neighbour ferromagnetic interaction:

$$\mathbf{H}_\ell^{[exc]} = -J \sum_{\ell' \text{ neighbour } \ell} \mathbf{s}_\ell(t) \cdot \mathbf{s}_{\ell'}(t) \tag{2.2}$$

where $J = 1$ is a positive constant. Secondly, the dipole–dipole interaction is written as

$$\mathbf{H}_\ell^{[d]} = \sum_{\substack{\ell' \\ (\ell' \neq \ell)}} \frac{d}{|\mathbf{r}_{\ell\ell'}|^3} \left(3 \frac{\mathbf{s}_\ell \cdot \mathbf{r}_{\ell\ell'}}{|\mathbf{r}_{\ell\ell'}|^2} \mathbf{r}_{\ell\ell'} - \mathbf{s}_{\ell'} \right) \tag{2.3}$$

where d is the strength coefficient of the dipole–dipole interaction and $\mathbf{r}_{\ell\ell'}$ is the vector connecting the locations of spins ℓ and ℓ'. With $J = 1$, the quantity d controls the relative strengths of the dipolar and exchange interactions: the usual exchange length in our case therefore becomes $l_{exc} = \sqrt{1/d}$.

Then, of course, we have

$$\mathbf{H}_\ell = \mathbf{H}_\ell^{[exc]} + \mathbf{H}_\ell^{[d]}$$

2.2.2 Method

The game is now to integrate equation (2.1) for a system large enough to provide the information of interest, at a finite temperature, on a time scale that will reproduce the dynamics we are interested in, this in a wall clock time compatible with the life expectancy of a standard researcher.

2.2.2.1 Numerical integration of the precession motion

Quite a number of methods can be used to solve the Landau–Lifshitz equation (2.1): established procedures are described in, e.g., [24]. Slightly different methods are used here to take advantage of the specifics of spin dynamics. It should first be noticed that the basic explicit Euler integration scheme will be quite unreliable: for a time integration step δt on spin ℓ, we have

$$\mathbf{s}_\ell(t + \delta t) = \mathbf{s}_\ell(t) + \dot{\mathbf{s}}_\ell(t)\,\delta t$$

However,

$$
\begin{aligned}
|\mathbf{s}_\ell(t + \delta t)|^2 &= |\mathbf{s}_\ell(t) + \dot{\mathbf{s}}_\ell(t)\,\delta t|^2 \\
&= |\mathbf{s}_\ell(t)|^2 + 2\mathbf{s}_\ell(t) \cdot \dot{\mathbf{s}}_\ell(t)\,\delta t + |\dot{\mathbf{s}}_\ell(t)\,\delta t|^2 \\
&= |\mathbf{s}_\ell(t)|^2 + |\dot{\mathbf{s}}_\ell(t)\,\delta t|^2
\end{aligned}
$$

results in a systematic and unphysical increase of $|\mathbf{s}_\ell(t)|$ with time. This could of course be taken care of by rescaling $\mathbf{s}_\ell(t)$ at every step, which is however a bit clumsy. The fourth-order Runge–Kutta integration [34] scheme does four estimates of the derivatives, but it is easily shown that, although better than the Euler scheme, it also will generate a systematic drift of $|\mathbf{s}_\ell(t)|^2$.

These general integration methods are in a way rather blind-folded and therefore poorly adapted: this can be avoided by noting that the instantaneous motion of a spin is a precession motion around the instantaneous local field $\mathbf{H}_\ell(t)$, as mentionned in Section 2.2.1. This can be done by using the Rodrigues equation (see, e.g., [32]) that allows to compute the rotation of a vector around another arbitrary vector by simply applying an appropriate rotation matrix:

$$\mathbf{s}_\ell(t + \delta t) = \mathbf{R}_\ell(t)\, \mathbf{s}_\ell(t) \tag{2.4}$$

with

$$\mathbf{R}_\ell(t) = \begin{pmatrix} h_x^2 u + \cos\omega & h_x h_y u - h_z \sin\omega & h_x h_z u + h_y \sin\omega \\ h_x h_y u + h_z \sin\omega & h_y^2 u + \cos\omega & h_y h_z u - h_x \sin\omega \\ h_x h_z u - h_y \sin\omega & h_y h_z u + h_x \sin\omega & h_z^2 u + \cos\omega \end{pmatrix}$$

with $u = 1 - \cos\omega$. h_x, h_y, and h_z are the coordinates of the unit vector $\mathbf{h} = \mathbf{H}_\ell / |\mathbf{H}_\ell|$, which is parallel to the local field and the precession angle is $\omega = |\mathbf{H}_\ell| \delta t$. The spin length is thus intrinsically conserved to numerical precision, as it should following equation (2.1). This, as pointed out in [33], where a similar method is used, has the additional advantage that it does not require the precession angle ω to be small, but only for $\mathbf{H}_\ell$ to vary slowly with respect to the integration step δt: in the extreme case in which the local field remains constant with time, the time step δt can be arbitrarily large. Reference [33] implements, as more efficient, quaternions instead of a full rotation matrix, since a quaternion has only four components instead of nine: however, most of the computation time will be taken by the dipolar interactions and the improvement should be negligible.

However, straightforward integration via equation (2.4) turns out to be disappointing as it requires unreasonably short time steps to obtain acceptable precision. As usual in such problems, one attempts to evaluate the derivatives of the differential equations at both ends of the integration step δt, instead of only at the beginning thereof as is implicit in equation (2.4). The "improved Euler method" or "Heun's method" for ordinary differential equations turns out to be efficient for multiplicative noise problems [24] and allows much better precision at the cost of two field computations per time step instead of one: the time step can therefore be increased and the additional cost thus absorbed. We first compute as before a "predicted" set of spins at time $t + \delta t$, as in equation (2.4)

$$\mathbf{s}_\ell^{[p]}(t + \delta t) = \mathbf{R}_\ell(t)\, \mathbf{s}_\ell(t)$$

These predicted spins yield a new "predicted" field estimate $\mathbf{H}_\ell^{[p]}(\{\mathbf{s}_k^{[p]}(t + \delta t)\})$ at the end of the time step $t + \delta t$. The "corrected" field used to compute the corrected rotation matrix $\mathbf{R}_\ell^{[c]}(t + \frac{\delta t}{2})$ estimated in the middle of the integration step is now the average of the fields at both ends of the integration step

$$\mathbf{H}_\ell^{[c]}\left(t + \frac{\delta t}{2}\right) = \frac{\mathbf{H}_\ell(t) + \mathbf{H}_\ell^{[p]}(t + \delta t)}{2}$$

and

$$\mathbf{s}_\ell(t + \delta t) = \mathbf{R}_\ell^{[c]}\left(t + \frac{\delta t}{2}\right) \mathbf{s}_\ell(t)$$

We thus have now both a quick and stable integration method of the Landau–Lifshitz equation (2.1).

2.2.2.2 The dipole–dipole interaction

While the exchange interaction is a nearest-neighbour interaction, and therefore computed at a reasonable cost, the dipole–dipole interaction decreases slowly as r^{-3} and will involve summing over the whole sample for each spin. For small samples and short times, brute force summation can be a rational solution, but for large, three-dimensional samples which are bound to contain many spins, this is not possible.

Using the convolution theorem and fast Fourier transforms [24] greatly improves speed. It the case of a 3D cubic lattice, the position of spin ℓ writes $\mathbf{r}_\ell = (i_\ell a,\, j_\ell a,\, k_\ell a)$, where a is the lattice constant, and i_ℓ, j_ℓ, and k_ℓ are integers. The dipolar field on spin ℓ thus, again, is written as

$$\mathbf{H}_\ell^{[d]} = d \sum_{\substack{\ell' \\ (\ell' \neq \ell)}} \frac{1}{|\mathbf{r}_{\ell\ell'}|^3}$$

$$\times \left(\mathbf{s}_{\ell'} - 3a \frac{s_{x\ell'}(i_{\ell'} - i_\ell) + s_{y\ell'}(j_{\ell'} - j_\ell) + s_{z\ell'}(k_{\ell'} - k_\ell)}{|\mathbf{r}_{\ell\ell'}|^2} \mathbf{r}_{\ell\ell'} \right)$$

where $|\mathbf{r}_{\ell\ell'}| = a\sqrt{(i_{\ell'} - i_{\ell})^2 + (j_{\ell'} - j_{\ell})^2 + (k_{\ell'} - k_{\ell})^2}$. This can be seen as the convolution of $\mathbf{s}_\ell$ (or equivalently $\mathbf{s}_{i_\ell, j_\ell, k_\ell}$) with the following matrix:

$$\mathbf{D}_{i_\ell, j_\ell, k_\ell} = \frac{d}{a^3 \left(i_\ell^2 + j_\ell^2 + k_\ell^2\right)^{\frac{3}{2}}}$$

$$\times \begin{pmatrix} 1 - 3\frac{i_\ell^2}{i_\ell^2 + j_\ell^2 + k_\ell^2} & -3\frac{i_\ell j_\ell}{i_\ell^2 + j_\ell^2 + k_\ell^2} & -3\frac{i_\ell k_\ell}{i_\ell^2 + j_\ell^2 + k_\ell^2} \\ -3\frac{i_\ell j_\ell}{i_\ell^2 + j_\ell^2 + k_\ell^2} & 1 - 3\frac{j_\ell^2}{i_\ell^2 + j_\ell^2 + k_\ell^2} & -3\frac{j_\ell k_\ell}{i_\ell^2 + j_\ell^2 + k_\ell^2} \\ -3\frac{i_\ell k_\ell}{i_\ell^2 + j_\ell^2 + k_\ell^2} & -3\frac{j_\ell k_\ell}{i_\ell^2 + j_\ell^2 + k_\ell^2} & 1 - 3\frac{k_\ell^2}{i_\ell^2 + j_\ell^2 + k_\ell^2} \end{pmatrix}$$

This matrix and its three-dimensional Fourier transform can be computed beforehand, once and for all.

The convolution itself thus uses fast Fourier transforms and zero padding for a finite sample [34]. If necessary, this is the part of the simulation that should be parallelized, which can be done at a moderate human time cost. Since the spin distribution is restricted to a discrete lattice, the numerical discrete Fourier transform introduces none of the usual approximations: this is an exact procedure.

2.2.2.3 Finite temperature simulations

In order to introduce temperature, a Langevin-type simulation will mimic thermal motion via a random force, the work of which is absorbed by damping: the temperature will result from the balance between these two terms. This can be easily adapted to magnetic motion [35, 36].

The Langevin "force" appears as a random or thermal field $\mathbf{H}_\ell^{[th]}$ with normal distribution, zero mean and variance

$$\left\langle \left(\mathbf{H}_\ell^{[th]}\right)^2 \right\rangle = 2\alpha k_B T \tag{2.5}$$

where α is a damping coefficient, k_B Boltzmann's constant (taken to equal 1 in reduced units) and T the temperature. The thermal field is added to $\mathbf{H}_\ell$ because spin $\mathbf{s}_\ell$ "feels" the thermal noise only via the local field. The equations of motion (2.1) are thus modified:

$$\dot{\mathbf{s}}_\ell = -\frac{\mathbf{s}_\ell \times \mathbf{H}_\ell' + \alpha \, \mathbf{s}_\ell \times (\mathbf{s}_\ell \times \mathbf{H}_\ell')}{1 + \alpha^2} \tag{2.6}$$

with

$$\mathbf{H}'_\ell = \mathbf{H}_\ell + \mathbf{H}^{[th]}_\ell$$

This is easily adapted to the Landau–Lifshitz equation in the Rodrigues formulation, as (2.6) can be rewritten as

$$\dot{\mathbf{s}}_\ell = -\mathbf{s}_\ell \times \mathbf{H}^{[eff]}_\ell \tag{2.7}$$

with

$$\mathbf{H}^{[eff]}_\ell = \frac{\mathbf{H}'_\ell + \alpha\,\mathbf{s}_\ell \times \mathbf{H}'_\ell}{1 + \alpha^2}$$

The damping parameter α controls the speed at which the system relaxes toward equilibrium after a temperature change but, if chosen small enough, should have no other effect. Indeed Langevin simulations generally have a reputation for being unreliable for dynamic properties as one must introduce a characteristic relaxation time that is connected with the damping coefficient. This is only partially true, as a given vibrational peak, assuming it is a delta function of frequency, will be broadened and slightly shifted by damping, but if the damping coefficient is chosen small enough, so that the broadening remains significantly less than the required frequency resolution, the obtained spectra can be considered reliable.

Simulation parameters are summarized in Table 2.1.

2.2.2.4 Seeking vortices

The vortex cores can be detected by simply computing the curl $\nabla \times \mathbf{s}(\mathbf{r})$; however, the accepted procedure is slightly different: the orientations φ_i, projected on the corresponding plane, of the four spins that surround a given square in the underlying cubic lattice (Fig. 2.1) are determined and their differences by rotation along the

Table 2.1 Simulation parameters

Time step	Exchange constant J	Dipole–dipole strength d
0.005	1	0.02 to 0.1
Temperature	Damping	System size
0.001	0.005	$64 \times 64 \times 4$ to $64 \times 64 \times 64$

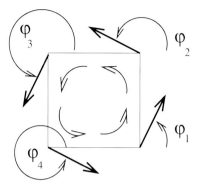

Figure 2.1 Vortex core detection: one square of the underlying lattice is selected. A counterclockwise rotation around that square is done and the sum of the four differences $\varphi_{i+1} - \varphi_i$ computed (Eq. (2.8)). If that sum equals 2π a vortex core is detected.

four edges of the square yield

$$\delta\varphi_i = \varphi_{i+1} - \varphi_i, \quad \delta\varphi_i \in [-\pi, \pi], \quad i \in [1, 4] \qquad (2.8)$$

The sum

$$\varphi_t = \sum_{i=1}^{4} \delta\varphi_i$$

yields the overall rotation about the square: if that rotation equals 2π, a vortex is detected on that square; if it equals -2π, then it is an antivortex; otherwise $\varphi_t = 0$. The vortex axis will be normal to the plane of the square.

2.3 Results

The system size was set to $n_x = n_y = 64$ and $n_z = 4, 8, 16, 32$ and 64 in succession, in order to compare the behaviors of a thin layer ($n_z = 4$) with that of a bulk sample ($n_z = 64$) for which the three directions x, y, and z are equivalent. The temperature was set to a relatively low value $T = 0.001$ (by comparison, with the same units the Kosterlitz–Thouless transition in a two-dimensional system [28] is for $T = 0.7$). The dipolar to exchange ratio d was varied from $d = 0.01$ to $d = 0.12$.

2.3.1 Initial Conditions: One Central Vortex Line, Normal to the (xOy) Plane

Initial conditions were set with one vortex line normal to the (xOy) plane as this is one of the stable states of the two-dimensional system [26]: a relaxation time is necessary as the spins are initially set in-plane, while the competition between exchange and dipolar

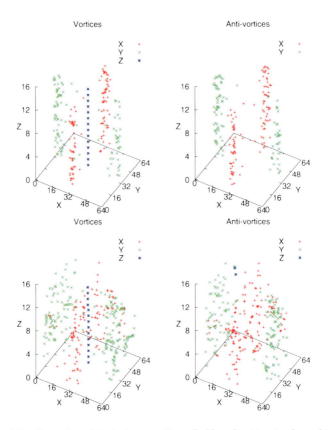

Figure 2.2 Snapshot of typical vortex lines (left) and antivortex lines (right) after relaxation for a $64 \times 64 \times 16$ sample and dipolar constant $d = 0.02$ (top) and 0.12 (bottom). The X, Y, and Z axes number the sites of the underlying cubic lattice. The blue stars labeled "Z" represent vortex cores or antivortex cores in the (xOy) plane, the red and green crosses, "X" and "Y" respectively, (anti)vortex cores in the (yOz) and (xOz) planes, respectively. The ratio between the vertical axis and the two others is not respected.

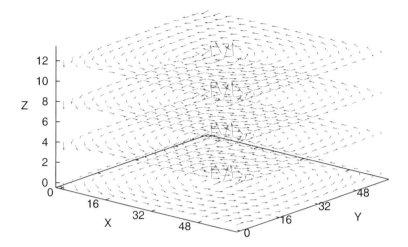

Figure 2.3 Snapshot of the spin configuration corresponding to Fig. 2.2. Four (xOy) planes out of 16 and 16 spins out of 64 in the x and y directions were selected for clarity.

interactions requires that the spins stick out of plane at the location of the vortex core. The transition from a "quasi-two-dimensional" system ($64 \times 64 \times 16$) to a fully volumic ($64 \times 64 \times 64$), cubic in fact, system is shown in Figs. 2.2–2.10, at the end of the relaxation period.

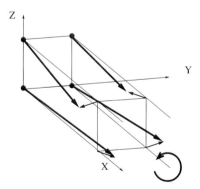

Figure 2.4 A spin configuration showing how a small motion of spins (long thick arrows) which are essentially parallel to the x axis can give rise to an X-type vortex (thick round arrow) through their projections in the yOz plane (short arrows).

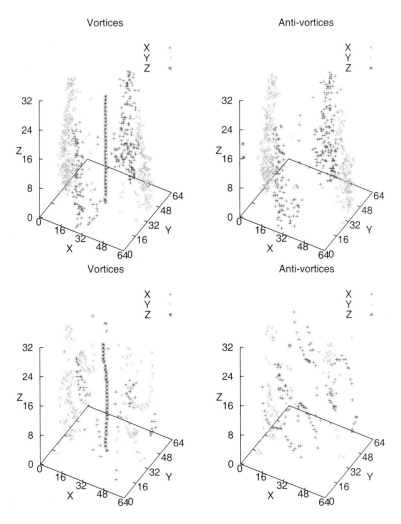

Figure 2.5 Typical vortex and antivortex lines for a 64 × 64 × 32 sample and $d = 0.02$ (top) and 0.06 (bottom). See caption of Fig. 2.2.

The configurations thus obtained are strikingly diverse:

- the "thin" slab (64×64×16) (aspect ratio = 1/4) retains the one-vortex line initial configuration for all the tested values of d (Fig. 2.2). In addition, a number of "X" and "Y" vortices and antivortices are present, which is a new feature with

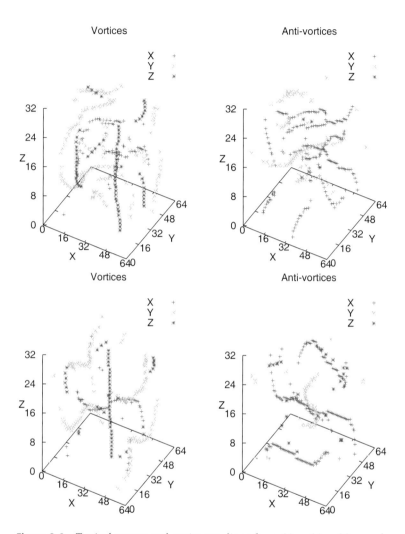

Figure 2.6 Typical vortex and antivortex lines for a $64 \times 64 \times 32$ sample and $d = 0.1$ (top) and 0.12 (bottom). See caption of Fig. 2.2.

respect to 2D systems [27]. This is a bit misleading, as the corresponding spin configuration (Fig. 2.3) only shows the central vortex line: small fluctuations around the average spin orientation can give rise to vortex (or antivortex) detection: Figure 2.4 shows four spins which are essentially

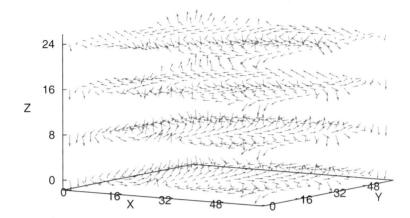

Figure 2.7 Snapshot of the spin configuration corresponding to Fig. 2.6 (top): 32 layers and $d = 1.0$. Four (xOy) planes out of 32 and 16 spins out of 64 in the x and y directions were selected for clarity.

parallel to the x axis, but their projections on the yOz plane generate a vortex that will be detected as an X-type vortex core. Indeed, these "clouds" of vortex cores move with time and can easily be attributed to the spin precession motion in connection with the eigenmodes (that would be magnons in an infinite system) of the system that have an effective wavelength of the order of the sample size. We can therefore distinguish between two different populations of vortex (and antivortex) cores: those organized as a line that denote the overall structure of the sample and those that are the signature of the system's vibrational modes. Thinner samples ($n_z = 4$ and 8) yield quite similar results.

- the "thick" slab ($64 \times 64 \times 32$) (aspect ratio $= 1/2$) shows a one-vortex state, similar to those for the thin slab, for small d ($d < 0.1$) but the vortex lines can be bent (Fig. 2.5), in a way reminiscent of what is observed in superfluids [37]. For the higher dipolar strengths $d = 0.1$ and $d = 0.12$ (Fig. 2.6) new patterns appear, denoting the onset of "3D behavior" with several vortex lines, some parallel to the z axis, some not. Observation of the actual spin configuration (Fig. 2.7) reveals an extremely complex situation, despite the fact that

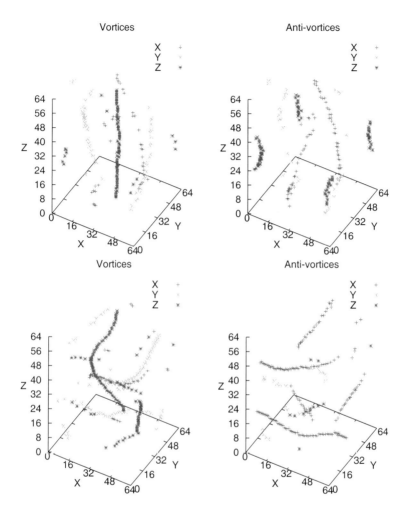

Figure 2.8 Typical vortex and antivortex lines for a 64 × 64 × 64 sample and $d = 0.02$ and 0.04. See caption of Fig. 2.2.

only a drastic selection of spins had to be made for minimal clarity...
- the cubic sample (64 × 64 × 64) (aspect ratio = 1) shows a great variety of behaviors. For $d = 0.02$ in Fig. 2.8 (top), we obtain one central vortex line with other vortex lines at the four corners of the slab. For $d = 0.04$ in Fig. 2.8

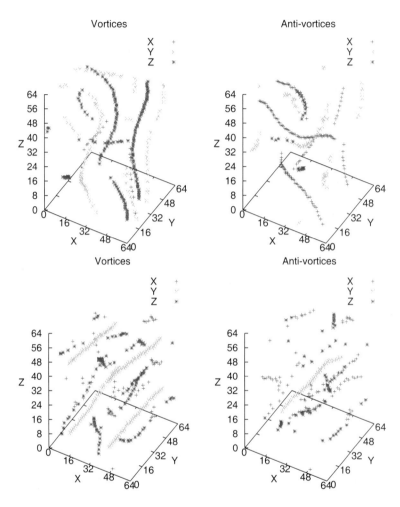

Figure 2.9 Typical vortex and antivortex lines for a 64 × 64 × 64 sample and $d = 0.06$ and 0.08. See caption of Fig. 2.2.

(bottom), twisted vortex lines without obvious order arise. For $d = 0.06$, Fig. 2.9 (top), some ordering takes place. For $d = 0.08$, Fig. 2.9 (bottom), $d = 01$ and $d = 0.12$, Fig. 2.10, the vortex lines tend to straighten up and form an array of parallel lines.

Results | 117

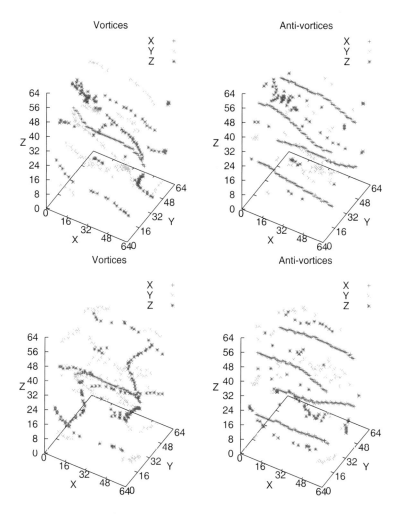

Figure 2.10 Typical vortex and antivortex lines for a 64 × 64 × 64 sample and $d = 0.1$ and 0.12. See caption of Fig. 2.2.

2.3.2 Ground State?

Confronted with this wealth of configurations, the reality of all this can be questioned. An attempt to sort things out was made by comparing the energies obtained after relaxation with initial conditions set either with one vortex line normal to the (xOy) plane

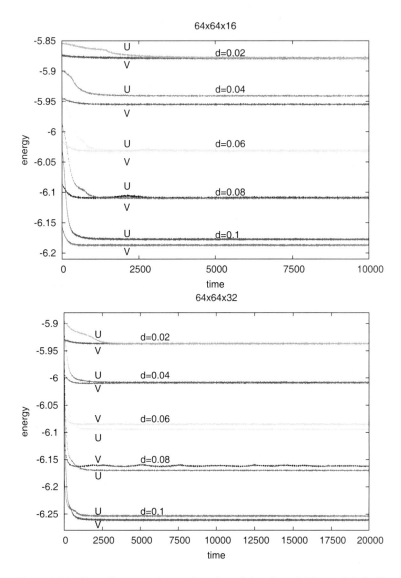

Figure 2.11 Energies per spin as a function of time for a 16-layer slab (left) and for a 32-layer slab (right) for $d \in [0.02, 0.1]$ and for two sets of initial conditions: 'U' stands for a uniform distribution of spin orientations, 'V' for a single vertical vortex core line.

as before, or with uniform magnetization over the sample, that is, with no vortex. In Fig. 2.11, for the thin slab (64 × 64 × 16) and for $d = 0.02$, after hesitating a bit ($t \simeq 2000$), the "uniform" energy converges toward the "vortex" energy, meaning the system spontaneously goes to the vortex state. For $d = 0.04$, the uniform energy remains higher than the vortex energy, meaning the uniform configuration is a metastable state (see, e.g., [26]). This goes on for the other values of d: the overall conclusion is that the vortex state is more stable than the uniform for the thin slab.

For the thick slab (64 × 64 × 32), the situation is more complex: for $d = 0.02$ and 0.04, the behavior is similar to the previous case, but for $d = 0.06$, 0.08 and 0.1 the uniform configuration yields lower energy! This is quite unusual as the dipolar interaction should favor the creation of vortices. The answer resides in Fig. 2.12, which shows the final configuration of the simulation started in the uniform configuration: we now have four parallel lines. This is consistent with Fig. 2.10, which shows arrays of parallel vortex lines for the bulk sample.

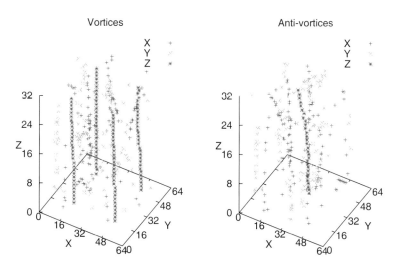

Figure 2.12 Typical vortex and antivortex lines for a 64 × 64 × 32 sample and $d = 0.1$ with initial conditions set with a uniform distribution of spin orientations. See caption of Fig. 2.2.

2.4 Concluding Remarks

Despite the simple geometry of our simulated system (cubic lattice, parallelepipedic samples) the competition between exchange and dipole–dipole interactions in a three-dimensional system generates a wealth of microscopic configurations with astounding variety. In two dimensions, although relatively complex from a dynamical point of view [25–27, 30, 31], the configurations roughly boil down to uniform, one central vortex, and an array of vortices on a surface. Now, the object we are looking at, the vortex core line, is much more complex: it can twist, form rings and arrays. The art of choosing initial conditions that are well suited to what is sought, therefore, is more crucial than ever.

References

1. Booth I., Isaac A. B., Whitehead J. P., and De'Bell K., *Phys. Rev. Lett.*, **75**, 951 (1995).
2. Sasaki J. and Matsubara F., *J. Phys. Soc. Japan*, **66**, 2138 (1997).
3. Vedmedenko E. Y., Ghazali A. and Lévy J.-C. S., *Surf. Sci.*, **402**, 391 (1998).
4. Cowburn R. P. and Welland M. E., *Phys. Rev. B*, **58**, 9217 (1998) and Cowburn R. P., Koltsov D. K., Adeyeye A. O., Welland M. E. and Tricker D. M., *Phys. Rev. Lett.*, **83**, 1042 (1999).
5. Ivanov B. A., Schnitzer H. J., Mertens F. G. and Wysin G. M., *Phys. Rev. B*, **58**, 8464 (1998).
6. Lévy J.-C. S., *Phys. Rev. B*, **63**, 104409 (2001).
7. Vaterlaus A., Stamm C., Maier U., Pini M. G., Politi P., and Pescia D., *Phys. Rev. Lett.*, **84**, 2247 (2000).
8. Usov N. A. and Kurkina L. G., *J. Magnetism and Magn. Mat.*, **242**, 1005 (2002).
9. Parkand J. P. and Crowell P. A., *Phys. Rev. Lett.*, **95**, 167201 (2005).
10. Aliev F. G., Sierra J. F., Awad A. A., Kakazei G. N., Han D.-S., Kim S.-K., Metlushko V., Ilic B. and Guslienko K. Y., *Phys. Rev. B*, **79**, 174433 (2009).
11. Rocha J. C., Coura P. Z., Leonel S. A., Dias R. A. and Costa B. V., *J. Appl. Phys.*, **107**, 053903 (2010).
12. Li J. and Rau C., *Phys. Rev. Lett.*, **97**, 107201 (2006).

13. Guslienko K. Yu., Ivanov B. A., Novosad V., Otani Y., Shima H. and Fukamichi K., *J. Appl. Phys.*, **91**, 8037 (2002), Guslienko K. Y., Heredero R. H. and Chubylako-Fesenko O., *Phys. Rev. B*, **82**, 014402 (2010), and Guslienko K. Y. and Slavin A. N., *J. Magnetism and Magn. Materials*, **323**, (2011) 2418, Guslienko K. Yu., *J. Appl. Phys.*, **89**, 022510 (2006).

14. Ivanov B. A. and Zaspel C. E., *J. Appl. Phys.*, **95**, 7444 (2004).

15. Park J. P., Eames P., Engebretson D. M., Berezovsky J. and Crowell P. A., *Phys. Rev. B*, **67**, 020403 (2003).

16. Puzic A., Van Waeyenberge B., Chou K. W., Fischer P., Stoll H., Schütz G., Tyliszczak T., Rott K., Brückl H., Reiss G., Neudecker I., Haug T., Buess M. and Back C. H., *J. Appl. Phys.*, **97**, 10E704 (2005), and Van Waeyenberge B., Puzic A., Stoll H., Chou K. W., Tyliszczak T., Hertel R., Fähnle, Brückl H., Rott K., Reiss G., Neudecker I., Weiss D., Back C. H and Schütz G., *Nature*, **444**, 461 (2006).

17. Kamionka T., Martens M., Drews A., Krüger B., Albrecht O. and Meier G., *Phys. Rev. B*, **83**, 224424 (2011).

18. Toscano D., Leonel S. A, Dias R. A., Coura P. Z., Rocha J. C. S. and Costa B. V., *J. Appl. Phys.*, **109**, 014301 (2011).

19. Villegas J. E., Li C. P. and Schuller I. K., *Phys. Rev. Lett.*, **99**, 227001 (2007).

20. Liu Y. and Du A. *J. Magnetism and Magn. Mat.*, **321**, 3493 (2009).

21. Tanase M., Petford-Long A. K., Heinonen O., Buchanan K. S., Sort J. and Nogués J., *Phys. Rev. B*, **79**, 014436 (2009).

22. Curcic M., Van Waeyenberge B., Vansteenkiste A., Weigand M., Sackmann V., Stoll H., Fähnle M., Tyliszczak T., Woltersdorf G., Back C. H. and Schütz G., *Phys. Rev. Lett.*, **101**, 197204 (2008), Kammerer M., Weigand M., Curcic M., Noske M., Sproll M., Vansteenkiste A., Van Waeyenberge B., Stoll H., Woltersdorf G., Back C. H., and Schütz G., *Nature Comm.*, 2:279 DOI: 10.1038/ncomms1277 (2011), and Kammerer M., Stoll H., Noske M., Sproll M., Weigand M., Illg C., Woltersdorf G., Fähnle M., Back C., and Schütz G., *Phys. Rev. B*, **86**, 134426 (2012).

23. Guslienko K. Y., Lee K.-S. and Kim S.-K., *Phys. Rev. Lett.*, **100**, 027203 (2008).

24. Nowak U. in *Annual Review of computational Physics IX*, Stauffer D. ed., World Scientific (2001) p. 105.

25. Depondt Ph. and Mertens F. G., *J. Phys.: Condens. Matter*, **21**, 336005 (2009).

26. Depondt Ph., Lévy J.-C. S. and Mertens F. G., *Phys. Letters A*, **375**, (2011) 628–632.

27. Depondt Ph. and Lévy J.-C. S., *Phys. Letters A*, **375**, (2011) 4085–4090.

28. Kosterlitz J. M. and Thouless J. D., *J. Phys. C: Solid State Phys.*, **6**, 1181–203 (1973), and Kosterlitz J. M., *J. Phys. C: Solid State Phys.*, **7**, 1146–1160 (1974).

29. Cuccoli A., Tognetti V. and Vaia R., *Phys. Rev. B*, **52**, 10221 (1995).

30. Mamica S., Lévy J.-C. S., Depondt Ph. and Krawczyk M., *J. Nanopart. Res.*, **13**, 6075 (2011).

31. Mamica S., Lévy J.-C. S., Krawczyk M. and Depondt Ph., *J. Appl. Phys.*, **112**, 043901 (2012).

32. Weisstein E. W. et al., *MathWorld* http://mathworld.wolfram.com/RodriguesRotationFormula.html

33. Visscher P. B. and Feng Xuebing, *Phys. Rev. B*, **65**, 104412 (2002).

34. Press W. H., Teukolsky S. A., Vetterling W. T., Flannery B. P., *Numerical Recipes in Fortran*, Cambridge University Press (1992).

35. Chubykalo O., Hannay J. D., Wongsam M., Chantrell R. W. and Gonzales J. M., *Phys. Rev. B*, **65**, 184428 (2002).

36. Kamppeter T., Mertens F. G., Moro E., Sanchez A. and Bishop A. R., *Phys. Rev. B*, **59**, 17, 11349 (1999).

37. Eltsov V. B., Finne A. P., Hänninen R., Kopu J., Krusius M., Tsubota M. and Thuneberg E. V., *Phys. Rev. Lett.*, **96**, 215302 (2006).

38. Garcia-Ripoll, J. J., Perez-Garcia, V. M., *Phys. Rev. A*, **64**, 053611 (2001).

39. Obukhov, S. P., Rubinstein, M., *Phys. Rev. Lett.*, **65**, 1279 (1990).

Chapter 3

In-Plane Magnetic Vortices in Two-Dimensional Nanodots

Sławomir Mamica

Faculty of Physics, Adam Mickiewicz University in Poznan,
ul. Umultowska 85, Poznan, Poland
mamica@amu.edu.pl

3.1 Introduction

Among others, two types of interaction are most common in magnetic systems, namely dipolar and exchange interactions, and these two differ significantly from each other. Dipolar interactions are long-range but weak while exchange ones are rather strong but very short-range, usually restricted to nearest or next nearest neighboring sites in the crystal lattice. Another difference rises from the preferred configuration of interacting magnetic moments. Exchange interactions, in the simplest case,[a] favor parallel or anti-parallel arrangement depending on the sign of the exchange integral

---

[a]There are several mechanisms responsible for exchange interactions (White, 2006) but for our purpose it is not necessary to distinguish them. Since our model is valid for ferromagnetic coupling we can assume just direct exchange.

Magnetic Structures of 2D and 3D Nanoparticles: Properties and Applications
Edited by Jean-Claude Levy
Copyright © 2016 Pan Stanford Publishing Pte. Ltd.
ISBN 978-981-4613-67-5 (Hardcover), 978-981-4613-68-2 (eBook)
www.panstanford.com

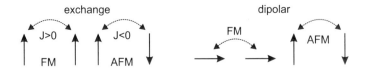

Figure 3.1 Competing interactions: Different preferred configuration of magnetic moments for exchange and dipolar interactions. In first case the sign of the exchange integral is crucial while in the second one the alignment of the moments.

positive (ferromagnetic, FM) or negative (antiferromagnetic, AFM), respectively (Morrish, 2001). In the case of dipolar interactions the situation is a bit more complex. If magnetic moments are alongside one another dipolar interactions are AFM but for magnetic moments one after another they are FM (see Fig. 3.1). The concurrence of these two interactions results in several specific effects in magnetic systems from which one of the best known is the existence of magnetic domains in ferromagnets. It is also responsible for a variety of possible magnetic structures of nanoparticles.

The circumstance of the interactions leads to novel effects in a number of systems, such as carbon nanotubes (Kotakoski et al., 2006), graphene (Kłos and Zozoulenko, 2010), elastic systems (Dubus et al., 2006), and the others. In magnetic nanomaterials this leads to the surface and subsurface localization in ultrathin films (Mamica et al., 1998, 2000; Puszkarski et al., 1998), causes the splitting of the spectrum into subbands in patterned thin films (Krawczyk et al., 2011; Pal et al., 2012), and is responsible for the complete band gap opening in magnonic crystals (Krawczyk et al., 2002; Mamica et al., 2012a,b). Additionally, the competition between exchange and dipolar interactions, each favoring a different magnetic configuration, results in specific phenomena. In small magnetic dots this competition effects in several possible stable (or meta stable) magnetic configurations (Bader, 2006). If a thickness of the dot is small enough with respect to its diameter one of these configurations is a vortex state (Chung et al., 2010; Metlov and Lee, 2008; Zhang et al., 2008).

The basic excitations in magnetic systems are called magnons in purely quantum approach, or spin waves if they are treated in semiclassical way. The last means spins act like classical vectors, i.e., they can be oriented at any angle with respect to the external (effective) field. In this picture we have magnetic moments tilted a bit out of the direction of the magnetic field so, as usually in such case, they have to precess. If this precession is collective, we call it spin wave. (The reader can find a very good explanation of magnons and spin waves in the book by Morrish (Morrish, 2001).) The precession of individual spin is elliptical,[a] in general; thus we have two amplitudes of the precession. If the amplitudes are small in comparison with the length of the spin, we can assume that its component along external field is approximately equal to its length and does not change in time. Finally, we have two components of the spin: static one, parallel to the external (effective) field, with fixed length, and dynamical one, rotating in the plane perpendicular to the field and changing its length except the special case of the circular precession.

In this chapter, we will speak about the stable magnetic configuration and spin-wave excitations in small two-dimensional (2D) magnetic dots. Both are highly affected by the competition of the exchange and dipolar interactions (Mamica et al., 2014).

In the vortex configuration the static components of the magnetic moments in the plane of the dot form a closed system (Fig. 3.2): a closure domain system (Landau state) in square dots and circular magnetization in circular dots (Park et al., 2003). As shown by simulations performed for square dots, the arrangement of the magnetic moments near the center of the vortex is circular also in square systems (Fig. 3.2c, see also (Depondt et al., 2013)). A detailed analysis leads to the observation that near the center of the vortex the magnetization is circular, distinct domains forming further outwards. Two effects compete here. On the one hand, the tendency to minimize the magnetic charges at the surface causes the magnetization to follow the edges of the dot, which in square dots results in the Landau state. On the other hand, the center of the vortex tends towards the circular magnetization as a result of minimizing the (local) exchange interaction at the meeting point of four domains. In large square dots, the center of the vortex is

---

[a]The origin of this ellipticity is an anisotropy, which could results, e.g., from the symmetry breaking by the surfaces (Lévy, 1981).

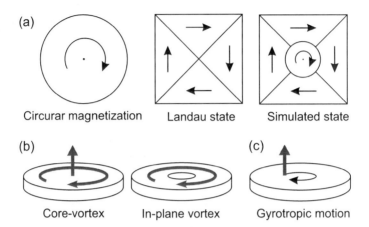

Figure 3.2 Schematic picture of different types of vortices: (a) Circular magnetization in circular dot, Landau state in square dot, and simulated state in square dot. The latest has circular magnetization around the center and Landau state near the edges. (b) Vortex with core vs. in-plane vortex. (c) Gyrotropic motion of the vortex core: If the core of the vortex is shifted from the dot center, it precesses around.

a relatively small region and the domain magnetization prevails in the major part of the dot. In contrast, in small dots only in minor corner regions the circular configuration fails to fit the geometry of the system.

Two types of vortices are observed in both cases: in-plane vortices and vortices with a core (Fig. 3.2). The latter have a nonzero out-of-plane component of static magnetization in the close vicinity of the center of the vortex (which does not necessarily coincide with the center of the dot, though is close to it in stable vortices); this nonzero out-of-plane magnetization region is referred to as a core (Shinjo et al., 2000). In-plane vortices, on the other hand, have no core: The static magnetization lies in the plane of the dot throughout its volume (Li et al., 2001).

Two quantities characterize the vortex state: the chirality, which defines the orientation of the magnetization in the plane of the dot clockwise or anticlockwise, and the polarity, defining the orientation of the core up or down. Each of these characteristic quantities has two possible values, ± 1 (obviously, with the exception of the in-plane vortex, in which the polarity is zero); thus, each represents a potential bit of information (Luo et al., 2008; Mani et al., 2004; Zhu et al., 2000). The vortex chirality and core polarity can be switched by external magnetic field (Jain and Adeyeye, 2008), electric current (Vavassori et al., 2007), or microwave radiation (Waeyenberge et al., 2006). A key role in magnetization switching is played by spin waves (Bauer et al., 2014; Kammerer et al., 2011). Moreover, they have a significant influence on the stability of the magnetic configuration (Mamica et al., 2011; Mozaffari and Esfarjani, 2007), also in systems smaller than the characteristic exchange length (Rohart et al., 2010). The applications of magnetic nanodots include also microwave generation using the gyrotropic motion of the core (Guslienko, 2012; Pribiag et al., 2007), and the generation of higher harmonics of microwave radiation by means of spin-wave excitations (Demidov et al., 2011). Circular rings were used as a building-blocs in magnetic logic gates (Bowden and Gibson, 2010; Rahm et al., 2005), and domain walls existing in square rings for single magnetic nanoparticle sensing and trapping (Donolato et al., 2009; Lagae et al., 2007; Miller et al., 2002; Vavassori et al., 2008).

The intention of this chapter is to describe quite simple and very effective method to explore the magnetic configuration and the spin-wave spectrum in nanodots without simulations. As an example, we will study the in-plane vortex state: its stability and spin-wave normal modes versus the dipolar-to-exchange interaction ratio, the size and shape of the system, and the symmetry of the lattice from which it has been cut.

3.2 Dynamical Matrix Method

Before we go to the calculation method let us introduce the model of the system under consideration. Our dots are cut out from a 2D discrete lattice with elementary magnetic moments in the lattice sites (let us call them spins, but the reader should keep in mind the difference). A system thus defined is naturally discrete, without recourse to artificial discretization applied to continuous systems, e.g., in micromagnetic simulations (Usov and Peschany, 1993). In the latter, we will use circular dots as an example so in Fig. 3.3 the model

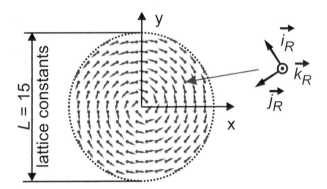

Figure 3.3 The model: Schematic plot of circular magnetic dot based on a 2D square lattice. Elementary magnetic moments (represented by the arrows) are arranged in all the lattice sites within the dot and form an in-plane vortex (circular magnetization). The size L is defined as the number of lattice constants in the diameter of the circle used for cutting out the dot (the dotted circle). To the right, local coordinate system associated with the magnetic moment in the lattice site indicated by the arrow.

of such dot is shown. In dots cut out from a discrete lattice the edges are not ideally circular; by a "circular" system we should understand a system cut out by means of circles. The edges cannot be smoothed; their smoothness is related to the size of the dot (measured in lattice constant units). Obviously, this problem does not arise in the case of square dots.

For each lattice site within the dot let us introduce a local Cartesian coordinate system in which a magnetic moment indicated by a position vector **R** is represented as $\mathbf{M_R} = \mathbf{M_{0,R}} + \mathbf{m_R}$, where $\mathbf{M_{0,R}}$ is the static component and $\mathbf{m_R}$ the dynamic component oscillating harmonically with an angular frequency ω ($\mathbf{m_R}(t) = \mathbf{m_R} \exp(i\omega t)$). In the case of the small precession amplitude one can use the linear approximation, assuming that $|\mathbf{m_R}| \ll |\mathbf{M_R}|$, $|\mathbf{M_{0,R}}| \simeq |\mathbf{M_R}|$, and $\mathbf{m_R} \perp \mathbf{M_R}$. The unit vectors of the local coordinate system are: $\mathbf{i_R}$, defining the direction of the static component of the magnetization; $\mathbf{j_R}$, lying in the plane of the dot and oriented towards its center; $\mathbf{k_R}$, the third unit vector of the right-handed coordinate system (see the inset in Fig. 3.3). Thus, the static component of the magnetic moment is given by $\mathbf{M_{0,R}} = g\mu_B S_R \mathbf{i_R}$, where g is the Landé g-factor, μ_B the Bohr magneton, and S_R the spin value.

Two coordinates of the dynamic component are defined: the in-plane coordinate $m_{j,\mathbf{R}}$, describing the oscillations of the magnetic moment in the plane of the system, and the "perpendicular" coordinate $m_{k,\mathbf{R}}$ related to the oscillations in the direction of the unit vector k; thus, $\mathbf{m_R} = m_{j,\mathbf{R}}\mathbf{j_R} + m_{k,\mathbf{R}}\mathbf{k_R}$. If magnetic configuration forms an in-plane vortex, the unit vector k is perpendicular to the plane of the dot for any lattice site; thus "perpendicular" coordinate means "out-of-plane" in this case. However, the method provided below is more general and works for any magnetic configuration.

For the description of the time evolution of each magnetic moment in the considered system we use the discrete damping-free Landau–Lifshitz (LL) equation:

$$\frac{\partial \mathbf{M_R}}{\partial t} = \gamma \mu_0 \mathbf{M_R} \times \mathbf{H_R^{eff}}, \tag{3.1}$$

where $\gamma = -56\pi \cdot 10^9$ [rad/sT] is the gyromagnetic ratio, μ_0 is the vacuum permeability, and $\mathbf{H_R^{eff}}$ the effective magnetic field acting on the magnetic moment at the position $\mathbf{R}$.

The effective field

Let us take into account the uniform and time-independent external field $\mathbf{H_0} = \mathbf{B_0}/\mu_0$, the dipolar field $\mathbf{H_R^d} = \mathbf{H_{0,R}^d} + \mathbf{h_R^d}$ and the exchange field $\mathbf{H_R^{ex}} = \mathbf{H_{0,R}^{ex}} + \mathbf{h_R^{ex}}$, where $\mathbf{H_{0,R}^d}$ and $\mathbf{H_{0,R}^{ex}}$ are the static components, and $\mathbf{h_R^d}$ and $\mathbf{h_R^{ex}}$ the dynamic components of the respective fields.

The dipolar field acting on a magnetic moment $\mathbf{M_R}$ is given by

$$\mathbf{H_R^d} = \frac{1}{4\pi} \sum_{\mathbf{R}'\neq\mathbf{R}} \left(\frac{3\,(\mathbf{R}' - \mathbf{R})\,(\mathbf{M_{R'}} \cdot (\mathbf{R}' - \mathbf{R}))}{|\mathbf{R}' - \mathbf{R}|^5} - \frac{\mathbf{M_{R'}}}{|\mathbf{R}' - \mathbf{R}|^3} \right),$$

where the summation $\sum_{\mathbf{R}'\neq\mathbf{R}}$ runs over all the lattice sites except that indicated by $\mathbf{R}$. If all the spins in the dot have the same value S and the position vectors are expressed in units of the distance a between nearest neighbors, the static component of the dipolar field becomes

$$\mathbf{H_{0,R}^d} = g\mu_B S\tilde{a} \sum_{\mathbf{R}'\neq\mathbf{R}} \left(\frac{3\,(\mathbf{R}' - \mathbf{R})\,(\mathbf{i_{R'}} \cdot (\mathbf{R}' - \mathbf{R}))}{|\mathbf{R}' - \mathbf{R}|^5} - \frac{\mathbf{i_{R'}}}{|\mathbf{R}' - \mathbf{R}|^3} \right),$$

and the dynamic component is

$$h_R^d = \tilde{d} \sum_{R' \neq R} \left(\frac{3 \, (R' - R) \, (m_{R'} \cdot (R' - R))}{|R' - R|^5} - \frac{m_{R'}}{|R' - R|^3} \right),$$

where we introduce $\tilde{d} = 1/\left(4\pi a^3\right)$. For the lattice types considered here the nearest-neighbor distance equals the lattice constant, hence its symbol a, and the interchangeability of these two terms.

Assuming that the exchange interaction is uniform and limited to nearest neighbors (NN), the exchange field is expressed by the following formula (Depondt and Mertens, 2009):

$$H_R^{ex} = \frac{2J}{\mu_0 \, (g\mu_B)^2} \sum_{R' \in NN(R)} M_{R'},$$

where J is the exchange integral and the summation runs over all the nearest neighbors of the spin indicated by R. Introducing $\tilde{J} = \frac{2J}{\mu_0 (g\mu_B)^2}$ the static component of the exchange field becomes

$$H_{0,R}^{ex} = g\mu_B S\tilde{J} \sum_{R' \in NN(R)} i_{R'},$$

and the dynamic component

$$h_R^{ex} = \tilde{J} \sum_{R' \in NN(R)} m_{R'}.$$

Finally, the effective magnetic field is a sum of described above static and dynamic components and could be written as

$$H_R^{eff} = H_0 + H_{0,R}^d + h_R^d + H_{0,R}^{ex} + h_R^{ex}. \tag{3.2}$$

Equations of motion

In the linear approximation, i.e., leaving only linear terms in view of dynamic components, and with the assumption of harmonic oscillation of the dynamic component of magnetic moment, the equation (3.1) takes the form:

$$i\frac{\omega}{\gamma \mu_0} m_R = M_{0,R} \times \left(h_R^d + h_R^{ex}\right) + m_R \times \left(H_0 + H_{0,R}^d + H_{0,R}^{ex}\right), \tag{3.3}$$

i being the imaginary unit. Dividing this equation by $g\mu_B S\tilde{J}$ and introducing following symbols:

$$\Omega = \frac{\omega}{\gamma \mu_0 g\mu_B S\tilde{J}} = \frac{g\mu_B}{2\gamma SJ} \omega,$$

$$\tilde{H}_R = \left(H_0 + H_{0,R}^d + H_{0,R}^{ex}\right)/\tilde{J},$$

$$\tilde{h}_R = \left(h_R^d + h_R^{ex}\right)/\left(g\mu_B S\tilde{J}\right),$$

one can obtain the equation of motion in the compact form:

$$i\Omega \mathbf{m_R} = \mathbf{i_R} \times \tilde{\mathbf{h}}_\mathbf{R} + \mathbf{m_R} \times \tilde{\mathbf{H}}_\mathbf{R}.$$

This vector equation can be rewritten as a set of equations of motion for three coordinates of the magnetic moment:

$$0 = m_{j,\mathbf{R}}\tilde{H}_{k,\mathbf{R}} - m_{k,\mathbf{R}}\tilde{H}_{j,\mathbf{R}},$$
$$i\Omega m_{j,\mathbf{R}} = -\tilde{h}_{k,\mathbf{R}} + m_{k,\mathbf{R}}\tilde{H}_{i,\mathbf{R}},$$
$$i\Omega m_{k,\mathbf{R}} = \tilde{h}_{j,\mathbf{R}} - m_{j,\mathbf{R}}\tilde{H}_{i,\mathbf{R}},$$

where the subscripts i, j, and k denote respective coordinates of the vectors. According to the assumption of small precession amplitudes, the component $\mathbf{M_i}$ is time-independent (static component); thus there is zero at the left-hand side of the first equation.

Finally, the system of equations of motion for the dynamic components of the magnetic moment is

$$i\Omega m_{j,\mathbf{R}} = m_{k,\mathbf{R}}\left(\sum_{\mathbf{R'}\in\mathrm{NN}(\mathbf{R})}\mathbf{i_R}\cdot\mathbf{i_{R'}} + d\sum_{\mathbf{R'}\neq\mathbf{R}}\right.$$
$$\times\left(\frac{3\left[(\mathbf{R'}-\mathbf{R})\cdot\mathbf{i_R}\right]\left[(\mathbf{R'}-\mathbf{R})\cdot\mathbf{i_{R'}}\right]}{|\mathbf{R'}-\mathbf{R}|^5} - \frac{\mathbf{i_R}\cdot\mathbf{i_{R'}}}{|\mathbf{R'}-\mathbf{R}|^3}\right)\right)$$
$$-\sum_{\mathbf{R'}\in\mathrm{NN}(\mathbf{R})}\mathbf{k_R}\cdot\mathbf{m_{R'}} - d\sum_{\mathbf{R'}\neq\mathbf{R}}$$
$$\times\left(\frac{3\left[(\mathbf{R'}-\mathbf{R})\cdot\mathbf{k_R}\right]\left[(\mathbf{R'}-\mathbf{R})\cdot\mathbf{m_{R'}}\right]}{|\mathbf{R'}-\mathbf{R}|^5} - \frac{\mathbf{k_R}\cdot\mathbf{m_{R'}}}{|\mathbf{R'}-\mathbf{R}|^3}\right)$$

$$i\Omega m_{k,\mathbf{R}} = -m_{j,\mathbf{R}}\left(\sum_{\mathbf{R'}\in\mathrm{NN}(\mathbf{R})}\mathbf{i_R}\cdot\mathbf{i_{R'}} + d\sum_{\mathbf{R'}\neq\mathbf{R}}\right.$$
$$\times\left(\frac{3\left[(\mathbf{R'}-\mathbf{R})\cdot\mathbf{i_R}\right]\left[(\mathbf{R'}-\mathbf{R})\cdot\mathbf{i_{R'}}\right]}{|\mathbf{R'}-\mathbf{R}|^5} - \frac{\mathbf{i_R}\cdot\mathbf{i_{R'}}}{|\mathbf{R'}-\mathbf{R}|^3}\right)\right)$$
$$+\sum_{\mathbf{R'}\in\mathrm{NN}(\mathbf{R})}\mathbf{j_R}\cdot\mathbf{m_{R'}} + d\sum_{\mathbf{R'}\neq\mathbf{R}}$$
$$\times\left(\frac{3\left[(\mathbf{R'}-\mathbf{R})\cdot\mathbf{j_R}\right]\left[(\mathbf{R'}-\mathbf{R})\cdot\mathbf{m_{R'}}\right]}{|\mathbf{R'}-\mathbf{R}|^5} - \frac{\mathbf{j_R}\cdot\mathbf{m_{R'}}}{|\mathbf{R'}-\mathbf{R}|^3}\right),$$

$$(3.4)$$

where $\Omega = \frac{g\mu_B}{\gamma SJ}\omega$ is the reduced frequency. Symbol d is the dipolar-to-exchange interaction ratio, defined as

$$d = \frac{(g\mu_B)^2\mu_0}{8\pi a^3 J}. \qquad (3.5)$$

Dynamical matrix

The system of equations 3.4 can be represented as a following eigenvalue problem:

$$\Omega \left[\begin{array}{c} m_j \\ m_k \end{array} \right] = \mathcal{M} \left[\begin{array}{c} m_j \\ m_k \end{array} \right], \qquad (3.6)$$

where m_j and m_k are column vectors containing in-plane and perpendicular amplitudes of magnetic moment precession, respectively, for every lattice site within the dot.

Matrix $\mathcal{M}$ is called *dynamical matrix*, the numerical diagonalization of which yields the spectrum of reduced frequencies Ω of the spin-wave modes and their profiles, i.e., the distribution of two amplitudes of the spins precession m_j and m_k. Its applicability is not limited to 2D magnetic vortices, though. The presented introduction is general enough for the obtained system of equations to apply to any two- or three-dimensional periodic structure with any configuration of the magnetic moments, and, if we neglect the exchange interaction, to nonperiodic structures as well.

The only material parameter in the adopted model is the dipolar-to-exchange interaction ratio d. In typical ferromagnets, such as cobalt, nickel or iron, strong exchange coupling results in very low values of d. For example, in the SPEELS experiments reported in Vollmer et al. (2004), the exchange integral in an ultrathin film of Co on a Cu(001) substrate is estimated at 15 meV. From the definition (3.5) we get $d = 0.00043$. By the definition d only depends on microscopic parameters. Thus, the stability of the assumed magnetic configuration will depend not only on d, but also on the structure of the system, i.e., its size and shape as well as the lattice from which it has been cut out.

It is worthy of notice that the approach in question does not involve any simulations; the magnetic configuration is assumed to be an in-plane vortex. Obviously, such a configuration may be unstable, in which case, however, zero-frequency modes, being nucleation modes responsible for magnetization reconfiguration, will occur in the spin-wave spectrum. In other words, the lack of zero-frequency modes is indicative of the stability (or metastability) of the assumed magnetic configuration (Skomski, 2008). On the other hand, if damping is neglected, only purely real solutions will be physical.

Thus, frequencies with a nonzero imaginary part will indicate that the assumed magnetic configuration is unstable (Rivkin et al., 2005). These two criteria are equivalent, which is reflected in the results: only frequencies with zero real part have a nonzero imaginary part. This approach allows relatively quick (in comparison with the simulation methods) exploration of various configurations, the stability of which is inferred from the obtained spin-wave spectrum.

Another advantage of the discussed theoretical approach is that the diagonalization of the dynamical matrix yields directly the spin-wave frequencies and profiles, without recourse to the Fourier transformation used in time-domain simulations. However, the assumption, rather than simulation, of the magnetic configuration is the main disadvantage of the method. Though reducing the calculation time, this approach can only be used for studying relatively simple magnetic configurations involving as few assumptions as possible. For example, the study of a core vortex will require additional assumptions regarding the shape and size of the core. In such cases a combined method seems to be the best solution, with simulations used for finding the stable configuration, and the dynamical matrix technique for calculating the corresponding spin-wave modes.

3.3 Normal Modes in Circular Dots

Dynamical matrix diagonalization gives the spectrum of so-called normal modes or standing waves. Normal modes are collective harmonic oscillations of the system about a local energy minimum, where all spins precess with the same frequency, the natural frequency of the system. From the wave point of view frequencies are quantized because of the spatial confinement of the system. On the other hand, they are linked with energy levels in quantum approach. Our model is semiclassical, i.e., the dot consists of discrete lattice of spins but every spin is treated as a classical vector. The number of normal modes of such system is equal to the number of spins within the dot.

Diagonalization of the dynamical matrix gives also the distribution of two coordinates of the dynamic component of the magnetic

134 | In-Plane Magnetic Vortices in 2D Nanodots

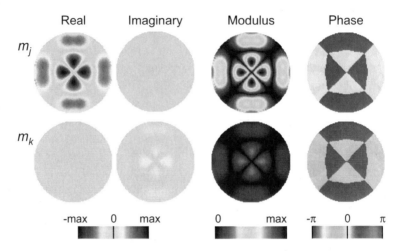

Figure 3.4 An example of a spin wave profile: Distribution of the in-plane (m_j, top profile) and out-of-plane (m_k, bottom profile) amplitude of precession of elementary magnetic moments for an exemplar normal mode in a dot of size $L = 101$ (8000 spins) in the in-plane vortex state for $d = 0.112$.

moment, i.e., spin-wave profiles. As an example we will study spin waves in circular dots. Their profiles are similar to those of vibrations of the circular membrane and the same terminology, based on the number of nodal lines in the radial and azimuthal direction, is useful to describe them. The mode type is given as (n, m), where the radial number n and the azimuthal number m specify the number of nodal lines in the respective directions. In Fig. 3.4 as an example the profile of one spin-wave mode is shown obtained for the circular dot in the vortex state. Since both coordinates of the dynamic magnetic moment are complex they can be expressed as real (Re) and imaginary part (Im) or as modulus and phase. The real part of the in-plane component (m_j) has well defined maxima and minima of the amplitude. Going from the dot center to the left (or right) we start in deep minimum then cross zero and finally reach maximum. Just antisymmetric situation we have traveling toward top (or bottom). Thus, we observe one nodal line in radial direction. In azimuthal direction, we have two maxima and two minima, i.e., there are two azimuthal nodal lines (across whole dot). The same symmetry has the imaginary part of the out-of-plane component m_k but the oscillation in this direction is 3.95 times smaller than the in-plane oscillation. $\text{Im}(m_j)$ and $\text{Re}(m_k)$ are just about zero within the whole dot. In the labeling described above this mode is a (1, 2) mode.

What is the physical meaning of the real and imaginary part of m_j and m_k? Let us go back to the time dependence of the dynamical component $\mathbf{m}$, which is given by $\mathbf{m}(t) = \mathbf{m} \exp i\omega t$. The phase shift between Re and Im, i.e., $\pi/2$, gives $T/4$ shift in time, where T is a period of oscillations for a given mode. For the mode in Fig. 3.4 it means for time $t = 0$ all spins lay purely in the plane of the dot (real part of m_k is zero) but after $t = T/4$ spins rise and m_k reaches the configuration given by its imaginary part. Meanwhile, the in-plane component changes from its maximum, given by its real part, to zero $(\text{Im}(m_j) = 0)$. In next $T/4$ spins go back to the plane of the dot with m_j having opposite sign than for $t = 0$. Finally, in $t = T$ the system goes back to the starting configuration and, since amplitudes of in-plane and out-of-plane oscillations are different, we obtain elliptical precession of spins.

Another way of presenting the profile is to show modulus and phase. The modulus reflects the distribution of the oscillation amplitude while the phase gives clear picture of the nodal lines; they coincide with lines at which the phase is shifted by π. For the mode in question (Fig. 3.4) there are three lines of such phase shift, two in azimuthal direction (along the dot diameter) and one in radial direction (almost circular), for both dynamic components.

As we can see in Fig. 3.4 for both dynamic components the profile is similar. Moreover, one part (real or imaginary) is sufficient to fully describe the oscillation amplitude distribution within the dot as well as the mode type. Therefore in the latter, if the situation is clear, we show only one part (Re or Im) of the one component (m_j or m_k).

3.4 Spin-Wave Spectrum

The frequency spectrum for the dot consists of 2032 spins (diameter $L = 51$ lattice constants) is shown in Fig. 3.5. An in-plane vortex is assumed as a magnetic configuration and the dipolar-to-exchange

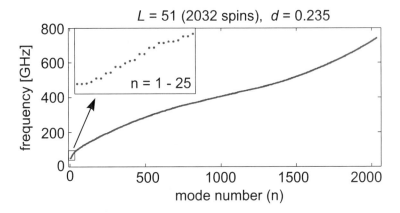

Figure 3.5 Spin wave spectrum: Frequencies of spin-wave normal modes versus the mode number for a 2D circular dot of size $L = 51$. The elementary magnetic moments are assumed to form an in-plane vortex and d is set to 0.235. The lack of zero-frequency modes is indicative of the stability of the assumed magnetic configuration. In the inset lowest 25 modes are shown.

interaction ratio d is set to 0.235. First of all, there are no zero-frequency modes in the spectrum, which, according to the above discussion (see Section 3.2), implies the stability of the assumed magnetic configuration. The overall shape of the spectrum is typical for dipolar-exchange systems. For low mode numbers (or for short wave vectors)[a] the shape is determined by dipolar interactions while by exchange interactions for high mode numbers (long wave vectors). In the inset frequencies of the 25 lowest modes are given and now it is clear that the spectrum is discrete. Another interesting feature is that some modes degenerate in pairs.

For the same dot spin-wave profiles of some selected modes are shown in Fig. 3.6: in panel (a) for 10 lowest modes in the spectrum, in panel (b) for few radial modes, and in panel (c) for some special modes.

Ten lowest modes in the dot under consideration are pairs of purely azimuthal modes with radial numbers zero and azimuthal

---

[a]Precisely speaking the wave vector is related rather to the azimuthal/radial number than to mode number in the spectrum, but as we will see later those are strongly related in this particular dot.

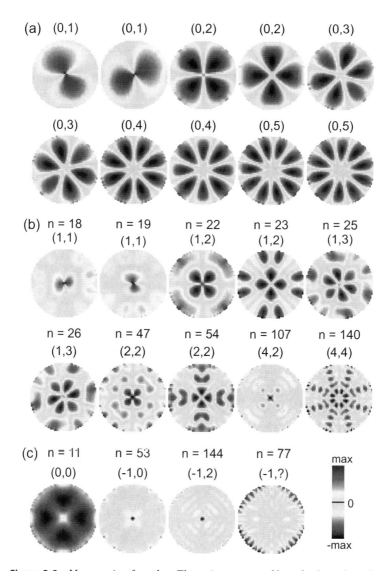

Figure 3.6 Menagerie of modes: The spin-wave profiles of selected modes obtained for a dot of size $L = 51$ in the in-plane vortex state for $d = 0.235$ (the same dot as in Fig. 3.5). (a) Ten lowest modes in the spectrum, all are purely azimuthal modes. (b) Few examples of radial modes. (c) Some special modes: the fundamental mode, two center-localized modes, and the edge mode. In brackets radial and azimuthal numbers are given. In panels (b) and (c) modes are also labeled with their mode number n.

ones being successive integers[a] (Fig. 3.6a). Among these modes six, namely those of odd order, are degenerate in pairs (compare the inset in Fig. 3.5): both (0, 1) modes have equal frequency, and the same can be said of the (0, 3) and (0, 5) modes. In contrast, the pairs of modes with an even azimuthal number, i.e., the second-order modes (0, 2) and the fourth-order modes (0, 4) are not degenerate. Their nondegeneracy is related to their symmetry, the same as that of the lattice from which the dot has been cut out. For example, one of the (0, 2) modes ($n = 3$) has nodal lines along the high spin density lines (x and y axes in Fig. 3.3). In contrast, in the other (0, 2) mode ($n = 4$) the high spin density lines coincide with anti-nodal lines. An analogical situation occurs in periodic structures, in which a band gap forms between two states at the boundary of the Brillouin zone if one state has nodes and the other anti-nodes in the potential wells. In the literature there are reports of lifted degeneracy of azimuthal modes in core vortices due to the coupling of the spin waves with the gyrotropic motion of the core (Guslienko et al., 2008; Hoffmann et al., 2007). Since we consider coreless vortices, coupling with the motion of the core is out of the question. The nondegeneracy is due to the fact that the dot has been cut out from a discrete lattice. It is the symmetry of the lattice that determines which modes have lifted degeneracy. For example in the case of hexagonal lattice the frequency splitting occurs for the modes with azimuthal number divisible by 3 (see our paper concerning hexagonal lattice based rings (Mamica, 2013a)).

The next group contains modes with nonzero radial number (Fig. 3.6b) and we will call them radial modes even if their azimuthal number is nonzero as well. Here the same rule holds as in the case of (purely) azimuthal modes: If the azimuthal symmetry of the profile matches the symmetry of the lattice the dot is based on modes of the same type are split. In other case their degenerate in pairs. Thus, both modes (1, 1) have the same frequency as well as the modes (1, 3), while pairs of modes (1, 2) and (2, 2) have lifted degeneracy. This is particularly strong manifested in the latter case where the

---

[a]In fact, for two azimuthal modes in the pair their azimuthal numbers differ in sign but for our purpose the sign of the azimuthal number is of no importance. Therefore, the second number should be understood as $|m|$.

lower of modes (2, 2) is 47th in the spectrum but the higher one is 54th.

Another feature we can see in Fig. 3.6 is dispersion relation, i.e., the dependence of the frequency versus wave vector. In the case of standing waves in confined system it is convenient to use the number of nodal lines instead of wave vector. More nodal lines means longer wave vector (shorter wave). If the radial number is fixed increasing the azimuthal number results in increasing the frequency (and the position of mode in the spectrum, Section 3.4). Exceptions are modes with azimuthal number zero, e.g., fundamental mode (0, 0) is 11th in the spectrum. Similarly, for fixed azimuthal number frequency increases with growing radial number. Thus, for the dot under the question, both dispersion relations are positive.

In Fig. 3.6c the profiles of some special modes are shown. First one ($n = 11$) is so-called fundamental mode, an analogue of the uniform excitation. In fact, its profile is not uniform but there are no nodal lines in any direction, azimuthal nor radial, so the mode is labeled as (0, 0). Its nonuniformity results from the fact that the circular symmetry of the dot is broken by the symmetry of the discrete lattice from which the dot has been cut out. A similar effect has been observed in micromagnetic simulations (Giovannini et al., 2007; Montoncello et al., 2010; Wang and Dong, 2012), in which, however, it stems from the artificial discretization of a continuous system into cubes (e.g., in the very popular OOMMF simulations). In the model used in our studies the lattice discreteness is an inherent feature of the system and the nonuniformity of the profile of the fundamental mode is its natural consequence. It is noteworthy that this effect should not occur in micromagnetic simulations based on the discretization into tetrahedra, e.g., in the NMAG approach (Fischbacher et al., 2007).

The second mode in Fig. 3.6c ($n = 53$) is strongly localized at the center of the vortex. This mode is labeled as $(-1, 0)$, where -1 refers to a complex wave number in the radial direction. Next mode ($n = 144$) is center-localized as well but additionally it has two nodal lines in azimuthal direction, $(-1, 2)$. The last mode in panel c is so-called edge-mode, strongly localized at the edge of the dot. In this case we observe several azimuthal nodal lines but their number is not clear. As we will see later, the localized modes gain in importance when the in-plane vortex loses its stability: They become soft modes responsible for the transition to new magnetic configuration.

3.5 Spin Waves vs. *d* and *L*

In the previous section, we have examined one particular example of a spin wave spectrum in a circular dot in the in-plane vortex state. The question arises: How will this spectrum change with the dipolar-to-exchange interaction ratio d? Figure 3.7a presents the spin-wave frequencies (in GHz) versus d in the same dot as in

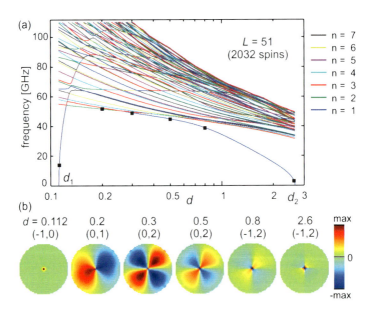

Figure 3.7 (a) Spin-wave spectrum versus d: The frequency of 80 lowest modes as a function of the dipolar-to-exchange interaction ratio d (in logarithmic scale) for a 2D circular dot of size $L = 51$ (2032 spins within the dot). The lack of zero-frequency modes is indicative of the stability of the assumed magnetic configuration (an in-plane vortex) for $d_1 < d < d_2$. The color assignment of the first seven mode lines is indicated at the right; the colors repeat cyclically for successive modes. (b) Spin-wave profiles of the lowest mode corresponding to the points marked with black squares in (a).

Figs. 3.5 and 3.6 ($L = 51$, 2032 spins). One of the characteristic features of this dependence is the existence of two critical values of d, dividing the presented dependence into three regimes. Below the critical value d_1 and above d_2 the spectrum includes zero-frequency modes (strictly speaking, modes with zero real part of the frequency); as discussed above, these indicate that the assumed magnetic configuration is unstable. Between the two critical values of d the in-plane vortex is a (meta)stable state, as evidenced by the fact that all the frequencies are greater than zero.

This reflects the source of the magnetic vortex state, namely the competition between the dipolar and exchange interactions. The exchange interaction favors the parallel alignment of the magnetic moments (in the case of ferromagnetic coupling), while the dipolar interaction attempts to align them in the plane of the dot to form a multi-domain or multi-vortex state (Depondt et al., 2011; Vedmedenko et al., 1999). Depending on the ratio between the two types of interaction, different magnetic configurations represent the compromise between these opposite tendencies and one of them is the single in-plane vortex, stable in some range of d. If d is too low, i.e., for strong enough exchange interaction, the out-of-plane component of the static magnetic moment rises at the vortex center and the magnetic configuration proceeds to the core-vortex (Mozaffari and Esfarjani, 2007). Since this spin reorientation is caused by the exchange interaction predominates over the dipolar interaction it will be referred to as the exchange-driven reorientation (transition). The reorientation of magnetic moments at $d = d_2$ results from the predomination of the dipolar interaction (too weak exchange interaction); thus, it will be referred to as the dipolar-driven reorientation: the dipolar interaction will lead to a multi-vortex state. (The stability of the in-plane vortex is studied in Section 3.6 of this chapter.)

In the in-plane vortex stability regime the frequencies of different spin-wave modes decrease at different rates with increasing d. As a consequence, the order of modes in the spectrum changes dynamically. In particular, the character of the lowest-frequency mode changes as a result of mode crossing. Also, two modes stand out in the presented dependence. One is the lowest mode in the spectrum for d near its critical values d_1 and d_2; with increasing d

its frequency increases steeply for $d \approx d_1$. For $d \approx d_2$ its frequency decreases much faster than frequencies of other modes. The other has an almost constant frequency in a certain range of d. As a consequence, in the first part of the dependence in Fig. 3.7a both modes ascend the spectrum by crossing or anti-crossing with modes the frequencies of which decrease with increasing d.

In Fig. 3.7a, in the frequency range above 80 GHz, a group of modes has frequencies that decrease with increasing d at a much lower rate than the frequencies of the other modes in this range. These slowly declining modes have the radial number 1 and the lowest of them is (1,1) mode. The frequencies of these modes decrease with increasing d as those of modes with radial number zero and the same azimuthal number. Thus, the influence of the dipolar-to-exchange interaction ratio on the frequency of a mode is mainly determined by its azimuthal number, the radial number being of little impact.

Lowest mode vs. d

To analyze the evolution of the lowest-frequency mode with d, in Fig. 3.7b we have plotted its profiles for selected values of d, marked with black squares in Fig. 3.7a. For d close to d_1 the lowest mode is a soft mode; its frequency reaches zero for $d = d_1$, and the mode becomes a nucleation mode, responsible for the magnetic reconfiguration of the system. In the case considered this mode is strongly localized at the center of the vortex (Fig. 3.7, $d = 0.112$). In this way the system manifests its tendency to form a vortex core. As d diverges from the critical value the in-plane vortex regains stability, and the mode in question is excited at increasing cost. This, in turn, results in a steep increase in its frequency (see Fig. 3.7a) and at $d \approx 0.1237$ the localized mode crosses two degenerate (0, 1) modes (Fig. 3.7b, $d = 0.2$). These first order azimuthal modes become the lowest modes while $(-1, 0)$ mode ascends the spectrum very fast.

Weak localization at the center of the dot, strengthening slowly with increasing d, occurs also for (0, 1) modes. The (0, 1) modes remain the lowest until they cross a (0, 2) mode at $d \approx 0.255$. The second-order azimuthal mode (only one due to the frequency splitting described in Section 3.4) remains the lowest mode in the spectrum to the end of the stability regime, i.e., until $d = d_2$, but its profile localizes gradually at the center of the dot ($d = 0.5 - 2.6$ in Fig. 3.7b). Finally, for larger values of d this mode becomes a $(-1, 2)$ mode. Weak localization of azimuthal modes at the center of the vortex, increasing as the dipolar interaction gains in importance, has been observed in micromagnetic simulations as well (Zhu et al., 2005).

Lowest mode vs. L

Figure 3.8a presents the frequency f of the lowest mode plotted versus the dipolar-to-exchange interaction ratio d for dots with a diameter L ranging from 27 (560 spins) to 101 (8000 spins); for successive curves the number of spins within the dot is approximately doubled. The character of presented curves is very similar regardless the size of the dot. In particular the $f(d)$ dependence for the localized $(-1, 0)$ mode is exactly the same in dots of any size in the considered range as long as this mode is the lowest one in the spectrum. The only difference is the point of the crossing with an azimuthal mode. This is due to the strong central localization of its profile and the local character of the exchange interaction, which predominates over the dipolar interaction for d close to d_1. The strongly localized mode does not "feel" the borders and consequently its behavior does not depend on the size of the dot. In fact, it does not depend on the shape of the dot as well, as we will see in Section 3.6 of this chapter.

In the second range of d an azimuthal mode has the lowest frequency and its dependence on the diameter of the dot is approximately $f \propto 1/\sqrt{L}$ (with the exception of the smallest dot), which is consistent with the results presented by Ivanov and Zaspel (2005) and Zivieri and Nizzoli (2005). However, the azimuthal number of this mode changes with both d and L. As we have already mentioned, in the dot of diameter $L = 51$ for smaller d (0, 1) mode is the lowest while for higher d (0, 2) mode. In Fig. 3.8 the point of intersection of these two azimuthal states is marked with a black circle for each considered size L. With the exception of the smallest dot, the intersection points follow a straight line (the dashed line in Fig. 3.8a), which crosses the frequency of the localized mode at

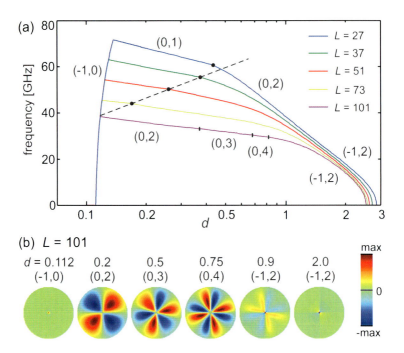

Figure 3.8 (a) Frequency of the lowest mode vs. dipolar-to-exchange interaction ratio d (in logarithmic scale) in dots of different size. From the top down, successive curves correspond to $L = 27$ (560 spins in the dot), 37 (1060 spins), 51 (2032 spins), 73 (4160 spins), and 101 (8000 spins). Black circles mark the points of intersection of first- and second-order azimuthal modes. (b) Evolution of the lowest mode profile with d in the dot of diameter $L = 101$ lattice constants.

ca. 39 GHz, corresponding to the frequency of the azimuthal (0, 1) mode for $L \approx 100$. This means that for dots with a diameter larger than 100 lattice constants the (0, 1) mode will not be the lowest in the spectrum for any value of d. This is confirmed by Fig. 3.8b, presenting the lowest mode profiles obtained for $L = 101$ (8000 spins): It is the (0, 2) mode that is the lowest in the spectrum after the crossing with the localized (−1, 0) mode. On the other hand, for $L = 101$ higher order azimuthal modes becomes the lowest in the spectrum with growing d (see successive profiles in Fig. 3.8b): (0, 3) at $d \approx 0.37$ and (0,4) at 0.69. For $d > 0.82$ the second-order azimuthal mode is the lowest again but now its profile is concentrated near the high spin density lines and further increase of d results in progressive localization of this mode at the center of the vortex similarly to the smaller dots.

For a given value of d (fixed material) increasing the size of the dot results in growing order of the lowest azimuthal mode. Moreover, for large enough dots and/or small enough exchange interactions (large d) the frequency of few lowest modes decreases with increasing azimuthal number, i.e., we have negative dispersion relation. Both effects were also observed in core-vortices experimentally by Buess et al. (2005) and in analytical calculations by Zivieri and Nizzoli (2005).

In the light of the results presented above, modes with increasing azimuthal number m fall successively to the bottom of the spectrum as the dipolar interaction gains in importance regardless of whether it is due to the material (d) or size (L) of the dot. On the other hand, the exchange interaction favors modes with $m = 1$. Thus, the competition between the dipolar and exchange interactions manifests itself not only in the magnetic configuration but also in the profile of the lowest-frequency spin-wave excitation.

Fundamental mode evolution

Evolution of the fundamental mode, another special mode in the spin-wave spectrum of a 2D circular dot, is studied in Fig. 3.9. In the panel (a) we present an enlargement of the relevant part of the spectrum shown in Fig. 3.7. For d up to 0.9 the frequency of (0, 0) mode is only slightly dependent on d, except few small ranges in which this mode anti-crosses with other modes in the spectrum. As we have already discussed, the profile of the fundamental mode is not uniform and its azimuthal nonuniformity reflects the symmetry of a lattice the dot is cut out (see Section 3.4). The nonuniformity of the profile results in two effects. First one is occurrence of weak dependence of the (0,0) mode frequency on d (for the uniform excitation its frequency should be the same for any d). The second one is the selection rule for the modes the fundamental mode is hybridizing with.

The most interesting hybridization is the multi-mode hybridization, which appears for d between d_1 and 0.145, where three modes

146 | *In-Plane Magnetic Vortices in 2D Nanodots*

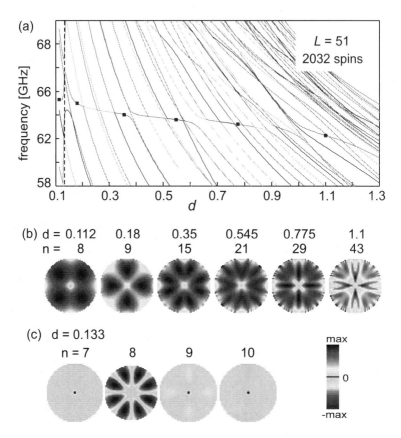

Figure 3.9 (a) Evolution of the fundamental mode as a function of the dipolar-to-exchange interaction ratio d for the dot of the diameter $L = 51$. (b) Profiles of the (0, 0) mode for points marked with black squares in (a). (c) Profiles of three hybridizing modes for $d = 0.133$: fundamental, center-localized, and one of two (0, 4) modes. Additionally, the profile of the second (0, 4) mode, which is not involved in the hybridization, is shown.

hybridize: the localized mode $(-1, 0)$, the fundamental mode $(0, 0)$, and one of the fourth-order azimuthal modes $(0, 4)$. The profiles of these modes are shown in Fig. 3.9c for $d = 0.133$, along with the profile of the second $(0, 4)$ mode, which is not involved in the hybridization. The profiles of hybridizing modes are completely mixed and highly localized at the dot center. The original profiles of $(0, 0)$ and $(0, 4)$ modes could be seen as a "shadow" in the rest of

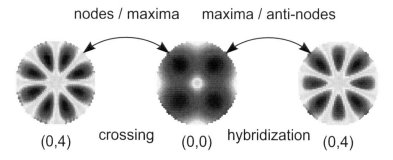

Figure 3.10 Selection rules for fundamental-azimuthal mode hybridization: the same symmetry and matching of maximums of the fundamental mode with in-phase anti-nodes of the azimuthal mode.

the dot. Comparing these "shadows" (or original profiles: $n = 8$ in Fig. 3.9b for (0, 0) and $n = 8$ in Fig. 3.6a for (0, 4)), one can infer the selection rule for fundamental-azimuthal mode hybridization (see Fig. 3.10). First of all the symmetry of the profiles has to be the same. In the considered case both profiles—for (0, 0) and (0, 4) modes—have fourfold symmetry (for the fundamental mode the symmetry is inherited from the lattice). The second condition is the matching of the in-phase anti-nodes in the azimuthal mode with the maximums in the fundamental mode. Only one of the two (0, 4) modes, namely that with anti-nodal lines in the same phase coinciding with the amplitude maximums in the fundamental mode, is involved in the hybridization, and consequently repulsion. The other (0, 4) mode is "ignored" by the fundamental mode, in spite of the same symmetry of the two. This is because the amplitude maximums of the fundamental mode match the nodal lines in the profile of this (0, 4) mode (please compare the profile $n = 8$ in Fig. 3.6c).[a]

---

[a] If the reader is a bit confused by our suggestion to compare the profiles for "the same" mode $n = 8$ in different figures please notice that these figures are obtained for different values of d, albeit for the same L. Since the mode order changes with d the modes under the question are, in fact, different modes. Moreover, azimuthal modes profiles are very little affected by d, except the regions of their hybridization and the lowest mode close to the critical value d_2. However, the fundamental mode profile changes a lot with d; thus it should be chosen carefully: for d just before the hybridization.

The hybridization of the (0, 4) and fundamental modes results in a sort of artifact at the center of the latter: localization and opposite phase with respect to the other parts of the dot (see $d = 0.112$ in Fig. 3.9b). It should be underlined that the phase shift in this case is an effect of the mode localization and does not mean the radial number is 1. A similar feature has been observed in core vortices in micromagnetic simulations (Giovannini et al., 2007; Wang and Dong, 2012).

After multi-mode hybridization ($d \approx 0.145$) the frequency of the localized mode, the lowest at first, becomes the highest in the end. The relative order of the other two modes does not change, the fundamental mode remaining below the azimuthal (0, 4) mode. As d continues to grow the frequency of the azimuthal mode decreases, while that of the fundamental mode is almost constant, hence another hybridization of these modes for d between 0.15 and 0.17. In this case again the hybridization requires both the same symmetry of the modes and the matching of the in-phase anti-nodes in the azimuthal mode with the maximums in the fundamental mode. At the end the (0, 0) mode profile is rotated by $\pi/4$ (Fig. 3.6b, $d = 0.18$) having maxima along the high spin density lines. Further increasing of d cause these maxima to split and the symmetry of the fundamental mode is doubled ($d = 0.35$ in Fig. 3.6b). In the same time, its frequency changes only little; thus it crosses descending azimuthal modes (0,5), (0,6), and (0,7), i.e., those whose profiles do not match its own symmetry. (Modes of 5th and 7th order degenerate in pairs, while modes (0,6) are split due to the coincidence with the symmetry of the lattice, according to the discussion in Section 3.4).

The symmetry of the fundamental mode matches the (0, 8) azimuthal modes. Therefore, next hybridization occurs for $d = 0.35 - 0.42$. Again, one of these modes is ignored and the other hybridize with (0, 0) mode. Afterwards the situation is repeated until $d \approx 1.0$: fundamental mode hybridizes with one of (0, 12) modes for $d = 0.60 - 0.70$ and one of (0, 16) for $d = 0.85 - 0.95$. For $d > 1.0$ the mode under the question loses its fundamental character; its profile has pronounced maximums and minimums ($d = 1.1$ in Fig. 3.6b), and its frequency noticeably depends on d. There is no more fundamental mode in the spectrum.

3.6 Stability of the In-Plane Vortex

As we already mentioned, even if we do not perform any simulations, we are able to deduce the stability of the in-plane vortex from the spin-wave spectrum: in Fig. 3.7 we have clear evidence of the existence of two critical values of d. In Fig. 3.11a, we show both critical values of d plotted versus the size of the dot (the number of spins within) for circular dots as well as for square ones. In the circular dot the critical value d_1 decreases at first, but quickly levels off at $d_1 = 0.1115$. To elucidate these effects let us refer to the profile of the lowest mode for $d \approx d_1$ (Figs. 3.7 and 3.8b, $d = 0.112$). The profile is strongly localized at the center, beyond which the magnetic moments do not precess, regardless the dot size. Note also that the spin reorientation at d_1 is caused by exchange interaction predomination. Since the exchange interaction has a local character, in larger dots the dynamic component, strongly localized at the center, fails to "feel" further increase in the size of the system, hence the size independence of d_1 in such systems. The situation is identical in square dots for the same reason: center localization of the lowest mode at $d \approx d_1$. The critical ratio d_1 is the same for both types of dot: square and circular, because strongly localized mode do not "feel" the shape of a dot as well. The results are consistent with those obtained in Monte Carlo simulations for circular dots, presented in Rocha et al. (2010), in both the size independence of d_1 and the value approached by it.

The reorientation of spins at $d = d_2$ results from the predomination of the dipolar interaction (too weak exchange interaction). The size dependence of d_2 is shown in Fig. 3.11a for dots of both shapes and this time, in contrast to d_1, different shape of a dot results in different size dependence. In square dots two regions can be distinguished in this size dependence, divided by an abrupt change in the slope for ca. 900 spins. In both regions the rate of decrease reduces as the dot grows in size. To explain this behavior in Fig. 3.11b,c, we have presented lowest mode profiles for square dots of different size. In small dots, i.e., in the first part of the size dependence of d_2, the lowest mode is strongly localized at the center of the vortex, and weakly at the corners (panel b). As the system

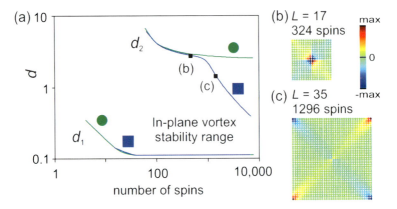

Figure 3.11 (a) Critical values d_1 and d_2 for in-plane vortex in circular and square dots versus the total number of spins within the dot. (b, c) Lowest mode profiles for $d = d_2$ in two square dots of different size indicated with black squares in (a).

grows in size, the corner localization increases at the cost of the center localization. When the amplitudes at the corners and at the center become comparable, the slope of the size dependence of d_2 changes abruptly. In the second part of the dependence the lowest mode is strongly localized at the corners with a residual localization at the center (panel c).

The size dependence of d_2 in circular dots is significantly different than that in square dots. The value of d_2 decreases at a rate decreasing with growing size of the dot (Fig. 3.11a). This is due to the strong localization of the lowest mode at the center of the vortex in circular dots of any size (see Figs. 3.7, $d = 2.6$ and 3.8b, $d = 2.0$), which in turn results from the lack of corners, playing the role of nucleation centers in square dots.

There is yet another major difference between the size dependences of d_2 in circular and square dots. In the former d_2 seems to saturate similarly to d_1. This suggests that even in very large circular dots there should be a range of the dipolar-to-exchange interaction ratio in which the in-plane vortex is a (meta)stable state. In contrast, in square dots the size dependence of d_2 suggests that in larger systems the circular magnetization is not stable for any value of d. This conclusion is consistent with the previous observation concerning circular magnetization in square dots. Increase in the size of the dot implies enlargement of the corner regions, in which the assumed magnetic configuration does not fit the geometry of the system. This mismatch causes also the localization of the lowest mode to change from central to corner as the dot enlarges.

Applying the obtained results to real materials, let us recall the value of d for Co on a Cu(001) substrate: 0.00043. It is much smaller than the limit 0.1115 obtained for d_1, which means in dots made of strong ferromagnetic materials the in-plane vortex is not stable. This is due to the strong center localization of the lowest mode and one could ask is it possible to stabilize the in-plane vortex in such materials by removing the central part of a dot? And the answer is yes, but it depends on the shape of a dot. It was shown experimentally in circular permalloy dots that introducing even very small hole cause the stabilization of the in-plane vortex (Hoffmann et al., 2007). In our study (Mamica et al., 2011), we show that removing even one single spin at dot center causes decrease of the d_1 down to 0.0149, but it is still too large. Thus, in typical ferromagnetic materials the introduction of the single-spin defect at the center of the dot, in spite of the significant reduction of the critical value d_1, will not stabilize the in-plane vortex configuration.

In circular rings making the hole bigger results in successive decreasing of d_1 (Mamica, 2013b). Moreover, the stability of the in-plane vortex increases (d_1 decreases) also as the external diameter of the ring grows. In sufficiently large rings d_1 is lower than the d value in Co/Cu(001), which means that in circular rings of this material the in-plane vortex will be a stable state. This increasing of the in-plane vortex stability is related to the change of the lowest mode profile from center-localized to fundamental. Its almost uniform profile is sensitive to both external and internal size of the ring. This result is in good agreement with experimental observations in cobalt rings (Li et al., 2001).

Completely different is the situation in square rings (Mamica et al., 2012). First of all, if we remove the central part of the dot the circular magnetization is no more justified and the Landau state is proper magnetic configuration. The elimination of four central magnetic moments destroys the central localization of the lowest mode but now it is moved to the corners, which play the role of nucleation centers instead of the vortex center. As a result, d_1 decreases but further enlargement of the hole in the resulting ring affects neither the profile of the lowest mode nor the value of d_1 as long as the ring is wide enough to maintain the strongly oscillating magnetic moments near the corners. Thus, enlargement of the hole in square rings in the Landau state will not stabilize the in-plane vortex in typical ferromagnets.

3.7 Final Remarks

The size range of the studied systems is rather far from that currently available to experimental studies. The method used allows to obtain numerical results in acceptable time for systems comprising up to a few thousand spins (corresponding to a diameter of tens of nanometers in 2D circular dots), whereas experimental studies tend to concern systems with a size ranging from a few hundred nanometers to a few micrometers, and 10-odd nanometer thick. Also, we consider the in-plane vortex magnetic configuration in rings and in full dots. In rings the in-plane vortex is a natural consequence of the removal of the center of the dot; however, full dots studied experimentally tend to be made of permalloy or cobalt, in which strong exchange interaction leads to the formation of a vortex core. On the other hand, the mismatch of spins at the center of the vortex in full dots plays the role of a nucleation center, similarly to the vortex core in dots studied experimentally. One of the major differences is the lack of coupling between the spin waves and the gyrotropic motion of the core; however, apart from this, our results should be, and are, in qualitative agreement with results for core-vortices: the experimental data as well as micromagnetic simulations and analytical calculations.

The use of the linear approximation limits the applicability of the discussed model to stable magnetic configurations and excitations with a moderate amplitude. However, even near the critical values of d, where strongly localized spin waves have a large amplitude, those of our results that can be compared with results obtained by other methods and reported in the literature, are in qualitative and quantitative agreement with the literature data.

Another disadvantage of the method used is, already mentioned, the lack of simulations. Therefore, it can be used for relatively simple magnetic configurations involving as few assumptions as possible, such as in-plane vortices.

Despite these limitations the dynamical matrix method is very useful approach for examining the normal modes in any discrete system as well as the stability of the magnetic configuration used.

Concerning the scientific results shown here, we can conclude that not only magnetic structure of the nanoparticle but also the lowest-frequency mode is indicative of the competition between exchange and dipolar interaction. Far from the critical values exchange interactions prefer lower azimuthal number m while dipolar interactions favor higher m, regardless of whether their predomination is due to the material or size of the dot. Close to the magnetic reorientation the lowest mode is center-localized (or uniform in the case of rings). Thus, the profile of the lowest mode carries information on the stability of the magnetic configuration.

Acknowledgments

The author acknowledges the support from the National Science Centre (NCN) of Poland, Project DEC-2-12/07/E/ST3/00538.

References

Bader, S. D. (2006). Colloquium: Opportunities in nanomagnetism, *Rev. Mod. Phys.* **78**, pp. 1–15.

Bauer, H. G., Sproll, M., Back, C. H., and Woltersdorf, G. (2014). Vortex core reversal due to spin wave interference, *Phys. Rev. Lett.* **112**, p. 077201.

Bowden, S. R., and Gibson, U. J. (2010). Logic operations and data storage using vortex magnetization states in mesoscopic permalloy rings, and optical readout, *J. Phys.: Conf. Ser.* **200**, p. 072033.

Buess, M., Knowles, T. P. J., Hollinger, R., Haug, T., Krey, U., Weiss, D., Pescia, D., Scheinfein, M. R., and Back, C. H. (2005). Excitations with negative dispersion in a spin vortex, *Phys. Rev. B* **71**, p. 104415.

Chung, S.-H., McMichael, R. D., Pierce, D. T., and Unguris, J. (2010). Phase diagram of magnetic nanodisks measured by scanning electron microscopy with polarization analysis, *Phys. Rev. B* **81**, p. 020403(R).

Demidov, V. E., Ulrichs, H., Urazhdin, S., Demokritov, S. O., Bessonov, V., Gieniusz, R., and Maziewski, A. (2011). Resonant frequency multiplication in microscopic magnetic dots, *Appl. Phys. Lett.* **99**, p. 012505.

Depondt, P., Lévy, J.-C. S., and Mamica, S. (2013). Vortex polarization dynamics in a square magnetic nanodot, *J. Phys.: Condens. Matter* **25**, p. 466001.

Depondt, P., Lévy, J.-C. S., and Mertens, F. G. (2011). Vortex polarity in 2-d magnetic dots by langevin dynamics simulations, *Phys. Lett. A* **375**, p. 628.

Depondt, P., and Mertens, F. G. (2009). Spin dynamics simulations of two-dimensional clusters with heisenberg and dipole–dipole interactions, *J. Phys.: Condens. Matter* **21**, p. 336005.

Donolato, M., Gobbi, M., Vavassori, P., Leone, M., Cantoni, M., Metlushko, V., Ilic, B., Zhang, M., Wang, S. X., and Bertacco, R. (2009). Nanosized corners for trapping and detecting magnetic nanoparticles, *Nanotechnology* **20**, p. 385501.

Dubus, C., Sekimoto, K., and Fournier, J.-B. (2006). General up to next-nearest-neighbour elasticity of triangular lattices in three dimensions, *Proc. R. Soc. A* **462**, pp. 2695–2713.

Fischbacher, T., Franchin, M., Bordignon, G., and Fangohr, H. (2007). A systematic approach to multiphysics extensions of finite-element based micromagnetic simulations: Nmag, *IEEE Trans. Magn.* **43**, pp. 2896–2898.

Giovannini, L., Montoncello, F., Zivieri, R., and Nizzoli, F. (2007). Spin excitations in nanometric magnetic dots: Calculations and comparison with light scattering measurements, *J. Phys.: Condens. Matter* **19**, p. 225008.

Guslienko, K. Y. (2012). Spin torque induced magnetic vortex dynamics in layered nanopillars, *J. Spintron. Magn. Nanomater.* **1**, p. 70.

Guslienko, K. Y., Slavin, A. N., Tiberkevich, V., and Kim, S.-K. (2008). Dynamic origin of azimuthal modes splitting in vortex-state magnetic dots, *Phys. Rev. Lett.* **101**, p. 247203.

Hoffmann, F., Woltersdorf, G., Perzlmaier, K., Slavin, A. N., Tiberkevich, V. S., Bischof, A., Weiss, D., and Back, C. H. (2007). Mode degeneracy due to vortex core removal in magnetic disks, *Phys. Rev. B* **76**, p. 014416.

Ivanov, B. A., and Zaspel, C. E. (2005). High frequency modes in vortex-state nanomagnets, *Phys. Rev. Lett.* **94**, p. 027205.

Jain, S., and Adeyeye, A. O. (2008). Probing the magnetic states in mesoscopic rings by synchronous transport measurements in ring-wire hybrid configuration, *Appl. Phys. Lett.* **92**, p. 202506.

Kammerer, M., Weigand, M., Curcic, M., Noske, M., Sproll, M., Vansteenkiste, A., Waeyenberge, B. V., Stoll, H., Woltersdorf, G., Back, C. H., and Schuetz, G. (2011). Magnetic vortex core reversal by excitation of spin waves, *Nat. Commun.* **2**, p. 279.

Kłos, J. W., and Zozoulenko, I. V. (2010). Effect of short- and long-range scattering on the conductivity of graphene: Boltzmann approach vs tight-binding calculations, *Phys. Rev. B* **82**, p. 081414.

Kotakoski, J., Krasheninnikov, A. V., and Nordlund, K. (2006). Energetics, structure, and long-range interaction of vacancy-type defects in carbon nanotubes: Atomistic simulations, *Phys. Rev. B* **74**, p. 245420.

Krawczyk, M., Mamica, S., Kłos, J. W., Romero-Vivas, J., Mruczkiewicz, M. and Barman, A. (2011). Calculation of spin wave spectra in magnetic nanograins and patterned multilayers with perpendicular anisotropy, *J. Appl. Phys.* **109**, p. 113903.

Krawczyk, M., Puszkarski, H., Lévy, J.-C., Mamica, S., and Mercier, D. (2002). Theoretical study of spin wave resonance filling fraction effect in composite ferromagnetic [A|B|A] trilayer, *J. Magn. Magn. Mater.* **246**, 1–2, pp. 93–100.

Lagae, L., Wirix-Speetjens, R., Das, J., Graham, D., Ferreira, H., Freitas, P. P. F., Borghs, G., and Boeck, J. D. (2007). On-chip manipulation and magnetization assessment of magnetic bead ensembles by integrated spin-valve sensors, *Appl. Phys. Lett.* **91**, p. 203904.

Lévy, J.-C. S. (1981). Surface and interface magnons: Magnetic structures near the surface, *Surf. Sci. Rep.* **1**, pp. 39–119.

Li, S. P., Peyrade, D., Natali, M., Lebib, A., Chen, Y., Ebels, U., Buda, L. D., and Ounadjela, K. (2001). Flux closure structures in cobalt rings, *Phys. Rev. Lett.* **86**, p. 1102.

Luo, Y., Du, Y., and Misra, V. (2008). Large area nanorings fabricated using an atomic layer deposition Al_2O_3 spacer for magnetic random access memory application, *Nanotechnology* **19**, p. 265301.

Mamica, S. (2013a). Spin-wave spectra and stability of the in-plane vortex state in two-dimensional magnetic nanorings, *J. Appl. Phys.* **114**, p. 233906.

Mamica, S. (2013b). Stabilization of the in-plane vortex state in two-dimensional circular nanorings, *J. Appl. Phys.* **113**, p. 093901.

Mamica, S., Józefowicz, R., and Puszkarski, H. (1998). The role of oblique-to-surface disposition of neighbours in the emergence of surface spin waves in magnetic films, *Acta Phys. Pol. A* **94**, pp. 79–91.

Mamica, S., Krawczyk, M., and Kłos, J. W. (2012a). Spin-wave band structure in 2D magnonic crystals with elliptically shaped scattering centres, *Adv. Cond. Mat. Phys.* **2012**, p. 161387.

Mamica, S., Krawczyk, M., Sokolovskyy, M. L., and Romero-Vivas, J. (2012b). Large magnonic band gaps and spectra evolution in three-dimensional magnonic crystals based on magnetoferritin nanoparticles, *Phys. Rev. B* **86**, p. 144402.

Mamica, S., Lévy, J. C. S., Depondt, P., and Krawczyk, M. (2011). The effect of the single-spin defect on the stability of the in-plane vortex state in 2D magnetic nanodots, *J. Nanopart. Res.* **13**, pp. 6075–6083.

Mamica, S., Lévy, J.-C. S., and Krawczyk, M. (2014). Effects of the competition between the exchange and dipolar interactions in the spin-wave spectrum of two-dimensional circularly magnetized nanodots, *J. Phys. D* **47**, p. 015003.

Mamica, S., Lévy, J.-C. S., Krawczyk, M., and Depondt, P. (2012). Stability of the landau state in square two-dimensional magnetic nanorings, *J. Appl. Phys.* **112**, p. 043901.

Mamica, S., Puszkarski, H., and Lévy, J.-C. S. (2000). The role of next-nearest neighbours for the existence conditions of subsurface spin waves in magnetic films, *phys. stat. sol. (b)* **218**, pp. 561–569.

Mani, A. S., Geerpuram, D., Domanowski, A., Baskaran, V., and Metlushko, V. (2004). Magnetic random access memory design using rings with controlled asymmetry, *Nanotechnology* **15**, p. S645.

Metlov, K. L., and Lee, Y. P. (2008). Map of metastable states for thin circular magnetic nanocylinders, *Appl. Phys. Lett.* **92**, p. 112506.

Miller, M. M., Prinz, G. A., Cheng, S.-F., and Bounnak, S. (2002). Detection of a micron-sized magnetic sphere using a ring-shaped anisotropic magnetoresistance-based sensor: A model for a magnetoresistance-based biosensor, *Appl. Phys. Lett.* **81**, p. 2211.

Montoncello, F., Giovannini, L., Nizzoli, F., Zivieri, R., Consolo, G., and Gubbiotti, G. (2010). Spin-wave activation by spin-polarized current pulse in magnetic nanopillars, *J. Magn. Magn. Mater.* **322**, pp. 2330–2334.

Morrish, A. H. (2001). *The Physical Principles of Magnetism* (Wiley-IEEE Press), ISBN 978-0-7803-6029-7.

Mozaffari, M. R., and Esfarjani, K. (2007). Spin dynamics characterization in magnetic dots, *Physica B* **399**, pp. 81–93.

Pal, S., Rana, B., Saha, S., Mandal, R., Hellwig, O., Romero-Vivas, J., Mamica, S., Kłos, J. W., Mruczkiewicz, M., Sokolovskyy, M. L., Krawczyk, M., and Barman, A. (2012). Time-resolved measurement of spin-wave spectra in CoO capped [Co(t)/Pt(7 Å)](n-1) Co(t) multilayer systems, *J. Appl. Phys.* **111**, p. 07C507.

Park, J. P., Eames, P., Engebretson, D. M., Berezovsky, J., and Crowell, P. A. (2003). Imaging of spin dynamics in closure domain and vortex structures, *Phys. Rev. B* **67**, p. 020403(R).

Pribiag, V. S., Krivorotov, I. N., Fuchs, G. D., Braganca, P. M., Ozatay, O., Sankey, J. C., Ralph, D. C., and Buhrman, R. A. (2007). Magnetic vortex oscillator driven by d.c. spin-polarized current, *Nat. Phys.* **3**, pp. 498–503.

Puszkarski, H., Lévy, J.-C. S., and Mamica, S. (1998). Does the generation of surface spin-waves hinge critically on the range of neighbour interaction? *Phys. Lett. A* **246**, pp. 347–352.

Rahm, M., Stahl, J., and Weiss, D. (2005). Programmable logic elements based on ferromagnetic nanodisks containing two antidots, *Appl. Phys. Lett.* **87**, p. 182107.

Rivkin, K., DeLong, L. E., and Ketterson, J. B. (2005). Microscopic study of magnetostatic spin waves, *J. Appl. Phys.* **97**, p. 10E309.

Rocha, J. C. S., Coura, P. Z., Leonel, S. A., Dias, R. A., and Costa, B. V. (2010). Diagram for vortex formation in quasi-two-dimensional magnetic dots, *J. Appl. Phys.* **107**, p. 053903.

Rohart, S., Campiglio, P., Repain, V., Nahas, Y., Chacon, C., Girard, Y., Lagoute, J., Thiaville, A., and Rousset, S. (2010). Spin-wave-assisted thermal reversal of epitaxial perpendicular magnetic nanodots, *Phys. Rev. Lett.* **104**, p. 137202.

Shinjo, T., Okuno, T., Hassdorf, R., Shigeto, K., and Ono, T. (2000). Magnetic vortex core observation in circular dots of permalloy, *Science* **289**, p. 930.

Skomski, R. (2008). *Simple Models of Magnetism* (Oxford University Press).

Usov, N. A., and Peschany, S. E. (1993). Magnetization curling in a fine cylindrical particle, *J. Magn. Magn. Mater.* **118**, p. L290.

Vavassori, P., Metlushko, V., and Ilic, B. (2007). Domain wall displacement by current pulses injection in submicrometer permalloy square ring structures, *Appl. Phys. Lett.* **91**, p. 093114.

Vavassori, P., Metlushko, V., Ilic, B., Gobbi, M., Donolato, M., Cantoni, M., and Bertacco, R. (2008). Domain wall displacement in py square ring for single nanometric magnetic bead detection, *Appl. Phys. Lett.* **93**, p. 203502.

Vedmedenko, E. Y., Ghazali, A., and Lévy, J.-C. S. (1999). Magnetic vortices in ultrathin films, *Phys. Rev. B* **59**, p. 3329.

Vollmer, R., Etzkorn, M., Kumar, P. S. A., Ibach, H., and Kirschner, J. (2004). Spin-wave excitation in ultrathin Co and Fe films on Cu(001) by spin-polarized electron energy loss spectroscopy (invited), *J. Appl. Phys.* **95**, p. 7435.

Waeyenberge, B. V., Puzic, A., Stoll, H., Chou, K. W., Tyliszczak, T., Hertel, R., Fahnle, M., Bruckl, H., Rott, K., Reiss, G., Neudecker, I., Weiss, D., Back, C. H., and Schutz, G. (2006). Magnetic vortex core reversal by excitation with short bursts of an alternating field, *Nature* **444**, pp. 461–464.

Wang, R., and Dong, X. (2012). Sub-nanosecond switching of vortex cores using a resonant perpendicular magnetic field, *Appl. Phys. Lett.* **100**, p. 082402.

White, R. M. (2006). *Quantum Theory of Magnetism* (Springer), ISBN 978-3540651161.

Zhang, W., Singh, R., Bray-Ali, N., and Haas, S. (2008). Scaling analysis and application: Phase diagram of magnetic nanorings and elliptical nanoparticles, *Phys. Rev. B* **77**, p. 144428.

Zhu, X., Liu, Z., Metlushko, V., Grutter, P., and Freeman, M. R. (2005). Broadband spin dynamics of the magnetic vortex state: Effect of the pulsed field direction, *Phys. Rev. B* **71**, p. 180408(R).

Zhu, J.-G., Zheng, Y., and Prinz, G. A. (2000). Ultrahigh density vertical magnetoresistive random access memory, *J. Appl. Phys.* **87**, p. 6668.

Zivieri, R., and Nizzoli, F. (2005). Theory of spin modes in vortex-state ferromagnetic cylindrical dots, *Phys. Rev. B* **71**, p. 014411.

Chapter 4

Magnetic Properties of Nanostructures in Non-Integer Dimensions

Pascal Monceau

Laboratoire Matière et Systèmes Complexes, UMR 7057 CNRS,
Université Denis Diderot-Paris 7, 10 rue A. Domon et L. Duquet,
75013 Paris Cedex, France
Université d'Evry-Val d'Essonne, France
pascal.monceau@univ-paris-diderot.fr

4.1 Introduction

Several physical systems [2, 5] (gels, polymers, magnetic domains, percolating structures, electrolytic deposits, …) and biological ones [25] (lungs, neuronal networks, …) exhibit laws involving a non-integer space dimension; the generic term "fractal" has been given to them. From a geometrical point of view, these systems are "irregular" in the sense that they lack the translation invariance symmetry specific to crystals. Nevertheless, a more detailed observation shows a hierarchical organization: A structuring appears at many scales, as if similar images were observed under a microscope when varying the magnification. Such a hierarchical structuring is characterized by scale invariance whose different versions will be

---

Magnetic Structures of 2D and 3D Nanoparticles: Properties and Applications
Edited by Jean-Claude Levy
Copyright © 2016 Pan Stanford Publishing Pte. Ltd.
ISBN 978-981-4613-67-5 (Hardcover), 978-981-4613-68-2 (eBook)
www.panstanford.com

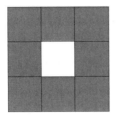 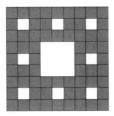 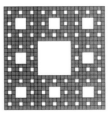

Figure 4.1 The three first iteration steps of the construction of a deterministic Sierpinski fractal $SP_a(3^2, 8, k)$.

described in a more detailed way below. The range over which the hierarchical organization can be seen enables to define the number of magnitude orders associated with the fractality. From a mathematical point of view, the scale invariance of a fractal covers an infinity of orders of magnitude while this number is finite for real physical systems; the right-hand pattern in Fig. 4.1 shows a fractal structure over three orders of magnitude. When dealing with nanostructures, fractals have mainly been obtained from growth process [6, 26], self-assembly process [7] or Diffusion Limited Aggregation, and the scale invariance covers a limited number of orders of magnitude.

4.2 Fractality and Scale Invariance

Fractals can be defined in a general way as objects that possess self-similarity properties. Topological properties of such objects have been studied as early as the nineteenth century by mathematicians as Cantor, Koch, Peano, Dedekind, Sierpinski... It is only around 1960 that the works of Benoit Mandelbrot enabled the concept of fractals to develop rapidly in many fields of science. Sierpinski carpets and their generalization to higher dimensions by Menger provide a generic model of fractals that has been widely used to model physical phenomena in non-integer dimensions. Let us firstly describe the two main categories of Sierpinski fractals we have to distinguish, namely the deterministic and random ones.

4.2.1 *Discrete Deterministic Scale Invariance*

Whereas discrete translations of an elementary cell enable to build Bravais lattices, generalized deterministic Sierpinski carpets are constructed from a generating cell according to an iteration process involving dilations. The generating cell is embedded in a d-dimensional space, and denoted $SP_g(n^d, N_{oc}, 1)$; let us describe the construction of Sierpinski fractals in the particular case where $d = 2$: A square, whose side has a size n unit length, is divided into n^2 subsquares and N_{oc} among them are occupied according to a given rule denoted g referring to the geometrical distribution of the occupied subsquares (center of the square, corner, ...). The fractal is constructed in the following way: A dilation by a factor n enlarges the generating cell, and each occupied subsquare is replaced by the original generating cell. This process can then be iterated as many times as wished in a deterministic way; no disorder is present here, since the rule g remains invariant all along the process. The lattice built up from a finite number k of iteration steps is denoted $SP_g(n^2, N_{oc}, k)$; its size is $L = n^k$ and the number of occupied subsquares is $N = (N_{oc})^k$. An example of such a construction is given in Fig. 4.1. The scale invariance of these fractal is discrete since they are exactly self-similar only at scales that are integer powers of n. It should be pointed out that such deterministic fractals nanostructures can be tailored by means of micro lithography or nanolithography [3].

4.2.2 *Discrete Random Scale Invariance*

The generating cell embedded in a two-dimensional space, denoted $SR_\epsilon(n^2, N_{oc}, 1)$ is still divided into n^2 subsquares but the N_{oc} occupied squares are chosen randomly among them. The iteration process differs from the deterministic case because the N_{oc} occupied squares in each occupied subsquare of an enlarged cell are once more chosen randomly. A lattice built up from a finite number k of iteration steps will be denoted $SR_\epsilon(n^2, N_{oc}, k)$, its lateral size is $L = n^k$, the number of occupied subsquares is $N = (N_{oc})^k$ and $\epsilon(L)$ designates the set $\{\epsilon_i\}$ of L^2 occupation numbers associated with such a process; $\epsilon_i = 0$ if a subsquare is deleted, $\epsilon_i = 1$ otherwise.

Figure 4.2 The three first iteration steps of the construction of a random Sierpinski fractal $SR_\epsilon(3^2, 8, k)$.

Figure 4.2 shows an example of the construction of such a random Sierpinski carpet. These structures are disordered because of the random character of the occupation process; the disorder is constrained by the structuring required by the hierarchical construction process, and each structure is self-similar only in a statistical way and at discrete scales, namely integer powers of n.

4.2.3 Random Scale Invariance

Fractal structures can be obtained in the framework of percolation [2]: Sites belonging to a square lattice or a cubic lattice of side L are randomly occupied with a probability equal to the percolation threshold p_c of such a lattice; $p_c \simeq 0.592746$ for a square lattice, $p_c \simeq 0.31161$ for a cubic one. The percolation cluster defined by the occurrence of at least one path connecting two edges through first neighbors is then extracted. Figure 4.3 shows a percolation cluster extracted from a square lattice. The disorder is no more constrained by a geometrical series of lengths involving the side of some generating cell and the statistical scale invariance is no more discrete.

4.2.4 Hausdorff Fractal Dimension

The global scale invariance of a fractal object can be quantified by its Hausdorff fractal dimension [2]. Without going into details of its rigorous mathematical definition, it is worth remembering that the "mass" contained in a box whose side is L behaves as L^{d_f}. When dealing with the fractals shown in Figs. 4.1 to 4.3, the number of

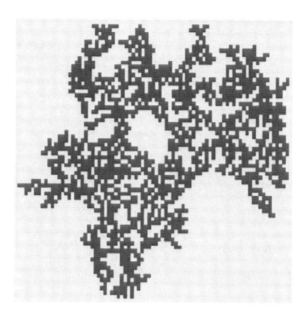

Figure 4.3 Percolation cluster on a square lattice occupied with a probability $p_c \simeq 0.592746$ (the small clusters have been removed).

occupied squares scales with the side L of the structure as $N \sim L^{d_f}$. In the case of discrete scale invariance, such a relation holds only for values of L that are integer powers of n and leads to $d_f = \ln(N_{oc})/\ln(n)$; when probing the lattice at scales that do not satisfy such a constraint, log periodic modulations around the leading power law behavior appear; log periodic oscillations bear the mark of discrete scaling invariance, and can be taken in account in an imaginary part of the associated exponent [24]. It is worth noticing that the Hausdorff dimension of the Sierpinski fractals depicted in Figs. 4.1 and 4.2, $d_f = \ln(8)/\ln(3) \approx 1.892789$ is very close to the fractal dimension $d_f \approx 1.89533$ of the percolation cluster from Fig. 4.3.

4.2.5 Configuration Entropy and Disorder Quantification

The remark noticed at the end of the last section suggests that it is necessary to go beyond the fractal dimension to characterize the disorder. This can be done by studying the configuration entropy per particle s; for a given size L, s is defined as $s = \frac{k_B \ln W}{L^{d_f}}$ where k_B is the Boltzmann constant and W the number of possible random configurations. For a given set $\{n, N_{oc}, k\}$, and provided that n and N_{oc} are chosen in such a way that $SR_\epsilon(n^2, N_{oc}, k)$ always overlaps the percolation cluster, the configuration entropy per site is equal to $s_{SR} = k_B \frac{1-1/(N_{oc})^k}{N_{oc}-1} \ln \binom{n^2}{N_{oc}}$, where $\binom{n^2}{N_{oc}}$ designates the binomial coefficients. In the case of the percolation cluster, the configuration entropy s_P scales as L^{β_p/ν_p} where β_p and ν_p are exponents of the percolation problem; $\beta_p/\nu_p = 5/48$ in the two-dimensional case [20]. Hence, the distinction between random fractality with or without discrete scale invariance is striking, since s_{SR} tends towards a finite limit as L tends to infinity while s_P grows with L as shown in Fig. 4.4: The disorder is strongly constrained by discrete scale invariance.

4.2.6 Geometrical Properties of Deterministic Sierpinski Fractals

Much work has been devoted to the characterization of geometrical properties of deterministic fractals beyond the fractal dimension;

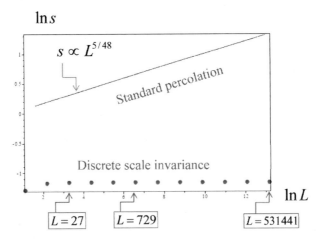

Figure 4.4 Evolution of the configuration entropy per site as a function of the size L according to the discrete or nondiscrete character of the scale invariance.

we will focus on two of them, relevant to the field of phase transitions.

4.2.6.1 Order of ramification

The order of ramification R at a given point of the fractal is equal to the number of bonds that must be cut in order to isolate an arbitrarily large set of sites connected to that point [4]. It can be easily checked that the order of ramification of the deterministic Sierpinski carpets described in Fig. 4.1 is infinite whereas it is finite in the case of the Sierpinski gasket shown in Fig. 4.5. One can convince oneself of the pertinence of the order of ramification by looking at the fractal whose generating cell is presented on the right side of Fig. 4.5: Its fractal dimension is $d_f = 2$, and R is finite; in the same way, a two dimensional square lattice can be considered as a fractal $SP_g(n^2, n^2)$, and has an infinite order of ramification. The question whether R is finite or not plays a central role in phase transitions: As a main result of the pioneering works of Gefen et al. dealing with discrete symmetry spin models in deterministic fractal structures [4], it has been found that second-order transitions at nonzero temperature occur only in fractals with an infinite order of ramification.

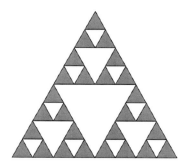 

Figure 4.5 Examples of fractals with finite ramification orders. Left: A Sierpinski gasket at the third iteration step with $d_f = \ln 3/\ln 2 \approx 1.58496$. Right: Elementary cell of a two-dimensional fractal.

4.2.6.2 Connectivity and mean number of links per site

Some topological features of the lattices built up from a given generating cell $SP_g(n^d, N_{oc}, k)$ in particular the mean number $z_s(n^d, N_{oc}, k)$ of links per site (with periodic boundary conditions), differs from an iteration step k to the next because of edge effects occurring all along the construction of the structure. It can be shown that

$$z_s(l^d, N_{oc}, k) = \frac{N_I}{N_{oc} - N_S} \left[1 - \left(\frac{N_S}{N_{oc}} \right)^k \right] + d \left(\frac{N_S}{N_{oc}} \right)^k, \quad (4.1)$$

where N_S is the number of occupied sites on each hypersurface of the generating cell and N_I the number of its internal bonds. It can clearly be seen from Eq. 4.1 that the convergence speed of $z_s(l^d, N_{oc}, k)$ towards its thermodynamical limit (that is when $k \to \infty$) is driven by the reason N_S/N_{oc} of a geometrical series involving a ratio of the hypersurface of the generating cell to its bulk. It should be pointed out that fractal structures encountered experimentally are not exactly scale invariant in the thermodynamical limit but only over a finite number k of order of magnitude; hence the variation of $z_s(l^d, N_{oc}, k)$ is expected to cause finite-size effects peculiar to discrete scale invariance. From a more general point of view, finite-size effects will play an important role in the field of nanostructures, since the fractality occurs over a limited or even small number of orders of magnitude.

4.3 Effects of Deterministic and Random Discrete Scale Invariance on Spin Wave Spectra

4.3.1 Spin Waves Eigenmodes Problem, Integrated Density of States and Spectral Dimensionality

Fractal features of a nanostructure are expected to induce peculiar, even unusual properties on the propagation of waves as it has been shown in recently in the case of plasmons [22]. The magnetic excitations spectra of a system are linked to the existence of spin waves. In the framework of the Heisenberg model, the evolution of the spins is described by a set of differential equations involving an exchange Hamiltonian [17]; a spin is placed at each point of a lattice, in the Oxy plane and the Hamiltonian reads

$$H = -\frac{1}{2}\sum_{f,l,u} J_{f,l} S_f^u S_l^u - g\mu_B H_e \sum_f S_f^z, \qquad (4.2)$$

where the spins are vectors and u designates one of the three space directions, namely x, y or z. The exchange interaction $J_{f,l}$ is assumed to be positive and nonzero only for first neighboring sites f and l, and to have the same value for each pair of nearest neighbors. The magnetic susceptibility is assumed to be uniform within the sample; the external field H_e is also assumed to be uniform over the sample, parallel to the z axis, neglecting the fluctuating dipolar part. μ_B is the Bohr magneton and g is the Landé factor. By linearizing the equations of evolution of a set of N spins located at the sites of a lattice, one obtains a set of N linear first-order coupled differential equations under the form (Tyablikov approximation)

$$i\hbar \frac{d}{dt} S_f^+ = \left(\sum_l J_{f,l}\right) S_f^+ - \sum_l J_{f,l} S_l^+ + g\mu_B H_e S_f^+, \qquad (4.3)$$

where spin excitations (or magnons) $S_f^+ = S_f^x + i S_f^y$ lie in the Oxy plane, and the sums over the index l run over the first neighbors of the spin indexed by f. A time Fourier-transform enables to derive a dynamical matrix equation from

$$S_f^+(t) = \int e^{-i\varpi t} S_f^+(\varpi) d\varpi \qquad (4.4)$$

In the following, we shall deal with the shifted frequency: $\omega = \hbar\varpi - g\mu_B H_e$. Hence, the frequencies are deduced from an eigenvalue problem:

$$\omega S_f^+(\omega) = \left(\sum_l J_{f,l}\right) S_f^+(\omega) - \sum_l J_{f,l} S_l^+(\omega) \qquad (4.5)$$

Eigenvalues and eigenvectors of the sparse matrix deduced from Eq. (4.5), i.e., spin wave eigenfrequencies and eigenmodes, can be calculated numerically with the help of a diagonalization QR algorithm [23]. A monodimensional spin wave can be represented as depicted in Fig. 4.6 where spins $S_f^+ = S_f^x + i S_f^y$ experiment precession motions around the z axis with a frequency ω. The phase difference

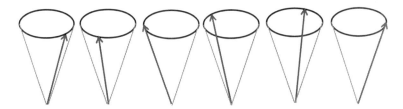

Figure 4.6 Representation of a spin wave along a ferromagnetic chain: spins experiment precession motions around a vertical axis Oz

between two successive spins is represented by the wavevector $\vec{k}$ associated to a wavelength λ. In the case of fully occupied square lattices with an elementary cell of side a, the eigenfrequencies and the quantified wavevectors are related by a dispersion relation under the form $\omega = \omega_0 \left(\sin^2 \left(\frac{k_x a}{2} \right) + \sin^2 \left(\frac{k_y a}{2} \right) \right)$; such a relation comes down to $\omega \sim k^2$ if the wavelengths are much larger than a, and it can easily be shown that the density of states $D(\omega)$ of low frequency ferromagnetic magnons is constant in the bidimensional case; more generally $D(\omega) \sim \omega^{d_s - 1}$, where the spectral dimensionality d_s is related to the space dimension d according to the relation $d_s = d/2$ when d is integer. The density of states of a system can be deduced from the computation of the eigenfrequencies by counting the number of modes in an interval $[\omega, \omega + \Delta\omega]$; in the following, we shall deal with the Normalized Integrated Density of States (NIDOS) $F(\omega) = \int_0^\omega D(u)du$, which enables us to compare more conveniently different geometry on a similar scale.

4.3.2 Discrete Deterministic Fractals

Figures 4.7 and 4.8 show spectra calculated for different lattices sharing the same size L. Whereas the NIDOS evolves regularly and continuously (at the scale ω_{max}/L^{d_f}) in the case of a translationally invariant lattice, singularity appear in the case of deterministic fractals. The strength and the number of these singularity increase as the fractal dimension is lowered, leading to "devil's staircase" spectra: Gaps open in the NIDOS and steep jumps linked to eigenfrequencies degeneracy appear. These singularity clearly prevent from

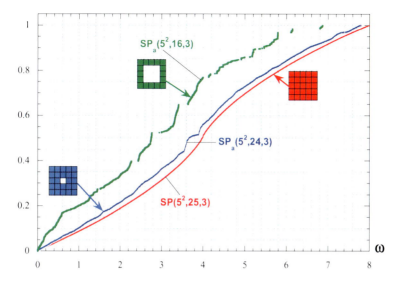

Figure 4.7 Evolution of the Normalized Integrated Density of States with the fractal dimension (1.72 for $SP_a(5^2, 16, 3)$ and 1.97 for $SP_a(5^2, 24, 3)$) for fourfold symmetrical deterministic fractal structures. The red curve is the translationally invariant lattice where a low frequency linear behavior of $F(\omega)$ can clearly be seen since $D(\omega) \sim \omega^{(\frac{d}{2}-1)}$.

defining a spectral dimension d_s of the low frequency magnons on deterministic fractals [21]. Moreover, the smoothing of the spectra associated with the increase in the number of eigenmodes with the system size $L = n^k$ does not remove the gaps and singularity, as shown in Fig. 4.9; these spectra are said to be singular continuous. It should be pointed out that the relative number of low frequency modes, expected to exhibit a long-range character, increases as d_f is lowered. Furthermore, Fig. 4.8 shows that, for a given fractal dimension, the spectra are strongly dependent on the topology of the generating cell. Let us notice that the gaps appearing in the NIDOS could be useful for magnonic applications since they induce unusual propagation properties [8, 22] and the improvement of surface techniques such as lithography [3] makes the realization of deterministic fractals possible nowadays. The possibility of designing such structures enables to control the position and width of the gaps, and to obtain magnonic filters.

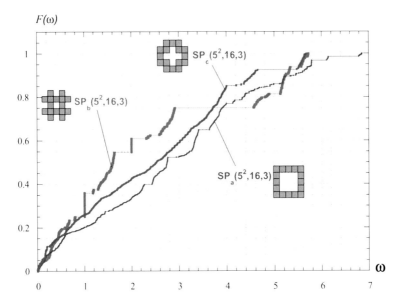

Figure 4.8 Effect of the topology of the generating cell on the Normalized Integrated Density of States for the fractal dimension $d_f \simeq 1.72$.

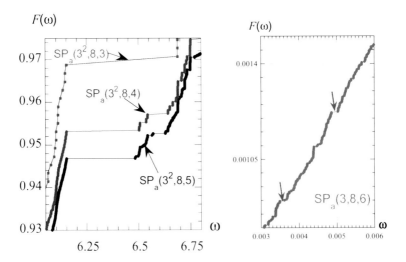

Figure 4.9 Effect of the size of the Sierpinski fractal structure on the Normalized Integrated Density of States for the fractal dimension $d_f \simeq 1.8927$; survival of the low frequency gaps as the size is increased.

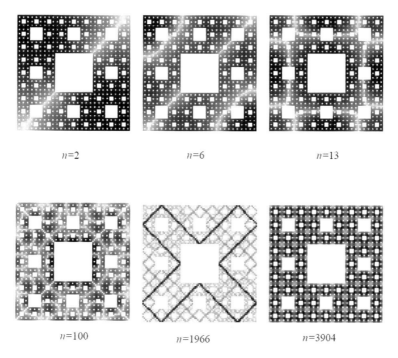

Figure 4.10 Some eigenmodes (among the 4096 ones) of the magnonic excitations on $SP_a(3^2, 8, 4)$. The gray levels represent the modules of the excitations amplitudes.

Several eigenmodes of a discrete scale invariance deterministic Sierpinski fractal are depicted in Fig. 4.10. The modulation of the amplitudes of magnetic excitations exhibits a complex structure; new geometry superposed on the underlying fractal network appears, underlined by nodal curves where the amplitude is very low. These patterns preserve some twofold or fourfold symmetry operations of the underlying network. It is worth noticing that a similar phenomena occurs in the case of vibrational properties of a Sierpinski membrane, and should be associated to some "fractal drum". The Fourier transform is less appropriate to the study of the geometrical properties of those modes than a wavelet one, since the lattices are not translationally invariant; the dispersion relation loses its relevance, and the main surviving general character, namely

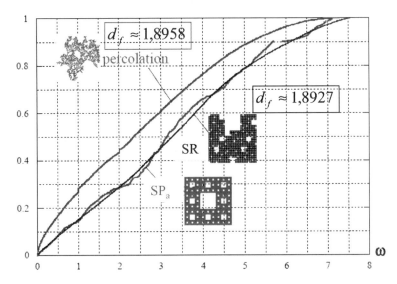

Figure 4.11 Effect of the nature of the scale invariance (discrete or not) on the Normalized Integrated Densities of States for $d_f \simeq 1.89$. Ten configurations are represented for the percolation clusters in order to show that the fluctuations are weak from a configuration to another (the NIDOS are self-averaging). A single configuration is shown for $SR_\epsilon(3^2, 5, 8)$; the fluctuations are of the same order as for percolation clusters.

the decrease in the typical size of spatial modulations as the frequency increases remains qualitative.

4.3.3 Random Scale Invariance

The NIDOS computed on a random fractal $SR_\epsilon(3^2, 8, 5)$ and percolation clusters sharing the same size are represented in Fig. 4.11 together with the NIDOS of a deterministic fractal $SP_a(3^2, 8, 5)$. These results call for two comments:

(i) By looking at the spectrum of a configuration $SR_\epsilon(3^2, 8, 5)$, it turns out that the random character of the discrete scale invariance has erased the singularity present in the deterministic case; it is worth noticing that such an effect is not an artifact linked to some averaging process, but is actually due to the intrinsic disorder of a given configuration [20].

(ii) Although the fractal dimensions of the percolation cluster and $SR_\epsilon(3^2, 8, 5)$ are very close, the NIDOS profiles are very different according to whether the scale invariance is discrete or not. A quantitative study shows that $F(\omega)$ follows low frequency power laws under the form $\omega^{0.65}$ for the percolation cluster and $\omega^{0.87}$ for $SR_\epsilon(3^2, 8, 5)$, leading to different spectral dimensions. In other words, the level of randomness quantified by the configuration entropy s_P and s_{SR} plays a crucial role in the spectra profiles [20].

4.4 Phase Transitions on Fractals

4.4.1 Critical Behavior of the Ising Model on Deterministic Fractals

4.4.1.1 Theoretical background

The magnetic behavior of a fractal structure with a strong ferromagnetic anisotropy can be described in the framework of the Ising model, where spins $1/2$ are located at the sites of percolating deterministic Sierpinski fractals. Thus, the Hamiltonian can be written $H = -J \sum_{<i,j>} s_i s_j - h \sum_i s_i$, where the spins s_i take on the values ± 1, the summation runs over all interacting first neighbors pairs and the coupling constant J is assumed to be positive and uniform all over the fractal; h denotes the external magnetic field. A second-order phase transition between a paramagnetic and a ferromagnetic phase is expected to occur at nonzero temperature T_c provided that the ramification order of the fractal is infinite. Just like translationally invariant systems, the order parameter, equal to zero in the less symmetrical phase, (paramagnetic) is the magnetization per spin m; it is worth noticing that the exchange interaction J responsible for the alignment of spins in the low temperature phase (ferromagnetic) can only be derived in the framework of quantum mechanics and has nothing to do with the dipolar interaction between magnetic dipoles.

Let us firstly recall the main features of such a transition in the framework of the standard critical phenomena theory [1] in the particular case of translationally invariant systems. The spontaneous symmetry breaking at the critical point lies in the heart of such a theory: The fundamental state of a system at zero temperature described by the Ising Hamiltonian is twice degenerated, since the spins are either all up or all down; an external field, regardless of how small it is, lifts the degeneracy since a potential barrier must be crossed in order to flip the whole set of spins. Such a symmetry breaking at the transition is closely linked to the characteristic size of the fluctuations. The main point is the emergence of a collective behavior when getting close to the critical temperature: The correlation length ξ, extracted from the two spins correlation function increases in such a way that fluctuations occur over larger and larger scales until ξ diverges right at the critical point. Hence, it is easy to understand that these fluctuations will drive the emergence of a macroscopic phenomenon, so that microscopic details are not relevant to the description of the critical behavior. Such an observation supports the striking notion of universality. In the thermodynamical limit, the singular behaviors of the order parameter m, the specific heat per spin C (thermal fluctuations of the energy), the susceptibility χ (thermal fluctuations of the order parameter), and the correlation-length ξ at the critical point are respectively described by the critical exponents $\beta, \alpha, \gamma, \nu$. Calling $t = \frac{T-T_c}{T_c}$ the relative deviation of the temperature T from the critical one T_c, they can be written in the following way: $|m(t)| \sim |t|^\beta$ with $t < 0$, $C(t) \sim |t|^{-\alpha}$, $\chi(t) \sim |t|^{-\gamma}$. $\xi(t) \sim |t|^{-\nu}$. The external field h is equal to 0, t is assumed to be small and the sign $\sim$ is understood as an asymptotic behavior when $t \to 0$. A universality class is characterized by a set of critical exponents $\{\alpha, \beta, \gamma, \nu\}$. In the case of translationally invariant systems, a universality class depends only upon the space dimension d, the symmetrical properties of the order parameter and the interaction range. It can be shown that these exponents are not independent but fulfill scaling laws [1]: The first one can be deduced from equilibrium thermodynamical properties ($\alpha + 2\beta + \gamma = 2$), and the second one, called hyperscaling law, involves the space dimensionality ($\gamma/\nu + 2\beta/\nu = d$).

The transposition of the theory recalled above to critical phenomena on fractals comes up against several conceptual difficulties [11]:

(i) Since a non-integer dimensional system is not translationally invariant, the question are to know what is going on with universality, and what role does the fractal dimension play.

(ii) Since voids are present over many scales, the definition of the two spins correlation function depends on the spin positions. As a result, it should be assumed that a characteristic length $\xi(t)$ can be brought out from the behavior of some position independent averaged two point correlation function in the vicinity of the critical point.

(iii) The generating cell is not a microscopic detail, since its topological properties are present at any discrete scale n^k. The question arises whether these multiscale properties of the underlying network leave a fingerprint on the critical behavior.

Moreover, the ϵ-expansions, based upon renormalization group perturbative methods allow the calculation of critical exponents in any non-integer dimensions. The question arises naturally whether there is a link between ε-expansions and critical phenomena on fractals.

4.4.1.2 Finite-size scaling and Monte Carlo simulations methods

Besides the trivial monodimensional case, a complete analytical solution of the statistical problem linked to the Ising model is hitherto available only in the bidimensional case. Such a solution involves a very complicated transfer matrix algebra. The difficulties in carrying out analytical calculations in non-integer dimensions make numerical simulations an alternative approach [9]. When computed through numerical simulations, i.e., for necessarily finite values of the size, L acts as a cut-off for $\xi(t)$. The finite system behaves as if L were its effective correlation length, so that the thermodynamical quantities are dependent on L. The main idea of the finite-size scaling analysis suggested by Fisher et al. is based upon an extrapolation of the results obtained from finite-size Monte Carlo simulations to the thermodynamical limit in order to compute the critical exponents. When dealing with fractal nanostructures, the simulation of finite-size systems is much more than an extrapolation, but the simulation of real nanostructures, since experiments are generally carried out far below the thermodynamical limit.

According to the finite-size scaling analysis canonical thermal averages of the system of size L read

$$m_L(t) \sim L^{-\beta/\nu} \mathcal{M}(tL^{1/\nu}) \tag{4.6}$$

$$C_L(t) \sim L^{\alpha/\nu} T \, \mathcal{C}(tL^{1/\nu}) \tag{4.7}$$

$$\chi_L(t) \sim L^{\gamma/\nu} \mathcal{X}(tL^{1/\nu}) \tag{4.8}$$

The scaling functions $\mathcal{C}(x)$, $\mathcal{K}(x)$, $\mathcal{M}(x)$, converge towards power laws of x when $x \to \infty$, and towards constant values when $x \to 0$. At the critical point, they are not dependent on L and take on fixed values, so that $C_L \sim L^{\alpha/\nu}$, $\chi_L \sim L^{\gamma/\nu}$, and $m_L \sim L^{-\beta/\nu}$. Finite-size effects replace the divergence of the susceptibility by a smoothed peak and shift $T_c^\chi(L)$ from its position T_c when $L \to \infty$:

$$T_c^\chi(L) = T_c + \mathcal{F}_\chi L^{-1/\nu}, \tag{4.9}$$

where $\mathcal{F}_\chi$ is a constant. Moreover, the maxima of $\chi_L(t)$ scale as $\chi_L^{max}(L) \sim L^{\gamma/\nu}$. The latter relation allows the computation of the ratio of exponents γ/ν, while the critical temperature T_c can be extracted from equation (4.9) provided that ν is known. ν is calculated through analyzing the size dependence of logarithmic derivatives $\Phi_L^n(t)$ of the magnetization M ($\beta_B = 1/k_B T$, E is the total energy and $\langle \rangle_T$ designates canonical thermal averages):

$$\Phi_L^n(t) = \frac{\partial \ln(M^n)}{\partial \beta_B} = \langle E \rangle_T \left\langle M^n \right\rangle_T - \left\langle E M^n \right\rangle_T \tag{4.10}$$

The maxima of $\Phi_L^n(t)$ scale as $\Phi_{max}^n(L) \propto L^{1/\nu}$, and their positions scale in a similar way as in equation (4.9): $T_c^{\Phi^n}(L) = T_c + \mathcal{F}_{\Phi^n} L^{-1/\nu}$. Hence the set of critical exponents and critical temperatures can be deduced from data obtained by simulating finite-sized fractals in the canonical situation. The consistency of the finite-size scaling analysis can be insured by calculating γ/ν and $1/\nu$ in two different ways. Let us notice that T_c can be calculated independently of ν with the help of the fourth-order magnetization cumulant $U_L(T) \equiv 1 - \left\langle M^4 \right\rangle_T / \left(3 \langle M^2 \rangle_T\right)$: Binder, showed that $U_L(T)$ should exhibit a fixed point at $T = T_c$ ($U_L(T_c) = U^*$ where U^* is a universal value that does not depend on the size of the system).

A necessary condition should be satisfied when generalizing the Finite-size analysis to the case of deterministic Sierpinski fractals:

The choice of L should be restricted to an integer power of the size n of the generating cell of the Sierpinski lattice in order to keep invariant the structure of the renormalized lattice [1]. Such a choice is closely linked to the discrete scale invariance [24], as already noticed when discussing the meaning of the Hausdorff dimension in Section 4.2.4. Such a constraint strongly hinders the finite-size analysis: Since the available sizes grow as a geometrical series, few points will be available, so that it will be necessary to carry out simulations on very large lattices; hence the computational cost expected to achieve such a task will be very high. Moreover, simulations come up against the fact that the convergence towards the thermodynamical limit, when increasing k, occurs at the same time that the fractal structures are constructed, leading to differences in topological properties of a fractal from an iteration step to the next (equation 4.1). One has to keep topological properties as close as possible to the thermodynamical limit in order to get rid of some drawback in the finite-size scaling analysis; hence one has to simulate sizes as large as possible.

Canonical Monte Carlo simulation are based upon a sampling of the phase space by building Markov chains, that is a series of spin configurations on a fractal lattice $SP_g(n^2, N_{oc}, k)$ drawn among the $2^{L^{d_f}}$ possible ones, according to a transition probability designed to sample correctly the behavior at thermal equilibrium and satisfying the criteria of accessibility and detailed balance [9]. The most popular algorithm is the Metropolis one. Canonical Monte Carlo simulations come up against a difficulty peculiar to second-order phase transition, the critical slowing down [9, 13]: Since the correlation length is very large, successive configurations generated in the critical region by randomly drawing spins on the network can be correlated over a large number of steps.

As a result of the critical slowing down, the statistical errors $\delta \langle A \rangle$ associated with the thermal averages of A (practically A is either the energy or the magnetization) extracted from a Monte Carlo run of N_s steps are enhanced by a term called statistical inefficiency; the stochastic kinetics associated with the Monte Carlo process enables to relate the statistical inefficiency to the integrated autocorrelation time τ_A associated with A [13]:

$$(\delta \langle A \rangle_T)^2 \simeq \frac{1}{N_s} \left(\langle A^2 \rangle_T - \langle A \rangle_T^2 \right) \left[1 + 2\frac{\tau_A}{\delta\theta} \right], \qquad (4.11)$$

where $\delta\theta$ is the unit time associated with one Monte Carlo step. It is worth noticing that, since the correlations between successive configurations increase with the lattice sizes, the number of Monte Carlo steps needed to obtain thermodynamical averages with a given accuracy can increase very quickly with the iteration step of the fractal lattice; τ_A can significantly be reduced by the use of nonlocal algorithms instead of the Metropolis one: The mapping between the percolation transitions and the magnetic ones, discovered by Fortuin and Kasteleyn, led Swendsen and Wang and Wolff to develop algorithms able to reduce significantly the critical slowing down [9]. Instead of flipping randomly chosen spins, one constructs one (or many) cluster at each Monte Carlo step. Moreover, as a result of thermal fluctuations, a canonical simulation at a given temperature T_0 brings information in the vicinity of T_0; an optimization of the data processing can be done with the help of the histogram method, which enables to calculate the thermodynamical averages over some range ΔT around the simulation temperature T_0. Although the reliability range ΔT decreases as the lattice size increases, it turns out to be very useful in looking for peaks in the thermodynamical quantities [9, 11, 13, 14].

4.4.1.3 Overall review of the numerical simulation results: weak universality

The whole set of results of canonical Monte Carlo simulations supports the occurrence of a second-order phase transition of the Ising model on deterministic Sierpinski fractals provided that the ramification order is infinite. It turns out from intensive simulation carried out on fourfold Sierpinski fractals whose dimension lies between 1 and 2 that the finite-size analysis suffers from scaling corrections all the more important that the fractal dimension is lowered [11, 12]. They are much less important when the fractal dimension lies between 2 and 3 [13]. These corrections sometimes prevent from calculating the critical exponents and only bounds can be provided. Nevertheless, $\chi_L^{max}(L)$ always follows a power law with a good reliability, enabling to calculate the ratio of exponents γ/ν

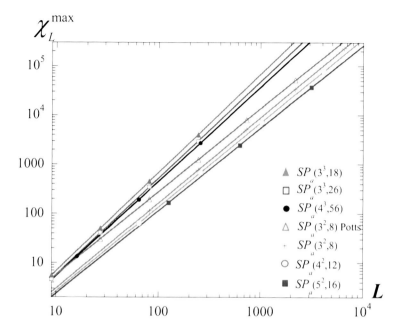

Figure 4.12 The maxima of the susceptibility $\chi_L^{max}(L)$ as a function of the size of the lattices for six fractals nanostructures investigated in the case of the Ising model. The open triangles show the values associated with the three-state Potts model on $SP_a(3^2, 8)$.

with a good precision as shown in Fig. 4.12. Scaling corrections affecting the behaviors of $m_L(T_c)$, $\Phi_L^n(T_c)$, $\Phi_n^{max}(L)$ and $U_L(T_c)$ when $1 < d_f < 2$ cannot be explained only by an usual finite-size effect, but also by a topological contribution to scaling corrections. The scaling corrections increase when the fractal dimension is lowered from 2 to 1 and the comparison between the results obtained for $SP_a(5^2, 24)$ and $SP_a(5^2, 16)$ confirms their topological nature, since the sizes investigated are the same. A contribution to scaling corrections arises from the slow convergence of the transition towards the thermodynamical limit as the size of the lattice increases. This convergence speed can be closely related to the behavior of the mean number of links per site $z_s(n^d, N_{oc}, k)$ with k discussed in Section 4.2.6.2 and driven by the reason N_S/N_{oc}. Scaling correction are almost not noticeable on $SP_a(5^2, 24)$ and it appears clearly that $z_s(5^2, 24, k)$ converges much faster towards its limit than $z_s(5^2, 16, k)$.

Despite the difficulties encountered in the calculation of exponents when the fractal dimension lies between 1 and 2, it turns out that the values of γ/ν and β/ν calculated from the whole set of simulations are always compatible with the hyperscaling relation $\gamma/\nu + 2\beta/\nu = d_f$ where the space dimension is the Hausdorff one.

The values of the ratio of exponents γ/ν calculated from the whole set of simulations are shown in Fig. 4.13. The results show evidence that, for a given fractal dimension, these values depend on the construction rule g of the fractal; hence the universality class is not the same if the removed squares of the generating cell are at its center or at the corners. These results provide a

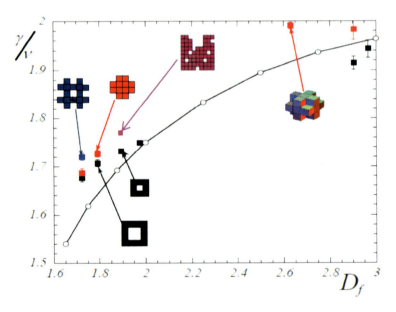

Figure 4.13 Ratio of exponents γ/ν calculated from Monte Carlo simulations as a function of the fractal dimension for 11 Deterministic fractals (and a random one). The open circles linked by a full black line are the values from ϵ-expansions. The removed squares are in the center of the generating cells for fractal represented by black squares, at the corners of the generating cells for the ones represented by red squares.

direct quantitative evidence that the usual statement of universality does not hold in the case of fractals, as already pointed out by Gefen et al. Although the hyperscaling relation is satisfied when the space dimension is replaced by the Hausdorff one, the set of critical exponents $\{\alpha, \beta, \gamma, \nu\}$ does not depend only upon the Hausdorff dimension, the symmetry of the order parameter and the interaction range but also upon topological features of the fractal, such as the geometrical properties of the generating cell. The critical behavior of the Ising model on fractals is said to be understood in the framework of weak universality [14]. It is worth noticing that such a result is in agreement with the fact that the values of γ/ν are significantly different from the results of ϵ expansions. The interpolation between integer dimension does not match a single hypothetical and somewhat universal fractal structure.

4.4.2 First-Order Magnetic Transitions on Deterministic Fractals: The Potts Model

4.4.2.1 Theoretical background

The Potts model generalizes the Ising one by taking into account the fact that spins can be in more than two discrete states. The Hamiltonian of the q-state Potts model reads $H = -J \sum_{\langle i,j \rangle} \delta(\sigma_i, \sigma_j)$, where σ_i and σ_j designate the spin states at the occupied sites i and j of the lattice and can take the integer values $1, 2, \ldots, q$. The sum runs over the nearest-neighbor spins pairs, $\delta(\sigma_i, \sigma_j)$ is equal to 1 if $\sigma_i = \sigma_j$, and 0 otherwise. $J > 0$ is the exchange ferromagnetic coupling uniform all over the lattice. For a given size L and a given spin configuration, the order parameter of the phase transition reads $m_L = \frac{q\rho_L - 1}{q - 1}$ where $\rho_L = \max\left\{N_1/N, \cdots, N_q/N\right\}$ and N_{q_0} is the number of spins whose state is q_0. Since the number of states q is related to the symmetrical properties of the order parameter, the critical behavior of the Potts model, in particular the universality class (excepted for $q = 2$), depends upon the value of this additional variable. Furthermore, one of its most striking features is the effect of q on the order of the transition as a function of the space dimension d: In the case of translationally invariant lattices, the phase transition in the bidimensional case is first order for $q > 4$ and second order for $q \leq 4$, whereas in the three-dimensional case, it is first order for $q \geqslant 3$ and second order for $q \leqslant 2$. This suggests the relevance of a phase diagram in the (d, q) plane [16], generalized to non-integer values of q and d, where some border should separate a first-order domain, where q is larger than a critical value $q_c(d)$, from a second-order one. The question of the physical meaning of the extrapolation to non-integer integer dimensions leads naturally to study phase transitions of the Potts model on fractal structures.

4.4.2.2 Multicanonical simulation of first-order transitions: the Wang Landau algorithm

Although a collective behavior occurs also in the case of first-order transitions, the correlation length remains finite and the two phases coexist at the transition temperature T_c, provided that the typical size of the ordered domains is smaller than the size of the sample. Since the energies of the ordered and disordered phases are separated by an interfacial energy, the probability density $\mathcal{P}_T(E)$ of the energy per occupied site is double peaked; it should be pointed out that the energy per spin is uniform inside a given domain while it varies along its interface according to the local topology (number of occupied first neighbors and local curvature of the domain wall). Hence, Monte Carlo simulations come up against difficulties linked to the existence of metastable states in the vicinity of T_c. Although the difficulties encountered in Monte Carlo simulations of first- and second-order phase transitions are quite different, they can be stated in terms of stochastic dynamics [15]. The characteristic time $\tau_A(L)$ extracted from the integrated autocorrelation time associated with a physical quantity A provides a measure of the statistical inefficiency in the vicinity of second-order phase transitions as recalled in Section 4.4.1.2. In the case of first-order transitions, calling $T_c(L)$ the effective temperature where the two peaks in $\mathcal{P}_T(E)$ have the same height, E_1 and E_2 the energies where these peaks occur, and E_0 the energy of the minimum between these peaks, a tunneling time $\tau_{\text{tun}}(L)$ is defined as the average number of Monte Carlo steps needed to go from a configuration where $E = E_1$ to another one where $E = E_2$, and back. $\tau_{tun}(L)$ is closely linked to the time needed by a stochastic Monte Carlo canonical dynamics to tunnel through the interfacial free energy barrier between the ordered and disordered phases when $\xi < L$. $\tau_A(L)$ usually grows as a power law of the system size while $\tau_{tun}(L)$ increases often exponentially, (or faster) with L. Hence, it is quite understandable that canonical simulations are inefficient in sampling properly the configuration space when two phases coexist. As an alternative method, the multicanonical random walk algorithm proposed by Wang and Landau [9], which enables to compute in an adaptative way the density of states $g(E)$ over the whole energy range $[E_{min}, E_{max}]$ of a finite system, turns out to be very efficient in overcoming this difficulties. Once $g(E)$ has been computed, the canonical partition function $\mathcal{Z}_T$ can be calculated whatever the value of the temperature T, according to its definition $\mathcal{Z}_T = \sum_{E_{min}}^{E_{max}} \exp(\ln g(E) - \beta_B E)$. Hence, for a given finite lattice $SP_g(n^d, N_{oc}, k)$, this method makes possible to calculate the canonical mean energy, the specific heat, the free energy and the entropy, without any a priori knowledge of the order of the transition. Unfortunately the Wang Landau algorithm does not enable to simulate large sizes. Let us point out that a bimodal density probability of the energy is not a sufficient condition for a phase transition to be a first-order one: The depth of the well separating the peaks associated with the coexistence of the two phases must increase with the system size L at the shifted transition temperature (where the two peaks have the same height). Figure 4.14 shows typical cases where the transition are unambiguously first-order ones.

4.4.2.3 Canonical simulations of first-order transitions: the Meyer-Ortmanns and Reisz criterion

If the correlation length ξ of the ordered domains is much larger than L, but remains finite, canonical simulations at a fixed temperature T may lead to the observation of a pseudo-critical behavior whereas the transition is a first-order one: As long as $\mathcal{P}_T(E)$ remains monomodal, the canonical sampling is correct, but its single peaked character does not enable to claim that the

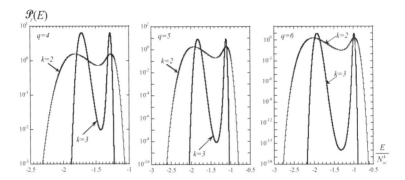

Figure 4.14 Probability densities $\mathcal{P}_T(E)$ of the energy per occupied site at the temperatures T where the two peaks have the same height at the second and third iteration step of $SP_a(3^3, 26)$; the fractal dimension of the nanostructures is $d_f \simeq 2.9656$. $\mathcal{P}_T(E)$ is represented in a logarithmic scale as a function of the energy per spin for $q = 4$, 5, and 6.

transition is a second-order one. For both first- and second-order transitions, $\chi_L(t)$ diverges at the transition temperature as $L \to \infty$. Near the transition points, finite-size effects round the divergence of thermodynamical averages and shift the rounded peaks away from T_c: $\chi_L(T)$ exhibits a peak at a maximum value $\chi_L^{max}(L)$ for a temperature $T_c^\chi(L)$. The criterion of Meyer-Ortmanns and Reisz [10] enables to discriminate between a first and a second-order transition; it rests only on the fact that the asymptotic behavior of $\chi_L(t)$ at $T_c^\chi(L)$ has a δ-function like singularity in the case of a first-order transition, and a power law type singularity in the case of a second-order one, as L tends to infinity. Hence, in the case of second-order transitions, the curves representing $\chi_L(t)$ do not intersect in their wings as L increases. Figure 4.15 shows the behavior of the susceptibility as a function of the iteration step in two typical cases of first and second-order transitions on deterministic Sierpinski and Menger fractals.

4.4.2.4 Phase diagram of the Potts model

The phase diagram of the Potts model in the (d, q) plane summarizing the set of simulations we carried out is shown in Fig. 4.16. The

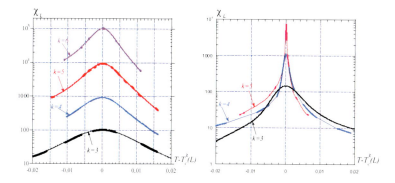

Figure 4.15 Left: Behavior of the susceptibility χ_L of the order parameter in the case of a second-order transition for the 10-state Potts model as a function of the shifted temperature on the fractal nanostructures $SP_a(4^2, 12, k)$, with a fractal dimension 1.7925 where L is equal to 64, 256, 1024, and 4096. Right: Case of a weakly first-order transition for the four-state Potts mode on the fractal nanostructures $SP_a(3^3, 18, k)$ with a fractal dimension 2.631 where L is equal to 27, 81, and 243.

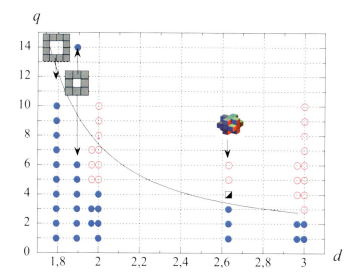

Figure 4.16 Phase diagram of the Potts model as a function of the space dimension d and the number of spin states q; the open red circles indicate a first-order transition whereas the blue ones indicate a second-order one and the square a weakly first-order transition. The full line is the Andelman–Berker approximation.

evolution of the border separating the first-order regime from the second-order one is in qualitative agreement with the approximate analytical result of Andelman and Berker if the fractal dimension is large enough. Furthermore, it should be pointed out that the lower bounds $q_c(1.8928) > 14$ and $q_c(1.7925) > 10$ [15] suggest that second-order transitions can occur for very large values of q in the case of hierarchically weakly connected systems, that is when the fractal dimension is significantly smaller than 2. Although a small decrease of the dimension from 2 to 1.9746 in the case of $SP_a(5^2, 24, k)$ implies a strong deviation from the translational invariance, it is not able to induce a second-order transition even for the 5-state Potts model. Moreover, in the cases where $d_f = 2.6309$, and $d_f = 2.9656$, the transitions are first order for small values of q. As a main conclusion, it turns out that the translational symmetry breaking linked to the discrete deterministic scale invariance of the fractals structures is not able to induce second-order transitions, unlike the case of bond randomness or aperiodic couplings.

4.4.3 Critical Behavior of the Ising and Potts Model on Random Fractals

In the case of random discrete scale invariance, besides thermo-dynamical averages, the physical quantities must be averaged over the disorder of fractal networks: Calling $\mathcal{O}$ an observable (energy, order parameter, ...) the thermodynamical average simulated over a given lattice $SR_\epsilon(n^2, N_{oc}, k)$ reads $\langle \mathcal{O} \rangle_T^{\epsilon(L)} = 1/N_s \sum_{i=1}^{N_s} \mathcal{O}_i^{\epsilon(L)}$, where $\mathcal{O}_i^{\epsilon(L)}$ is the value of an observable associated to one generated spin configuration and N_s is the steps number of a Monte Carlo run. The average value of $\mathcal{O}$ over the random fractality can be written:

$$\overline{\langle \mathcal{O}_L \rangle_T} = \sum_{\{\epsilon(L)\}} \sum_{i=1}^{N_s} \mathcal{O}_i^{\epsilon(L)} \mathcal{P}(\epsilon(L)), \qquad (4.12)$$

where $\mathcal{P}(\epsilon(L))$ is the weight of the random fractal defined by the set $\epsilon(L)$; an estimator for $\overline{\langle \mathcal{O}_L \rangle_T}$ is computed from an average over a sample of $\mathcal{N}$ different random fractals, so that $\mathcal{P}(\epsilon(L)) = 1/\mathcal{N}$. The computational cost at fixed values of T and L is proportional to $\mathcal{N}.N_s$. The variability of $\langle\mathcal{O}\rangle_T^{\epsilon(L)}$ can be very important from a random lattice to another one. Hence, self-averaging properties of disorder averaged quantities is of main interest and can be conveniently characterized by the relative variance $R_{\mathcal{O}}(L) = \left(\overline{[\langle\mathcal{O}_L\rangle_T]^2} - \overline{[\langle\mathcal{O}_L\rangle_T]}^2\right) / \overline{[\langle\mathcal{O}_L\rangle_T]}^2$. A quantity is said to be strongly self-averaging if $R_{\mathcal{O}}(L) \sim L^{-d}$ where d is the space dimension and weakly self-averaging if $R_{\mathcal{O}}(L) \sim L^{-a}$ where $0 < a < d$. $\mathcal{O}$ is said to exhibit lack of self-averaging if $R_{\mathcal{O}}(L)$ tends towards a nonzero constant as $L \to \infty$. The main results of the simulations carried out on $SR_\epsilon(3^2, 8, k)$ for $q = 2$ and $q = 14$ are the following [18, 19]:

(i) The transitions are second-order ones: This result suggests that random discrete scale invariance on low-dimensional systems (above a critical dimension still remaining to be calculated) can exhibit long-range order even for high values of the number of states of the Potts model.

(ii) Although the disorder of the fractals is strongly constrained by the discrete random scale invariance as already noticed in Section 4.2.5, the magnetic susceptibility lacks self-averaging.

(iii) Scaling corrections are much weaker than in the deterministic case. These corrections were attributed to a topological character of the fractal. Such an interpretation is strengthened by the present result, since topological constraints induced by random fractality are much weaker than in the deterministic one; it is as if the averaging process over randomness had reduced these topological scaling corrections.

(iv) Critical exponents can be calculated with a precision enabling to evidence new universality classes satisfying the hyperscaling relation; these exponents are clearly different from the ones estimated by the ϵ-expansions where $d_f \simeq 1.8928$. Averaging over the disorder does not restore any hypothetical translational invariance underlying the ϵ-expansions.

At last, the comparison between the deterministic fractal $SP_a(5^2, 24, k)$ and the random one $SR_\epsilon(5^2, 24, k)$ where the transition are first order when $q = 7$ shows that the disorder constrained by fractality does not induce a second-order transition. Let us recall that an infinitesimal amount of disorder can induce

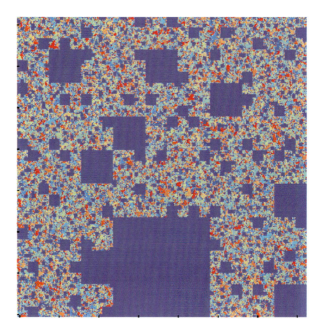

Figure 4.17 Configuration of spin states computed by Monte Carlo simulation of the 14-state Potts model on $SR_\epsilon(3^2, 8, 6)$ in the ferromagnetic phase at $T = 0.54 > T_c$; each color is associated to a spin state except the blue-violet showing the unoccupied squares at different scales.

a second-order transition of the Potts model from a transition exhibiting a first-order one ($q > 4$) in the translationally invariant two-dimensional case.

Figures 4.17 to 4.20 show the evolution of some spin configurations simulated at equilibrium for a given temperature T when T is decreased from the paramagnetic to the ferromagnetic phase in the case of the second-order transition associated to $q = 14$ on the low dimensionality fractal nanostructures $SR_\epsilon(3^2, 8, 6)$ where $d_f \simeq 1.89533$. The emergence of a collective behavior when approaching the critical point can clearly be seen when looking at clusters of spins sharing the same value of q whose mean characteristic size ξ increases as T is decreased. Moreover, the presence of hierarchically structured voids has a strong effect on the shape of clusters: It is as if there were a coupling between the statistical scale invariance

Figure 4.18 Configuration of spin states computed by Monte Carlo simulation of the 14-state Potts model on $SR_\epsilon(3^2, 8, 6)$ in the paramagnetic phase at $T = 0.5 > T_c$.

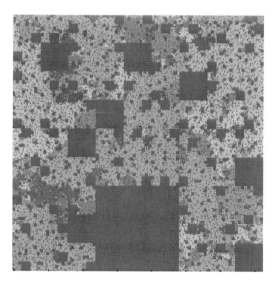

Figure 4.19 Configuration of spin states computed by Monte Carlo simulation of the 14-state Potts model on $SR_\epsilon(3^2, 8, 6)$ in the vicinity of the critical temperature at $T = 0.496$. The effect of the discrete scale invariance on the clusters geometry can easily be seen.

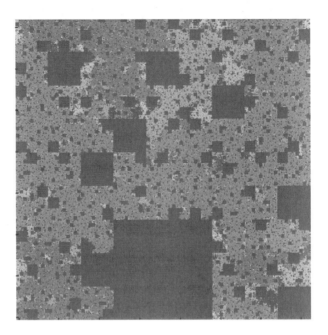

Figure 4.20 Configuration of spin states computed by Monte Carlo simulation of the 14-state Potts model on $SR_\epsilon(3^2, 8, 6)$ in the ferromagnetic phase $T = 0.493 < T_c$; each color is associated to a spin state.

associated to the emergence of the collective behavior until the scale ξ and the random discrete scale invariance of the underlying Sierpinski network associated to the scales n^k. In the case of discrete spin models on deterministic fractals such a coupling is reflected in the statement of weak universality. Furthermore, when dealing with the spin waves described in the framework of the Heisenberg model, localization patterns emerging from the properties of the eigenmodes (Fig. 4.10) can be seen as a coupling between scale invariance of the underlying network and the set of wavevectors that characterize a given spin wave.

References

1. Binney J. J., Dowrick N. J., and Fisher A. J. (1992), *The Theory of Critical Phenomena*, Oxford University Press.

2. Bunde A. and Havlin S. (1996), *Fractals and Disordered Systems*, Springer, Berlin Heidelberg.

3. Cerofolini G. F., Narducci D., Amato P., and Romano E. (2008), Fractal nanotechnology, *Nanoscale Res. Lett.* **3**, 381–385.

4. Gefen Y., Aharony A., and Mandelbrot B. B. (1984), Phase transitions on fractals. III. Infinitely ramified lattices, *J. Phys. A: Math. Gen.* **17**, 1277–1289.

5. Gouyet J. F. (1992), *Fractales et structures Fractales*, Masson, Paris.

6. Gracheva I. E, Moshnikov V. A., Maraeva E. V., Karpova S. S., Alexsandrova O. A., Alekseyev N. I., Kuznetsov V. V., Olchowik G., Semenov K. N., Startseva A. V., Sitnikov A. V., and Olchowik J. M. (2012), Nanostructured materials obtained under conditions of hierarchical self-assembly and modified by derivative forms of fullerenes, *J. Non-Crystalline Solids* **358**, 433–439.

7. Kannan P., Maiyalagan T., and Opallo M. (2013), One-pot synthesis of chain-like palladium nanocubes and their enhanced electrocatalytic activity for fuel-cell applications, *Original Nano Energy* **2**, 677–687.

8. Kruglyak V. V., Demokritov S. O., and Grundler D. (2010), Magnonics, *J. Phys. D-Appl. Phys.* **43**, 264001–264015.

9. Landau D. P. and Binder K. (1996), *A Guide to Monte Carlo Simulations in Statistical Physics*, Cambridge University Press.

10. Meyer-Ortmanns H. and Reisz T. (1998), The monotony criterion for a finite size scaling analysis of phase transitions, *J. Math. Phys.* **39**, 5316–5323.

11. Monceau P., Perreau M., and Hébert F. (1998), Magnetic critical behavior of the Ising model on fractal structures, *Phys. Rev. B* **58**, 6386–6393.

12. Monceau P. and Perreau M. (2001), Critical behavior of the Ising model on fractals structures in dimensions between 1 and 2: Finite-size scaling effects, *Phys. Rev. B* **63**, 184420-1–10.

13. Monceau P. and Hsiao P. Y. (2002), Cluster Monte-Carlo distributions in fractal dimension between two and three: Scaling properties and dynamical aspects for the Ising model, *Phys. Rev. B* **66**, 104422-1–5.

14. Monceau P. and Hsiao P. Y. (2004), Direct evidence for weak universality on fractal structures, *Physica A* **331**, 1–9.

15. Monceau P. (2006), Critical behavior of the ferromagnetic q-state Potts model in fractal dimensions: Monte Carlo simulations on Sierpinski and Menger fractal structures, *Phys. Rev. B* **74**, 094416-1–12.

16. Monceau P. (2007), First order phase transitions of the Potts model in fractal dimensions, *Physica A* **379**, 559–568.

17. Monceau P. and Lévy J. C. S. (2010), Spin waves in deterministic fractals, *Phys. Lett. A* **374**, 1872–1879.

18. Monceau P. (2011), Critical behavior of the Ising model on random fractals, *Phys. Rev. E.* **84**, 051132-1-10.

19. Monceau P. (2012), Effects of random and deterministic discrete scale invariance on the critical behavior of the Potts model, *Phys. Rev. E* **86**, 061123-1-7.

20. Monceau P. and Lévy J. C. S. (2012), Effects of deterministic and random scale invariance on spin wave spectra, *Physica E* **44**, 1697–1702.

21. Nakayama T., Yakubo K., and Orbach R. L. (1994), Dynamical properties of fractal networks: Scaling, numerical simulations and physical realizations, *Rev. Mod. Phys.* **66**, 381–443.

22. Ng K. F., Leung C. W., Jim K. L. (2014), Influence of center fractal patterns on the transmission spectrum and electric field intensity enhancement in gold/glass plasmonic nanostructures, *Microelectronic Engineering* **119**, 79–82.

23. Press W. H., Flannery B. P., Teukolsky S. A., and Vetterling W. T. (1986), *Numerical Recipes, the Art of Scientific Computing*, Cambridge University Press.

24. Sornette D. (1998), Discrete scale invariance and complex dimensions, *Phys. Rep.* **297**, 239–270.

25. Werner G. (2010), Fractals in the nervous system: Conceptual implications for theoretical neuroscience, *Frontiers in Physiology*, vol. **1**, 15, 1–28.

26. Zhu Y. Q., Hsu W. K., Zhou W. Z., Terrones M., Kroto H. W., and Walton D. R. M. (2001), Selective co-catalysed growth of novel MgO fishbone fractal nanostructures, *Chem. Phys. Lett.* **347**, 337–343.

Chapter 5

Magnetic Anisotropy and Magnetization Reversal in Self-Organized Two-Dimensional Nanomagnets

Vincent Repain

Matériaux et Phénomènes Quantiques, Université Paris Diderot et CNRS,
10 rue A. Domon et L. Duquet, 75205 Paris Cedex 13, France
vincent.repain@univ-paris-diderot.fr

5.1 Introduction

The study of magnetism at the nanoscale is mandatory for the scalability of all non-volatile memories based on magnetic materials and still remains a challenging task at the fundamental level. Experimentally, different routes are explored, from the single nanoparticle study to the properties of bulk nanocomposite materials. The measurement of magnetism in single particles is generally not easy due to the small related signal. Pioneering micro-SQUID measurements have provided among the first data on the magnetization reversal (Wernsdorfer et al., 1997) and magnetic anisotropy (Jamet et al., 2004) at the nanoscale but the experiments are still challenging and scarce, limited to very specific

---

Magnetic Structures of 2D and 3D Nanoparticles: Properties and Applications
Edited by Jean-Claude Levy
Copyright © 2016 Pan Stanford Publishing Pte. Ltd.
ISBN 978-981-4613-67-5 (Hardcover), 978-981-4613-68-2 (eBook)
www.panstanford.com

samples. More recently, scanning probes microscopies such as spin-polarized scanning tunneling microscopy have been used to study the magnetic reversal of single epitaxial magnetic clusters (Krause et al., 2009; Ouazi et al., 2012b) with intriguing results. However, in these last cases, the possible interaction with the magnetic tip makes the interpretation of the results somehow delicate and the technique itself remains of complex use. Averaging methods such as SQUID, vibrating sample magnetometer, and magneto-optical effects are much more widely used to characterize large ensembles of nanoparticles, in either bulk or thin films forms. The size, shape, and structural distributions among the nanoparticles together with interactions generally leads to a complex averaged signal that has been widely studied by applied magnetism methods (Blanco-Mantecon and O'Grady, 2006). Meanwhile, the physics of self-organized surfaces and ordered growth of epitaxial nanostructures was developed with an impressive control on the size distribution (Brune et al., 1998) and homogeneity (Repain et al., 2002) of the samples. The study of nanomagnetism with such model samples is tempting and appears finally very fruitful. In this chapter, I will review the results that we have obtained over the last few years on the physics of magnetization switching at the nanoscale and the magnetic anisotropy in pure, core–shell and alloyed nanoparticles, all self-organized on crystalline surfaces.

5.2 Self-Organized Magnetic Nanocrystals Supported on Crystalline Surfaces

The discovery of self-organization on crystalline surfaces at the nanometer scale has arisen in the nineties with the growing number of scanning probe microscopes around the world. One of the most famous patterns is the herringbone reconstruction of the Au(111) surface (cf. Fig. 5.1a) (Barth et al., 1990). To minimize the surface energy, this surface increases its surface density by 4% and the extra atoms rearrange in a long-range ordered pattern with a herringbone shape. As a general result for such self-organized systems, the period of this reconstruction is driven by the competition between an elastic energy release and a boundary energy cost (Narasimhan and

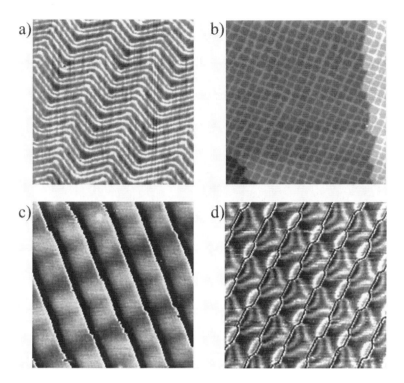

Figure 5.1 STM images of self-organized surfaces with periodic nanometer scale patterns. (a) Herringbone reconstruction of the Au(111) surface. The typical periodicity is 30 nm. (b) CuN islands on a Cu(100) surface. The periodicity is around 5 nm. (c) Reconstruction of the Au(788) vicinal surface that defines a two dimensional pattern with step edges of dimensions 7.2 nm by 3.8 nm. (d) Reconstruction of an Au vicinal surface with 8 nm-wide terraces. The interplay between step edges and the strain-relief pattern induces a complex morphology.

Vanderbilt, 1992). One of the drawbacks of such pattern is that it is very sensitive to any elastic field like dislocations or step edges, changing its periodicity and coherency over the sample. The use of pure well-oriented good single crystals is therefore mandatory to get a homogeneous self-organization. A two-dimensional square lattice with a period of 5 nm can also be obtained by the chemisorption of atomic nitrogen on a Cu(100) surface at around 600 K (Ellmer et al., 2001), as can be seen in Fig. 5.1b. Copper nitride islands, imaged as depletions by STM due to their insulating character, self-order with some bare Cu area in between. The important change of sticking coefficient between the bare metal and the nitride makes the latter an ideal mask for further growth of ordered nanoparticles (Ellmer et al., 2002). Unfortunately, most of these self-organized crystalline surfaces have a natural periodicity and symmetry that is generally very difficult to modify. Few tries to control the period and coherency of such lattices have been done using vicinal surfaces, i.e., showing a regular atomic steps array. By cutting the crystal with different misorientation angles, it is therefore possible to change one distance, i.e., the terrace width between two step edges. In the other direction, Au vicinal surfaces display a surface reconstruction that makes a two-dimensional pattern. Examples are given in Figs. 5.1c,d that show the surface pattern of an Au(788) and a facet of Au(12,11,11) surface, respectively (Rousset et al., 2003). Changing the miscut angle has modified the period of the self-organized lattice from 3.8×7.2 nm on Au(788) to 8×14 nm on the facet of Au(12,11,11). It has also to be noticed that these vicinal surfaces generally display self-organized patterns that are coherent over a macroscopic scale in contrary to flat surfaces, what is of great interest for the study of physical properties by averaging techniques (Ortega et al., 2002).

Beyond the fundamental physics of such unusual nanoscale patterns, these substrates have been proposed very soon as templates for ordered growth of nanoparticles. However, the transfer from a surface pattern to a lattice of nanoparticles of a given material is not always obvious. The subtle variations of the diffusion coefficient or place exchange energy barrier among a surface dislocations lattice like in Au(111) leads nevertheless to ordered growth in a certain temperature range, including room temperature in this case for elements such as Fe, Co, and Ni (Chambliss et al., 1991; Stroscio et al., 1992; Voigtlander et al., 1991)(cf. Fig. 5.2a for a STM image of ordered Co nanoparticles on this surface). However, as already noticed, the coherency of the lattice of nanoparticles on this surface is typically limited to around 100 nm, what can be an issue if one wants to study for example the magnetic ordering between particles with averaging techniques. It also slightly enlarges the size distribution. In order to overcome these drawbacks, we have proposed to use vicinal surfaces where the symmetry breaking

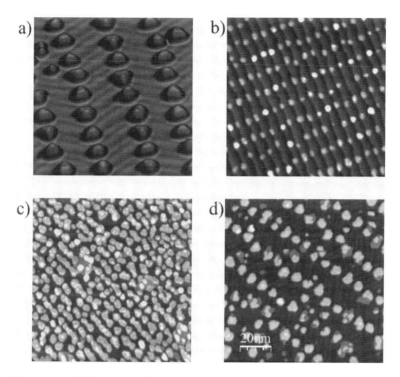

Figure 5.2 STM images of ordered growth of magnetic nanoparticles on surfaces. (a) 60 nm 3D image of Co bilayer islands on an Au(111) surface. (b) Array of Co nanodots on Au(788) (60 × 60 nm). (c) Quasi-unidimensional Co nanostructures on strained Au(111) (100 × 100 nm). (d) CoPt alloy nanostructures on Au(111) (100 × 100 nm).

induced by the step array naturally increases the coherency of the lattice to a macroscopic scale. In the case of Co on Au(788), a deposition at 130 K was needed to obtain the best ordered growth of Co nanoparticles (Repain et al., 2002) (cf. Fig. 5.2b). Fe nanoparticles can be also ordered on this surface with a similar size distribution, although with different growth conditions (Rohart et al., 2008). It is worth noting that more recently, the use of Moiré patterns between a graphene sheet and metallic substrates has also lead to ordered growth with a very good long-range order (N'Diaye et al., 2009). Quasi-unidimensional Co nanostructures can also be grown with such patterns. The most famous example is the realization of Co atomic wires along the step edges of a Pt vicinal surface (Gambardella et al., 2002). It is also possible to induce a one-dimensional coalescence of clusters by using an anisotropic self-organized surface. This is the case, for example, of the Au(111) reconstruction on a strained substrate that displays a linear pattern over the whole sample. Once again, choosing the good deposition temperature or using heterogeneous nucleation, one can obtain self-assembled epitaxial Co nanowires (Campiglio et al., 2011b) (cf. Fig. 5.2c). Finally, this method not only is limited to pure elements but also can be generalized to alloys of interest for magnetism, although not obviously. The range of temperature where the ordered growth can take place is likely to be narrower than for single species as the preferential nucleation has to occur for both species simultaneously. Figure 5.2d shows an example of CoPt nanoparticles made on Au(111) at room temperature (Moreau et al., 2014). Although the ordering is clear, we can observe that it is less perfect than in the case of pure Co (cf. Fig. 5.2a). This is due to the Pt atoms that are known to perturb the Au(111) reconstruction, leading to imperfections in the final lattice (Nahas et al., 2010).

5.3 A Model System for the Study of Ultra-High Density Magnetic Recording Media

Those self-organized systems can be exploited to study the physics of magnetism at very small cluster size and with state-of-the-art particles areal densities, typically from 10 to 100 Tbits/inch^2. Beyond the fundamental challenge, such studies are required if one wants to continue increasing the density of non-volatile devices such as hard disk drives and magnetic random access memories. It is known in the magnetic recording industry that reducing the size distribution and going to ordered bits should help significantly to increase the areal density limit. A huge effort has been made on so-called self-organized magnetic arrays (SOMA), starting exclusively from a wet chemistry colloidal approach (Sun et al., 2000). Although the periodicity and the size distribution of these SOMA are still unchallenged nowadays (typically less than 30% of full width at half maximum for the volume distribution), the distribution of

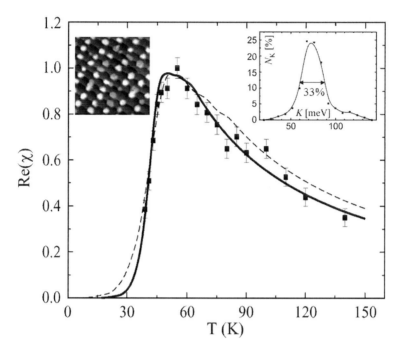

Figure 5.3 Real part of the magnetic susceptibility of an array of Co nanodots grown on Au(788) (STM image in inset) as a function of the temperature. The fitting of this curve gives access to the distribution of magnetic anisotropy shown in inset.

magnetic anisotropy was always found to be far larger (Woods et al., 2001), certainly due to the random orientation of structural axis in such samples. This drawback naturally disappears in epitaxial self-organized systems, what greatly improves their magnetic properties and mimics better the actual physical vapor deposition technologies of magnetic memories. Taking advantage of the long-range order growth of Co nanoparticles on a gold vicinal surface (cf. Fig. 5.2b), it has been shown by measuring the magnetic susceptibility as function of the temperature (cf. Fig. 5.3) that the distribution of magnetic anisotropy in such a sample was indeed very narrow (cf. inset of Fig. 5.3), close to the volume distribution (Weiss et al., 2005).

This out-of-plane magnetic bits array is still at the challenging density of 26 Tbits/inch^2, although not magnetically stable at room temperature. From this measurement, it was also possible to determine that dipolar interactions were negligible in the switching field distribution and to extract a magnetic anisotropy around 0.4 meV/atom, which is far beyond the Co bulk value. It is worth noting that at smaller cluster size, magneto-elastic effects have to be taken into account to explain a broadening of the magnetic anisotropy distribution, even in these epitaxial systems (Rohart et al., 2006).

5.4 The Thermal Stability of Nanomagnets: Beyond the Néel–Brown Model

A key issue for magnetic recording is also to have a reliable law of the magnetic thermal stability with the particle size. Up to now, the famous Néel–Brown model (Brown, 1963) has been widely used to predict the scalability of magnetic memories. However, it appeared recently that its domain of application was maybe overestimated and that even few nanometers nanoparticles can behave very differently than predicted. Although measurements on single particles would lead to the best understanding of such a physic (Krause et al., 2009; Ouazi et al., 2012b), it is not yet clear if techniques like spin-polarized scanning tunneling microscopy is not biasing the measurement due to interactions with the probe. Instead, we have chosen to study the variation of the thermal magnetic stability with size using self-organized Co nanoparticles and an averaging magneto-optical probe. Although the deconvolution of the anisotropy distribution can be delicate in general, it appears to be rather straightforward in such samples due to the narrowness of the size distribution. Concretely, we have measured the magnetic susceptibility as function of temperature for 11 different cluster diameters, from 1 to 9 nm. The fitting procedure of such curves gives access, as in the case of Section 5.3, to the energy barrier for the thermal reversal with size (cf. Fig. 5.4a). The Néel–Brown model (black dotted line) states that this energy barrier should be the magnetic anisotropy energy and therefore scale linearly with the number of atoms in the particle, which is clearly deviating from the measurements. Although such a behavior could be explained by introducing a specific anisotropy for less coordinated atoms (Rusponi et al., 2003), we have shown either that

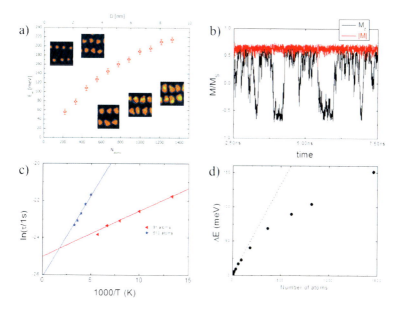

Figure 5.4 (a) Experimental determination of the activation energy barrier for thermal reversal as function of the size of Co bilayer nanodots self-organized on Au(111). The dashed line corresponds to the expected value in a standard model. 10 nm STM images of the Co nanodots of different sizes are shown in inset. (b) Simulation of the dynamics of the out-of-plane total magnetic moment (black) and of the magnetic moment norm (red) of an island composed of 613 atomic spins at a temperature of 200 K. Few thermal reversal event are visible on this time scale. (c) Arrhenius plot of typical switching time for the magnetization reversal. The slope gives an effective energy barrier for this process. (d) Energy barrier for the thermal magnetic switching as function of the island size (in number of atoms). The magnetic anisotropy energy is the dotted line.

it was due to a deviation of the Néel–Brown model at such a small scale (Rohart et al., 2010).

By using atomic scale magnetic modeling, we have simulated the thermal reversal of circular islands with increasing number of spins from 1 to 1500. The input parameters of such a simulation are the local magnetic anisotropy and exchange. Dipolar interactions are calculated considering the experimental value of the atomic magnetic moments and the temperature is introduced following Brown's formalism (Brown, 1963). One snapshot of magnetization reversal with time is shown in Fig. 5.4b for a particle of 613 spins at 200 K. The analysis of such a telegraph noise gives a typical time for the switching event. The dependence of this time with temperature, displayed in an Arrhenius plot in Fig. 5.4c for two different cluster sizes, gives directly access to the energy barrier for the magnetization reversal. This key quantity is plotted versus the island size in Fig. 5.4d. Once again, the black dotted line shows the variation of the magnetic anisotropy energy, which is an input parameter. The results of this simulation obviously show that the energy barrier deviates strongly from the magnetic anisotropy for particles as small as 3 nm in diameter. A very likely explanation for such an effect is the occurrence of spin-wave modes inside the magnetic particles (Rohart et al., 2010) that are not taken into account by the Néel–Brown model and that could help the magnetization reversal through energy transfer between excited modes and the uniform precessional one.

5.5 Magnetic Anisotropy in Core–Shell and Alloys Nanomagnets

One interest of self-organized epitaxial clusters is the versatility of the physical vapor deposition to modify easily the chemical surrounding of magnetic nanoparticles using various capping, what is an important issue of nanomagnetism. Considering the easy oxidation under ambient conditions of Co and Fe, it is indeed mandatory for applications to protect them with less reactive capping layer. However, such overlayers can have a strong impact on the magnetic properties, especially at the nanometer scale through interface effects. Although many studies have been done in the past 30 years on the influence of capping layers on the magnetism of ultrathin films, there is much less work on nanoparticles. We have used the Co on Au(111) self-organized system to explore the change of magnetic anisotropy when capped with Au and Pt, which are among the most studied interfaces in magnetism. Following the magnetic properties directly under ultra-high vacuum, it is therefore possible to study step by step how a non-magnetic shell modifies the anisotropy of the Co core. Figure 5.5a shows a 3D STM image

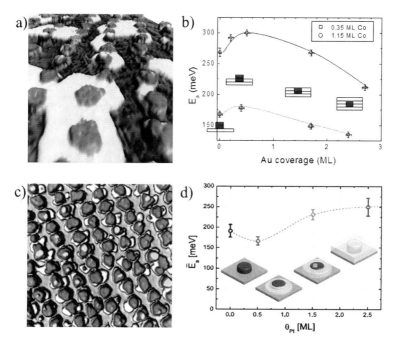

Figure 5.5 (a) 3D STM view of self-organized Co nanodots on Au(111) surrounded by an Au layer. (b) Energy barrier for the thermal magnetic reversal as function of the gold capping. The drawings in inset show the geometric configurations of the Co core (black)-Au shell (gray) for different Au depositions. (c) 100 nm STM image of self-organized Co nanodots on Au(111) surrounded by a Pt layer. (d) Energy barrier for the thermal magnetic reversal as a function of the platinum capping. The 3D drawings in inset show the progressive capping of the Co core.

of ordered Co nanodots surrounded by Au in the submonolayer regime. At this stage, mainly the perimeter Co atoms have Au neighbors, whereas a subsequent deposition will also cover the top Co atoms. The drawings in the inset of Fig. 5.5b summarize how the black Co core is progressively surrounded by Au. For every configuration, we have measured the magnetic susceptibility with temperature and extracted the energy barrier for magnetization reversal. As shown in Fig. 5.5b for two different Co cluster sizes, we observe an increase of the barrier when perimeter atoms are capped and a significant decrease during and after the capping of the top Co layer. This behavior is in strong contrast with the ultrathin film properties where the Au capping increases the Co out-of-plane anisotropy. Molecular dynamic calculations of the Co relaxations during the Au capping process gives an answer to this paradox (Nahas et al., 2009) by showing that huge in-plane and out-of-plane dilatations occurs, increasing the out-of-plane anisotropy through a magneto-elastic coupling. This result also indicates that the influence of electronic hybridizations between Co and Au on the interfacial magnetic anisotropy is certainly negligible, in contrast to what is often claimed.

We have repeated roughly the same experiment with Pt instead of Au in order to highlight similarities and differences between the Co/Au and Co/Pt systems that are generally believed to be magnetically rather similar. Although Co and Pt have a strong tendency to mix, the STM image of Fig. 5.5c, complemented by EXAFS measurements, shows that it is possible to obtain a clean Co–Pt core–shell geometry by deposition at room temperature. The change of energy barrier for magnetization reversal is now reversed as compared to the Au case (cf. Fig. 5.5d). After a slight decrease during the lateral capping, the out-of-plane magnetic anisotropy increases with the top capping (Campiglio et al., 2011a). It is worth noting that a similar behavior has been found with a Pd capping of Co islands on Pt(111) (Ouazi et al., 2012a). In these two latter cases, it is very likely that electronic hybridizations are now the key driving force to explain the anisotropy change, with a simple pair rule favoring the magnetic moment along the axis of the chemical bond. Moreover, the experimental study of Co relaxations during Pt capping shows that they are less than with Au (Campiglio et al., 2011a), the remaining part counterbalancing the interfacial anisotropy and explaining the weak experimental change we observe during the Pt capping.

Finally, we have studied the magnetic properties of CoPt alloys self-organized nanoparticles, as the one shown in Fig. 5.2d. The CoPt $L1_0$ ordered phase is one of the strategic materials for the magnetic recording industry and has been widely studied in ultrathin films and in three-dimensional nanoparticles. There are few studies on epitaxial two-dimensional nanoparticles, as naturally obtained by the physical self-organization method. By codepositing Co and Pt on the Au(111) surface at room temperature, one obtains an array of

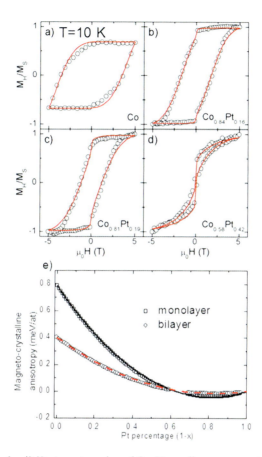

Figure 5.6 (a–d) Hysteresis cycles of Co_xPt_{1-x} alloys nanoparticles grown on $Au(111)$ (cf. Fig. 5.2d) (a) $x = 1$, (b) $x = 0.84$, (c) $x = 0.81$, (d) $x = 0.58$, (e) variation of the mean magnetic anisotropy of a 631-atom island, either one atomic layer high (square) or two atomic layers high (circle) as a function of the Pt percentage inside the island. The squares and the circles come from a simulation based on a pair model. The full and dotted lines come from analytical expressions in a mean field model.

islands that are made of a disordered CoPt alloy. By changing the relative flux of Co and Pt, it is possible to tune the Co concentration and study its influence on the magnetic properties. Figures 5.6a–d show hysteresis cycles obtained at 10 K using x-ray dichroism for 600 atoms nanoparticles with an increasing concentration of

Pt (pure Co, $Co_{0.84}Pt_{0.16}$, $Co_{0.81}Pt_{0.19}$, and $Co_{0.58}Pt_{0.42}$, respectively). The decrease of the coercive field indicates a decrease of out-of-plane anisotropy that can be quantitatively determined by the fitting of such cycles (red lines of Figs. 5.6a–d) (Moreau et al., 2014).

Starting from an out-of-plane anisotropy around 0.3 meV/atom for pure Co, one finally obtains very soft magnetic nanoparticles for $Co_{0.58}Pt_{0.42}$. This result is in very good agreement with the pair rule that the magnetic moments lie more favorably along the Co–Pt bonds. In order to quantify this rule, we have developed a pair model that considers different local anisotropy for the Co–Co and Co–Pt bonds. This phenomenological model, introduced a long time ago by Néel, uses as input parameters the magnetic anisotropy of the CoPt $L1_0$ phase and of the Co nanoparticles on Au(111). With these values, one can calculate numerically (open symbols of Fig. 5.6e) or analytically using a mean-field model (full lines of Fig. 5.6e) the variation of the magneto-crystalline anisotropy in CoPt islands (single or bilayer) (Moreau et al., 2014). The comparison between experimental data and this prediction is rather good, strengthening the idea that magnetic anisotropy in CoPt nanoparticles is mainly driven by geometrical considerations regarding the Co–Pt bonds' orientations.

5.6 Conclusion

Among the variety of experimental systems that can help to understand the magnetism at the nanoscale, the epitaxial self-organized magnetic nanoparticles are certainly a very fruitful approach. Their magnetic measurements by averaging techniques do not suffer (too much) from the deconvolution of a broad distribution that can hinder the important physic. These samples are also rather close to applicative ones using epitaxy and physical vapor depositions and can therefore be considered as model systems to study the future limits of two-dimensional recording. The results presented in this chapter raise a lot of exciting physics on nanomagnetism. The development of ab initio methods coupled to atomic scale magnetic simulations will certainly help to understand more deeply the experimental observations in a near future.

Acknowledgment

I first thank my collaborators in the STM team (C. Chacon, Y. Girard, J. Lagoute, S. Rousset and every Ph.D students and post-docs) for their work and stimulating discussions on this particular topic. I also thank H. Bulou, H. Brune, O. Fruchart, C. Goyhenex, S. Rohart, and A. Thiaville for long-standing fruitful collaborations. I finally thank IUF for support.

References

Barth, J. V., Brune, H., Ertl, G., and Behm, R. J. (1990). Scanning tunneling microscopy observations on the reconstructed Au(111) surface: Atomic structure, long-range superstructure, rotational domains, and surface defects, *Physical Review B* **42**, 15, pp. 9307–9318, doi:10.1103/PhysRevB.42.9307, URL http://link.aps.org/doi/10.1103/PhysRevB.42.9307.

Blanco-Mantecon, M., and O'Grady, K. (2006). Interaction and size effects in magnetic nanoparticles, *Journal of Magnetism and Magnetic Materials* **296**, 2, pp. 124–133, doi:10.1016/j.jmmm.2004.11.580, URL http://www.sciencedirect.com/science/article/pii/S03048853050\01927.

Brown, W. F. (1963). Thermal fluctuations of a single-domain particle, *Journal of Applied Physics* **34**, 4, pp. 1319–1320, doi:10.1063/1.1729489, URL http://scitation.aip.org/content/aip/journal/jap/34/4/10.1063\/1.1729489.

Brune, H., Giovannini, M., Bromann, K., and Kern, K. (1998). Self-organized growth of nanostructure arrays on strain-relief patterns, *Nature* **394**, 6692, pp. 451–453, doi:10.1038/28804, URL http://www.nature.com/nature/journal/v394/n6692/abs/394451a0.\html.

Campiglio, P., Moreau, N., Repain, V., Chacon, C., Girard, Y., Klein, J., Lagoute, J., Rousset, S., Bulou, H., Scheurer, F., Goyhenex, C., Ohresser, P., Fonda, E., and Magnan, H. (2011a). Interplay between interfacial and structural properties on the magnetism of self-organized core-shell Co/Pt supported nanodots, *Physical Review B* **84**, 23, p. 235443, doi:10.1103/PhysRevB.84.235443, URL http://link.aps.org/doi/10.1103/PhysRevB.84.235443.

Campiglio, P., Repain, V., Chacon, C., Fruchart, O., Lagoute, J., Girard, Y., and Rousset, S. (2011b). Quasi unidimensional growth of Co nanostructures on a strained Au(111) surface, *Surface Science* **605**, 1314, pp. 1165–1169, doi:10.1016/j.susc.2011.03.019, URL http://www.sciencedirect.com/science/article/pii/S00396028110\0121X.

Chambliss, D. D., Wilson, R. J., and Chiang, S. (1991). Nucleation of ordered Ni island arrays on Au(111) by surface-lattice dislocations, *Physical Review Letters* **66**, 13, pp. 1721–1724, doi: 10.1103/Phys RevLett.66.1721, URL http://link.aps.org/doi/10.1103/PhysRevLett.66.1721.

Ellmer, H., Repain, V., Rousset, S., Croset, B., Sotto, M., and Zeppenfeld, P. (2001). Self-ordering in two dimensions: Nitrogen adsorption on copper (1 0 0) followed by STM at elevated temperature, *Surface Science* **476**, 1–2, pp. 95–106, doi: 10.1016/S0039-6028(00)01121-3, URL http://www.sciencedirect.com/science/article/pii/S00396028000%11213.

Ellmer, H., Repain, V., Sotto, M., and Rousset, S. (2002). Pre-structured metallic template for the growth of ordered, square-based nanodots, *Surface Science* **511**, 1–3, pp. 183–189, doi: 10.1016/S0039-6028 (02)01440-1, URL http://www.sciencedirect.com/science/article/pii/S00396028020\14401.

Gambardella, P., Dallmeyer, A., Maiti, K., Malagoli, M. C., Eberhardt, W., Kern, K., and Carbone, C. (2002). Ferromagnetism in one-dimensional monatomic metal chains, *Nature* **416**, 6878, pp. 301–304, doi:10.1038/416301a, URL http://www.nature.com/nature/journal/v416/n6878/abs/416301a.h\tml.

Jamet, M., Wernsdorfer, W., Thirion, C., Dupuis, V., Molinon, P., Porez, A., and Mailly, D. (2004). Magnetic anisotropy in single clusters, *Physical Review B* **69**, 2, p. 024401, doi:10.1103/PhysRevB.69.024401, URL http://link.aps.org/doi/10.1103/PhysRevB.69.024401.

Krause, S., Herzog, G., Stapelfeldt, T., Berbil-Bautista, L., Bode, M., Vedmedenko, E. Y., and Wiesendanger, R. (2009). Magnetization reversal of nanoscale islands: How size and shape affect the Arrhenius prefactor, *Physical Review Letters* **103**, 12, p. 127202, doi:10.1103/PhysRevLett.103.127202, URL http://link.aps.org/doi/10.1103/PhysRevLett.103.127202.

Moreau, N., Repain, V., Chacon, C., Girard, Y., Lagoute, J., Klein, J., Rousset, S., Scheurer, F., and Ohresser, P. (2014). Growth and magnetism of self-organized Co_xPt1_x nanostructures on Au(111), *Journal of Physics D: Applied Physics* **47**, 7, p. 075306, doi:10.1088/0022-3727/47/7/075306, URL http://iopscience.iop.org/0022-3727/47/7/075306.

Nahas, Y., Repain, V., Chacon, C., Girard, Y., Lagoute, J., Rodary, G., Klein, J., Rousset, S., Bulou, H., and Goyhenex, C. (2009). Dominant role of the epitaxial strain in the magnetism of core-shell Co/Au self-organized nanodots, *Physical Review Letters* **103**, 6, p. 067202, doi:10.1103/PhysRevLett.103.067202, URL http://link.aps.org/doi/10.1103/PhysRevLett.103.067202.

Nahas, Y., Repain, V., Chacon, C., Girard, Y., and Rousset, S. (2010). Interplay between ordered growth and intermixing of Pt on patterned Au surfaces, *Surface Science* **604**, 9–10, pp. 829–833, doi:10.1016/j.susc.2010.02.008, URL http://www.sciencedirect.com/science/article/pii/S00396028100\00506.

Narasimhan, S., and Vanderbilt, D. (1992). Elastic stress domains and the herringbone reconstruction on Au(111), *Physical Review Letters* **69**, 10, pp. 1564–1567, doi:10.1103/PhysRevLett.69.1564, URL http://link.aps.org/doi/10.1103/PhysRevLett.69.1564.

N'Diaye, A. T., Gerber, T., Busse, C., Mysliveek, J., Coraux, J., and Michely, T. (2009). A versatile fabrication method for cluster superlattices, *New Journal of Physics* **11**, 10, p. 103045, doi:10.1088/1367-2630/11/10/103045, URL http://iopscience.iop.org/1367-2630/11/10/103045.

Ortega, J. E., Mugarza, A., Repain, V., Rousset, S., Porez-Dieste, V., and Mascaraque, A. (2002). One-dimensional versus two-dimensional surface states on stepped Au(111), *Physical Review B* **65**, 16, p. 165413, doi:10.1103/PhysRevB.65.165413, URL http://link.aps.org/doi/10.1103/PhysRevB.65.165413.

Ouazi, S., Vlaic, S., Rusponi, S., Moulas, G., Buluschek, P., Halleux, K., Bornemann, S., Mankovsky, S., Minor, J., Staunton, J. B., Ebert, H., and Brune, H. (2012a). Atomic-scale engineering of magnetic anisotropy of nanostructures through interfaces and interlines, *Nature Communications* **3**, p. 1313, doi:10.1038/ncomms2316, URL http://www.nature.com/ncomms/journal/v3/n12/abs/ncomms2316.ht\ml.

Ouazi, S., Wedekind, S., Rodary, G., Oka, H., Sander, D., and Kirschner, J. (2012b). Magnetization reversal of individual Co nanoislands, *Physical Review Letters* **108**, 10, p. 107206, doi:10.1103/PhysRevLett.108.107206, URL http://link.aps.org/doi/10.1103/PhysRevLett.108.107206.

Repain, V., Baudot, G., Ellmer, H., and Rousset, S. (2002). Two-dimensional long-range ordered growth of uniform cobalt nanostructures on a Au(111) vicinal template, *EPL (Europhysics Letters)* **58**, 5, p. 730, doi:10.1209/epl/i2002-00410-4, URL http://iopscience.iop.org/0295-5075/58/5/730.

Rohart, S., Campiglio, P., Repain, V., Nahas, Y., Chacon, C., Girard, Y., Lagoute, J., Thiaville, A., and Rousset, S. (2010). Spin-wave-assisted thermal reversal of epitaxial perpendicular magnetic nanodots, *Physical Review Letters* **104**, 13, p. 137202, doi:10.1103/PhysRevLett.104.137202, URL http://link.aps.org/doi/10.1103/PhysRevLett.104.137202.

Rohart, S., Girard, Y., Nahas, Y., Repain, V., Rodary, G., Tejeda, A., and Rousset, S. (2008). Growth of iron on gold (7 8 8) vicinal surface: From nanodots to step flow, *Surface Science* **602**, 1, pp. 28–36, doi:10.1016/j.susc. 2007.09.036, URL http://www.sciencedirect.com/science/article/pii/ S00396028070\09648.

Rohart, S., Repain, V., Tejeda, A., Ohresser, P., Scheurer, F., Bencok, P., Ferré, J., and Rousset, S. (2006). Distribution of the magnetic anisotropy energy of an array of self-ordered Co nanodots deposited on vicinal Au(111): X-ray magnetic circular dichroism measurements and theory, *Physical Review B* **73**, 16, p. 165412, doi:10.1103/PhysRevB.73.165412, URL http://link.aps.org/doi/10.1103/PhysRevB.73.165412.

Rousset, S., Repain, V., Baudot, G., Garreau, Y., and Lecoeur, J. (2003). Self-ordering of Au(111) vicinal surfaces and application to nanostructure organized growth, *Journal of Physics: Condensed Matter* **15**, 47, p. S3363, doi:10.1088/0953-8984/15/47/009, URL http://iopscience. iop.org/0953-8984/15/47/009.

Rusponi, S., Cren, T., Weiss, N., Epple, M., Buluschek, P., Claude, L., and Brune, H. (2003). The remarkable difference between surface and step atoms in the magnetic anisotropy of two-dimensional nanostructures, *Nature Materials* **2**, 8, pp. 546–551, doi:10.1038/nmat930, URL http://www. nature.com/nmat/journal/v2/n8/abs/nmat930.html.

Stroscio, J. A., Pierce, D. T., Dragoset, R. A., and First, P. N. (1992). Microscopic aspects of the initial growth of metastable fcc iron on Au(111), *Journal of Vacuum Science & Technology A* **10**, 4, pp. 1981–1985, doi:10.1116/1. 578013, URL http://scitation.aip.org/content/avs/journal/jvsta/10/ 4/10.11\16/1.578013.

Sun, S., Murray, C. B., Weller, D., Folks, L., and Moser, A. (2000). Monodisperse FePt nanoparticles and ferromagnetic FePt nanocrystal superlattices, *Science* **287**, 5460, pp. 1989–1992, doi:10.1126/science.287.5460. 1989, URL http://www.sciencemag.org/content/287/5460/1989.

Voigtlander, B., Meyer, G., and Amer, N. M. (1991). Epitaxial growth of thin magnetic cobalt films on Au(111) studied by scanning tunneling microscopy, *Physical Review B* **44**, 18, pp. 10354–10357, doi:10.1103/ PhysRevB.44.10354, URL http://link.aps.org/doi/10.1103/PhysRevB. 44.10354.

Weiss, N., Cren, T., Epple, M., Rusponi, S., Baudot, G., Rohart, S., Tejeda, A., Repain, V., Rousset, S., Ohresser, P., Scheurer, F., Bencok, P., and Brune, H. (2005). Uniform magnetic properties for an ultrahigh-density lattice of noninteracting Co nanostructures, *Physical Review Letters* **95**, 15, p. 157204, doi:10.1103/PhysRevLett.95.157204, URL http://link.aps.org/doi/10.1103/PhysRevLett.95.157204.

Wernsdorfer, W., Orozco, E. B., Hasselbach, K., Benoit, A., Barbara, B., Demoncy, N., Loiseau, A., Pascard, H., and Mailly, D. (1997). Experimental evidence of the Néel-Brown model of magnetization reversal, *Physical Review Letters* **78**, 9, pp. 1791–1794, doi:10.1103/PhysRevLett.78.1791, URL http://link.aps.org/doi/10.1103/PhysRevLett.78.1791.

Woods, S. I., Kirtley, J. R., Sun, S., and Koch, R. H. (2001). Direct investigation of superparamagnetism in Co nanoparticle films, *Physical Review Letters* **87**, 13, p. 137205, doi:10.1103/PhysRevLett.87.137205, URL http://link.aps.org/doi/10.1103/PhysRevLett.87.137205.

Chapter 6

High-Aspect-Ratio Nanoparticles: Growth, Assembly, and Magnetic Properties

Frédéric Ott,[a] Jean-Yves Piquemal,[b] and Guillaume Viau[c]

[a]*Laboratoire Léon Brillouin, CEA CNRS UMR12, IRAMIS, CEA-Saclay, 91191 Gif sur Yvette, France*
[b]*Université Paris Diderot, Sorbonne Paris Cité, ITODYS, CNRS UMR 7086, 15 rue J.-A. de Baïf, 75205 Paris Cedex 13, France*
[c]*Université de Toulouse, INSA CNRS UPS, UMR 5215 LPCNO, 135 avenue de Rangueil, 31077 Toulouse Cedex 4, France*
frederic.ott@cea.fr, jean-yves.piquemal@univ-paris-diderot.fr, gviau@insa-toulouse.fr

6.1 Introduction

Nanorods (NRs) and nanowires (NWs) are a particular class of nanoparticles, the properties of which are directly related to their shape anisotropy and can strongly differ from those of the isotropic particles. It is particularly true when the magnetic properties are considered. Nanorods and nanowires have a mean diameter in the nanometre range, 1–100 nm, and an aspect ratio higher than 3 for the rods, higher than 20 for the wires, although no official classification was clearly established. Ferromagnetic nanoparticles

Magnetic Structures of 2D and 3D Nanoparticles: Properties and Applications
Edited by Jean-Claude Levy
Copyright © 2016 Pan Stanford Publishing Pte. Ltd.
ISBN 978-981-4613-67-5 (Hardcover), 978-981-4613-68-2 (eBook)
www.panstanford.com

with such a high-aspect-ratio exhibit a strong magnetic anisotropy that has strong effects on the magnetic configuration of their remanent state and on their magnetization loop. As hard magnetic material ferromagnetic NRs and NWs may find several applications in the field of magnetic recording [1] and permanent magnets [2]. Their particular magneto-rheological behaviour makes them also interesting for biosensors [3].

Several methods of elaboration of ferromagnetic NRs and NWs have been developed in the last decade that give various mean diameters and aspect ratios, for a large range of chemical composition. In Section 6.2, we describe the most developed methods to grow high-aspect-ratio ferromagnetic particles emphasizing the morphological and structural properties because these are the key point to understand the resulting magnetic properties.

Besides these experimental works, theoretical studies and modelling have been carried out to describe the magnetic behaviour of high-aspect-ratio ferromagnetic particles. In Section 6.3, we report on studies on the magnetic properties of single elongated ferromagnetic particles comparing modelling and experimental results.

Most of the applications of elongated nanoparticles involve dense assemblies. In Section 6.4, we describe the magnetic properties of assemblies of elongated ferromagnetic particles in which the dipolar interactions can play an important role.

6.2 Elaboration of High-Aspect-Ratio Magnetic Nanoparticles: Relation between Growth and Structural Properties

Several methods were developed to grow ferromagnetic particles with high-aspect-ratio, liquid-phase chemical synthesis, electrochemical growth using solid host template and physical deposition. The resulting magnetic properties of the particles depend, first, on their magnetization through magnetostatic effect. We will focus in the following on growth method of pure metal, Co, Fe, Ni and alloys. The mean diameter and mean length of these elongated particles and also their structure and crystallinity are very dependent on the growth process. All these parameters have a strong influence on the shape and magnetocrystalline anisotropy of the particles. The remanent state and the magnetization reversal process are thus highly dependent on the elaboration process.

6.2.1 *Chemical Synthesis*

The chemical syntheses of metal nanoparticles consist in the reduction or decomposition of a metal precursor in a liquid solvent. The control of the metal nucleation and particle growth allows in a lot of cases the production of monodisperse particles with various shapes.

There are basically two classes of chemical methods which differ according to the nature of the metal precursor and the solvent, the polyol process, on one hand, the organometallic approach and the thermodecomposition, on the other hand. The polyol process consists of the reduction of a metal salt in a liquid α-diol. In this method, the diol acts as both the solvent for the metal precursor and the reducing agent. The α-diols like 1,2 propanediol or 1,2 butanediol exhibit high dielectric constant that make them good solvents for inorganic salts and present high boiling points. The liquid polyols are mild reducing agents, but at high temperature they can reduce metals like nickel, cobalt or iron. The organometallic approach uses organometallic complexes as precursors. They are dissolved in an aprotic and apolar solvent like mesitylene and are reduced under the hydrogen atmosphere. The thermodecomposition route consists in heating a solution of metal carbonyl complexes such as $Fe(CO)_5$ and $Co_2(CO)_8$, in the same kind of solvent as the organometallic route. The oxidation state of the metal atom in the carbonyl compounds is zero. The decomposition at high temperature leads to the growth of metal particles with the evolution of carbon monoxide.

In all of these processes, the particle growth is controlled by the addition of surfactants. These molecules are generally carboxylic acids $(C_nH_{2n+1}CO_2H)$, amines $(C_nH_{2n+1}NH_2)$ or mixtures of both. The polar head of these molecules has a strong affinity with the metal surface and the long-chain ensures a steric repulsion. In most cases, a non-specific adsorption of the surfactants at the metal surface leads to isotropic particles. The growth of non-isotropic particles is realized when a specific adsorption of the long-chain molecules on particular (hkl) crystallographic planes is favoured. Cobalt with the hexagonal close packed (hcp) structure is much more favourable than nickel and iron, which crystallize with the fcc and bcc structures, respectively, to grow high-aspect-ratio nanoparticles, like nanorods and nanowires, because of its unique six-fold c-axis.

6.2.1.1 Polyol process

The polyol process was developed by Fiévet and co-workers, first for the synthesis of isotropic particles of cobalt and nickel [4, 5], then for CoNi [6] and FeCoNi [7, 8] spheres. FeCo nanocubes were also synthesized by a modified polyol process [9]. Figlarz et al. described the first example of anisotropic particles by the polyol process: the synthesis of silver nanorods by the reduction of silver nitrate in a solution of PVP in 1,2 ethanediol [10]. According to Xia et al. [11], the first seeds exhibit a decahedral shape that grow parallel to their five-fold axis thanks to the selective grafting of the PVP on the Ag {100} planes.

More recently, the polyol process was developed for the synthesis of anisotropic magnetic particles such as cobalt and cobalt–nickel nanorods and nanowires. Cobalt NRs and NWs were obtained using a long-chain carboxylate as cobalt precursor, such as cobalt laurate $Co(C_{11}H_{23}CO_2)_2$, and the 1,2 butanediol as solvent [12]. The protocol is easy to carry out. The metal precursor is dispersed in a 1 eq. NaOH solution of polyol containing a small amount of ruthenium chloride (Ru/Co = 2.5%). The mixture is heated at 175°C for 30 min. The suspension turns black indicating the formation of the metal particles. The electron microscopy images reveals high-aspect-ratio nanoparticles (Fig. 6.1). Depending on the experimental conditions it is possible vary the mean diameter in the range 7–40 nm and aspect ratio in the range 3–30 (Fig. 6.1).

The fcc and hcp phases of cobalt are very close in energy, and even if the hcp phase is supposed to be more stable at low temperature, the cubic phase is often obtained with the liquid-phase processes. This anisotropic structure is more favourable than the

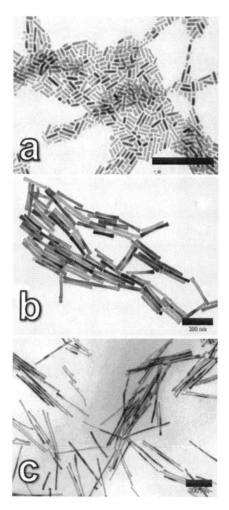

Figure 6.1 Electron microscopy images of elongated cobalt nanoparticles prepared by the polyol process exhibiting different mean diameter, D_m, and mean length, L_m : (a) $D_m = 7$ nm, $L_m = 28$ nm; (b) $D_m = 16$ nm, $L_m = 160$ nm; (c) $D_m = 20$ nm, $L_m = 350$ nm. Scale bar denotes 200 nm.

fcc to obtain non-spherical particles. Nevertheless, the final size and shape of the particle depends on the nucleation and growth steps. It was evidenced that the cobalt NRs and NWs prepared by the polyol process crystallize with the hexagonal closed-packed

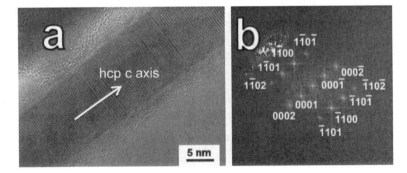

Figure 6.2 (a) High-resolution image of a cobalt rod showing the hcp structure in the [11–20] zone axis; (b) the corresponding numerical electron diffraction pattern (adapted from [12]).

structure. They are nearly single crystals with the long axis parallel to the hcp c-axis (Fig. 6.2). Thus, the formation of elongated particles is fulfilled when the growth along the crystallographic c-axis is favoured with respect to a growth of the basal plane.

Theoretical calculations showed that the adsorption of the carboxylate ions on the hcp cobalt nanocrystal controls the growth step [13]. On the one hand, it was calculated that the adsorption energy of carboxylate was higher on the {10-10} facets than on the (0001) basal compact planes. On the other hand, it was also found that, for a given surface coverage value, the surface ligand concentration is higher on basal facets than on lateral ones. This subtle balance was appreciated by plotting the surface energy of the decorated (0001) and {10-10} surfaces as a function of the chemical potential of the carboxylate ligand, hence its concentration. The results have shown that the final hcp Co particle shape depends on the carboxylate ion concentration: For high concentration the higher surface coverage of the (0001) basal plane favours the growth of platelets, while for lower ligand concentrations, the rod shape is obtained [13]. This prediction was verified experimentally: With the same experimental protocol it is possible to switch for rods to platelets by increasing the concentration of long-chain carboxylate in the medium. Moreover, an intermediate concentration may lead to the formation of dumbbells (Fig. 6.3a). These original shapes are explained by two growth steps: A cobalt rod is formed at first, the tips of which progressively enlarge as the free carboxylate concentration increases in the medium.

The final shape is also strongly dependent on the nucleation step. The respective nucleation and growth rate must also be carefully adjusted to produce rods and wires. The role of ruthenium chloride is to produce in situ ruthenium tiny nanoparticles that act as seeds for the cobalt growth. If the number of seeds inside the medium is not high enough or if the growth rate is too high compared to the nucleation rate cobalt stars are produced rather than isolated wires [12]. The variation of the number of seeds in the growth medium permits the variation in the aspect ratio (Fig. 6.1).

Bimetallic cobalt-nickel elongated particles were also synthesized by the polyol process using the reduction of mixtures of cobalt and nickel acetates in 1,2 propanediol [14,15]. $Co_{80}Ni_{20}$ NWs with a mean diameter of 7 nm and a mean length in the range 100–300 nm were obtained (Fig. 6.3b). High-resolution electron microscopy showed the same hcp structure and the same crystallographic orientation as the pure Co NRs. Nevertheless, local chemical analysis evidenced Ni enrichment at the tips showing that nickel is reduced slower than the cobalt. For the $Co_{50}Ni_{50}$ composition, particles with dumbbell shape were obtained (Fig. 6.3c) that consisted in a Co-rich nanorod with the hcp structure capped by two Ni-rich hexagonal platelets exhibiting the fcc structure [16].

6.2.1.2 Organometallic chemistry

The reduction of an organometallic precursor dissolved in a non-polar solvent was developed for the synthesis of metal magnetic nanoparticles [18–20]. For iron, cobalt and nickel the particle growth is controlled by the addition of long-chain carboxylic acid and long-chain amine in the medium. The case of cobalt is particularly interesting for the large variety of shapes that was obtained. Cobalt nanorods with a mean diameter lower than 10 nm were obtained by the reduction of $[Co(\eta_3\text{-}C_8H_{13})(\eta_4\text{-}C_8H_{12})]$ under 3 bars of dihydrogen gas in anisole containing an equimolar mixture of octadecylamine and oleic acid [21]. The self- assembly in solution of cobalt nanorods prepared by organometallic chemistry into hexagonal superlattices was described [22]. Similar superlattices

220 | High-Aspect-Ratio Nanoparticles

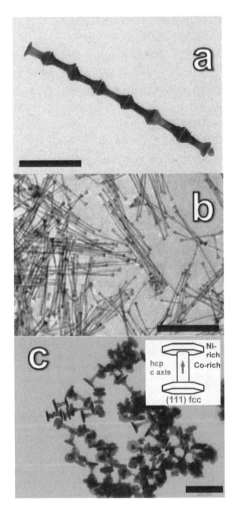

Figure 6.3 Transmission electron microscopy image of particles prepared by the polyol process: (a) Co dumbbells (adapted from [17]); (b) $Co_{80}Ni_{20}$ elongated particles (adapted from [17]); (c) $Co_{50}Ni_{50}$ dumbbells. Inset: scheme of a $Co_{50}Ni_{50}$ dumbbell (adapted from [16]). Scale bar denotes 200 nm.

with improved magnetic properties were obtained with cobalt nanorods prepared using [Co{N(SiMe$_3$)$_2$}$_2$] complex as precursor [23] (Fig. 6.4a). Like for the rods prepared by the polyol process described above, the cobalt rods prepared by the organometallic

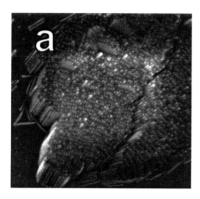 

Figure 6.4 Electron microscopy images of elongated nanoparticles prepared by the organometallic chemistry approach (a) Self-assembled Co nanorods (length 100 nm, diameter 5 nm) (reproduced with permission from [23]; (b) Co stars (adapted with permission from [25], copyright 2012 American Chemical Society).

route crystallized with the hcp structure with the c-axis parallel to the long axis. These rods are generally single crystals.

The cobalt particle shape was found to depend strongly on the amine/acid ratio and on the protocol of addition. Cobalt particles with other shapes such as spheres, wires, and stars (Fig. 6.4b) were also obtained [24, 25]. In the case of the nanorods, the mean aspect ratio of the cobalt nanorods was found to depend on the nature of the long-chain acid and amine. Bimetallic particles can also be prepared by the organometallic route. The growth of metal iron on cobalt nanorods led to dumbbells with iron cubes at the tips of the cobalt nanorods [26]. Recently, an extension of the organometallic approach has been carried out for the growth of Co NWs on substrates leading to arrays of parallel cobalt rods [1]. These very thin rods grow with hcp structure in epitaxy on (111) platinum surface.

Another application of the cobalt rods prepared by the organometallic approach is the elaboration of liquid crystalline magnetic materials by dispersing the rods in polymers [27, 28].

The thermodecomposition of metal carbonyl in high boiling point solvent is a very well-known method to grow spherical particles. However, only a few examples of anisotropic magnetic nanoparticles were obtained by this method. The synthesis of metal particles by thermodecomposition consists of heating a metal carbonyl compound dissolved in a high boiling point solvent containing surfactants. Since the metal centre in this family of precursors is already at the zero-valent state, no additional reducing agent is needed. Cobalt nanodisks were obtained by the thermodecomposition of cobalt carbonyl $Co_2(CO)_8$ [29]. The thermodecomposition of $Fe(CO)_5$ in presence of didodecyldimethylammonium bromide (DDAB) leaded to bcc-Fe nanorods with a mean diameter of 2 nm and a length varying between 10 and 30 nm [30]. FePt nanowires were obtained by the co-decomposition of $Fe(CO)_5$ and reduction of platinum acetylacetonate $Pt(acac)_2$ [31].

6.2.2 *Electrochemical Synthesis in Solid Templates*

Many works have been devoted to the growth of metal nanowires in the uniaxial pores of alumina or track-etched polycarbonate membranes, even if other types have also been successfully utilized, such as radiation track-etched mica membranes (TEPC) [32].

The porous alumina elaboration involves first the deposition of an aluminium thin film, which is then electrochemically oxidized to form a porous alumina membrane. Note that besides flat membranes, cylindrical porous alumina was recently reported: It was obtained after the two-step anodization of a 100 μm diameter Al wire [33]. During the anodization process, under peculiar conditions, hexagonally ordered pores lying perpendicularly to the substrate can be generated [34]. Depending on the anodization conditions, the diameter of the pores can be varied in the range *ca.* 5–300 nm, while very high pore densities, about 10^{11} pores·cm^{-2}, can be obtained [34,35]. With TEPC membranes, pores as small as 10 nm can be obtained with a pore density of about 10^9 pores·cm^{-2}, but, due to the preparation procedure, pores are randomly distributed within the material.

About a decade ago, the different synthetic procedures for the preparation of metallic and non-metallic nanomaterials inside the porosity of these solid templates were reviewed [36]. As regards metal nanowires, the reported methods are (i) electrodeposition,

(ii) chemical deposition which requires a reducing agent and (iii) chemical vapour deposition requiring the use of volatile precursors.

Electrodeposition is considered to be the most employed strategy [36]. The first step consists of depositing a thin metal film at the backside of the membrane, acting as a cathode. Then, the electrochemical growth of different metals is subsequently performed. An obvious advantage of this method is that arrays of parallel nanowires can be obtained (Fig. 6.5a). The wire length ranges from less than 20 nm to about 1 μm [37], while the wires' diameter is dictated by the pore diameter, which in turn depends on the anodization process of the membrane (see above). To the best of the authors' knowledge, the first report for the preparation of ferromagnetic metal nanowires was published in 1975 with AAO as the template [38]. Pure Fe, Co and Ni wires were obtained, as well as Fe–Co, Fe–Ni and Co–Ni alloys [38]. Since then, an impressive number of studies has been devoted to the preparation of Fe, Co and Ni (and their alloys) nanowires and nanotubes following this approach. The NWs are generally polycrystalline [39, 40]. Recently, hcp Co NWs with different textures, [101], [002] and [110], were obtained by this method by playing upon the electroplating conditions [41]. FeCo alloyed nanowires with high saturation magnetization were also obtained by this method [41]. High coercivity was measured on $Fe_{28}Co_{67}Cu_5$ nanowires arrays grown by this method and annealed at $500°C$ [42].

Review on the magnetism of parallel self-assembled nanowire array was already published: the maximum coercivity measured at room temperature for Fe, Co and Ni nanowires parallel assemblies in host matrices were 3.0, 2.6 and 0.95 kOe, respectively [37]. The observed coercivity is the sum of two contributions: (i) shape anisotropy and (ii) magnetocrystalline anisotropy. However, for Fe and Ni particles, the shape anisotropy contribution overcomes the magnetocrystalline one and tends to align the magnetization along the NW axis, the situation is a little more complex for Co. Indeed, for this metal, the two contributions can be on the same order of magnitude since hcp Co exhibits the highest magnetocystalline anisotropy constant of the three ferromagnetic metals. Depending on the experimental conditions such as the electrodeposition parameters (DC or AC, voltage, current density, pH of the electrolytic

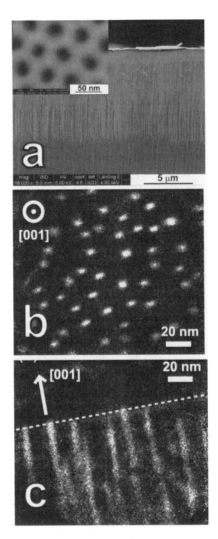

Figure 6.5 Scanning electron microscopy image of a cross section of FeCoCu nanowires array (diameter of 18 nm, inter-wire distance of 55 nm, length around 8 μm) grown electrochemically in a AAO membrane, Inset: top view of the membrane (reproduced with permission from [42]); (b) top view and cross section; (c) EFTEM images, taken at the Ni L-edge, of a Ni containing CeO_2 film grown on $SrTiO_3(001)$ by pulse laser deposition; the arrow indicates the growth direction along [001] of $SrTiO_3$; reprinted with permission from [56], copyright (2013) American Chemical Society.

bath, etc.) and the diameter of the template pores, it was found that not only a different crystallographic phase of Co (cubic, hexagonal or a mixture of both allotropic varieties) can be obtained but also different textures were observed [43–47]. For instance, with hcp Co, the c-axis (which corresponds to the easy magnetization axis) was found to range from an orientation in the substrate plane (perpendicular to the wire axis) to an orientation perpendicular to the substrate plane, i.e. parallel to the wire axis. Note, however, that for hcp Co, the easy magnetization axis is very often found to lie perpendicular to the wire axis [48]. In such a case, the shape anisotropy and magnetocrystalline anisotropy contributions are competing, which results in reduced coercivity and squareness of the hysteresis cycle [43]. It has been shown that it is possible to control, to some extent, the orientation of the easy magnetization axis relative to the NW axis by using an external magnetic field during electrodeposition: with $\mu_0 H_{ext} = 5$ T, the c-axis moved from 90 to 70° [49].

6.2.3 *Pulsed Laser Deposition*

Metal and oxide ultrathin ferromagnetic nanowires can be grown in an oxide matrix using pulsed laser deposition. This approach does not require any template, and it is based on the heteroepitaxy and the controlled reduction of perovskite-type oxides in ultrahigh vacuum or under a reducing atmosphere. The reduction of the oxide matrix leads to a segregation at nanometre scale that makes possible the growth of well-separated nanowires. The crystallographic orientation of the wires is generally dependent on the lateral epitaxy conditions between the wires and the matrix.

Fe nanowires with bcc structure were formed by pulsed-laser deposition of $La_{0.5}Sr_{0.5}FeO_{3-x}$ on single-crystal $SrTiO_3$ (001) substrate in a reducing atmosphere. The single crystal α-Fe nanowires are embedded in a crystalline $LaSrFeO_4$ matrix. The mean diameter of the nanowires was varied in the range 4–50 nm by varying the growth temperature in the range $T = 560$ to 840°C [50, 51]. With the same approach $CoFe_2O_4$ nanowires were grown in a $BaTiO_3$ matrix [52] or in a $BiFeO_3$ matrix on a $SrTiO_3$ (001) substrate [53]. Recently, Vidal and co-workers developed a similar approach for the growth of cobalt nanowires assemblies in a CeO_2 matrix on $SrTiO_3$ (001) [54, 55]. The diameter of the cobalt wires could be varied from 3 to 5 nm. The 3 nm diameter nanowires are made of hcp Co grains with the c-axis pointing along one of the four <111> directions of the CeO_2 matrix, separated by fcc Co regions. In the 5 nm diameter nanowires, the grains are smaller and the density of stacking faults within the wire is much higher. A combinatorial method was also used successfully to grow Ni and alloyed cobalt-nickel nanorods in $CeO_2/SrTiO_3$ (001) [56] (Fig. 6.5b,c). Cobalt nanowires were also grown in yttria-stabilized zirconia matrix [57].

6.2.4 Summary

The polyol process and the organometallic route are very efficient to produce cobalt nanorods and nanowires with a mean diameter in the range 5–40 nm. The great interest of these methods is that the growth axis is the crystallographic c-axis and that the particles are generally single crystals. These properties make that these particles model objects for the understanding of the magnetic properties of single nanorods and nanowires. Cobalt-nickel elongated nanoparticles were also grown by the polyol process. Nevertheless, the chemical methods are less efficient for other metal like iron and nickel that crystallize with a cubic structure.

The electrochemical growth of metal particles in template membranes and the pulse laser co-deposition methods are very efficient to produce in one-step ferromagnetic nanowire arrays perpendicular to the substrate. The mean diameter of the electrochemically grown wires is directly controlled by that of the pores of the template. While it is generally hard to decrease the mean diameter of the NWs below 15 nm with the template method, the pulse laser deposition method is very efficient to grow ferromagnetic nanowires with mean diameters lower than 5 nm. The nanowires prepared by both methods are generally polycrystals, but recent improvements have been realized for the control of crystallinity.

6.3 Magnetism of High-Aspect-Ratio Nanoparticles

6.3.1 Description of the Magnetism of Nanometric and Micrometric Scaled Objects

In the field of magnetism, atomistic calculations cannot yet be performed to predict the behaviour of systems with a volume larger than a few 1000 nm^3. Thus a mesoscopic description of the magnetic material in which the material is considered as a continuous medium is used. It is referred to as *micro-magnetic modelling*. Note that this description is presently also limited for technical reasons to systems with sizes on the order of 1–10 μm^3. An exchange stiffness constant A (expressed in J/m) is introduced. It characterizes the local interaction strength. Its expression is related to the microscopic parameters by the relation:

$$A = \frac{S^2 a^2 J \, N_V}{2},\tag{6.1}$$

where N_V, S, a and J are, respectively, the number of nearest-neighbour atoms per unit volume, the electronic spin of the atom, the distance between spins and the exchange interaction. The exchange stiffness constant is between 1×10^{-11} and 2×10^{-11} J·m^{-1} for most ferromagnets.

The exchange energy is expressed as $E_{ex} = A \, (\partial m / \partial x)^2$. The larger the change of orientation between adjacent spins, the larger the exchange energy.

The magnetic material is further described by a magneto-crystalline anisotropy constant K (expressed in J·m^{-3}), which describes the anisotropy of the material with respect to its crystallographic axes. For cubic crystals, the magneto-crystalline energy is expressed as $E_{mc} = K_1^2 (m_x^2 m_y^2 + m_y^2 m_z^2 + m_z^2 m_x^2)$, where m_x, m_y and m_z are the direction cosine of the magnetization with respect to the crystallographic axes. For cubic materials the magneto-crystalline anisotropy constants at room temperature are low: $K_1(\text{bcc-Fe}) = 4.8 \times 10^4$ J·m^{-3}, K_1 (fcc-Ni) $= -4.5 \times 10^3$ J·m^{-3}. In the case of hexagonal materials, the magneto-crystalline energy is expressed as $E_{mc} = K_1 \sin^2 \theta$, where θ is the angle between the magnetization and the crystallographic axis [0001]. In the case of hcp-Co, $K_1 = 4.1 \times 10^5$ J·m^{-3} at room temperature which is 10 times larger than bcc-Fe. This magneto-crystalline anisotropy tends to align the magnetization along the [0001] crystallographic axis.

Since we are dealing with finite-size objects, a new magnetic effect arises which is the interaction of the magnetization within the object with the dipolar field created by the object itself. This is referred to as the self-dipolar energy. The magnetic moments within the object create a field $H_D(r)$, which is named *demagnetizing field* since its orientation is opposite to the average magnetization of the object. This demagnetizing field has a complex spatial distribution, which depends on the distribution $M(r)$ within the object and on the object geometry. $H_D(r)$ is usually not homogeneous within the magnetic object except in the very specific case of ellipsoidal objects and their limiting cases, thin films and infinite cylinders. In the particular case of ellipsoid, it is possible to demonstrate that the homogeneous demagnetizing field within the magnetic objet can then be expressed as

$$\begin{pmatrix} H_{dx} \\ H_{dy} \\ H_{dz} \end{pmatrix} = \begin{bmatrix} N_x & 0 & 0 \\ 0 & N_y & 0 \\ 0 & 0 & N_z \end{bmatrix} \begin{pmatrix} M_x \\ M_y \\ M_z \end{pmatrix},$$

where N_x, N_y and N_z are the demagnetizing factors.

For a very long wire with the (Oz) direction parallel to the wire axis, $N_x = N_y = 1/2$, $N_z = 0$. A zero demagnetizing field is obtained when the magnetization lies along the long axis of the ellipsoid. On the other hand, the demagnetizing field is maximized when the magnetization is perpendicular to the long axis of the ellipsoid. This effect is referred to as shape anisotropy. Note that this anisotropy is very large since it is on the order of the magnetization in the material. In the case of finite-size objects the demagnetizing factors depend on the ellipsoid aspect ratio (i.e. the ratio between the length of the long axis and the short axis). For an aspect ratio of 10, the demagnetization coefficient is already very small, $N_z = 0.017$. In a zero-order approximation, cylinders may be modelled by ellipsoids of the same aspect ratio.

From these various magnetic energy terms, it is possible to introduce two mesoscopic length-scales: the exchange length l_{ex} and the wall width parameter δ. The exchange length l_{ex} is the minimal

Table 6.1 Values of the exchange length and wall width parameters for some common magnetic materials (hard and soft)

Material	l_{ex} (nm)	δ (nm)
Fe	1.5	40
Co	2	14
Ni	3.4	82
$SmCo_5$	4.9	3.6
$Nd_2Fe_{14}B$	1.9	3.9

length-scale over which the direction of the magnetic moments can change. It defines the threshold below which atomic exchange interaction prevails on the magneto-static fields [58]:

$$l_{ex} = \sqrt{\frac{A}{\mu_0 M_S^2}},$$

where A is the exchange length, M_S is the saturation magnetization and μ_0 is the vacuum permittivity. The exchange length is typically on the order of a few nanometres (see Table 6.1).

The second micromagnetic length is the wall width parameter δ, which accounts for the competition between the exchange energy and the magnetocrystalline anisotropy. The wall width parameter characterizes the wall width which separates magnetic domains and is expressed as

$$\delta = \sqrt{\frac{A}{K_1}}$$

It is possible to define a relation between these mesoscopic length-scales and the magnetic energy competition. This is illustrated in Fig. 6.6. For very small objects with sizes of a few l_{ex}, the exchange energy is prevalent and one may consider that the objects are mostly monodomain particles (blue particle). This is typically the case of magnetic nanoparticles with diameters smaller than 10–20 nm. For magnetic objects with sizes larger than a few δ, the object is able to accommodate magnetic domains (green particle) which will strongly minimize the dipolar energy. In the intermediate regime (10–200 nm) there is a subtle balance of competition between

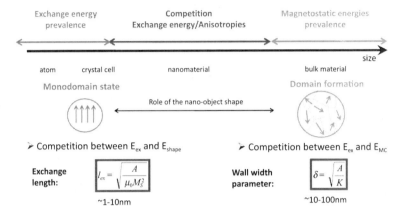

Figure 6.6 Magnetic energy competition in nanomaterials.

the exchange, anisotropy and shape energies which leads to a magnetic behaviour which is not single domain but where well-defined magnetic domains can neither be accommodated.

6.3.2 Description of the Magnetization Reversal Processes of Magnetic Nanoparticles within the Stoner–Wohlfarth Model

In a first approximation, in order to describe the reversal of magnetic nano-objects, it is often assumed that the magnetization of the nanoparticle remains homogeneous within the particle during the reversal process. The Stoner–Wohlfarth model thus simply consists of minimizing the magnetic energy:

$$E = E_{MC} + E_{shape} + E_Z = [(K_{MC} + K_{shape})\sin^2\theta - \mu_0 M_S H \cos(\phi - \theta)]V,$$

where E_{MC} is the magneto-crystalline energy, E_{shape} is the shape anisotropy and E_Z is the Zeeman energy which corresponds to the interaction of the externally applied field H and the magnetization M_S; ϕ is the angle between the applied magnetic field and the easy axis of the particle and θ the angle between the magnetization and the easy axis of the particle (Fig. 6.7). Beware that this expression only applies to single domain ellipsoidal objects in which the demagnetizing field and the magnetization are homogeneous within the object. This equation, nevertheless, enables to give a

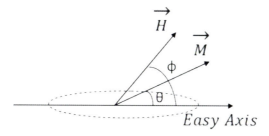

Figure 6.7 The Stoner–Wohlfarth geometry: definition of θ and ϕ.

qualitative description of the behaviour of the reversal of elongated nanoparticles.

Figure 6.8 illustrates the hysteresis cycles that are expected for an ellipsoid for different orientations of the magnetic field with respect to the long axis of the ellipsoid (varying ϕ) assuming a zero magneto-crystalline anisotropy. When the external magnetic field is applied along the long axis of the ellipsoid (easy axis), i.e. $\phi = 0$, blue curve, a square cycle is calculated. The coercive field is $H_c = M_s/2$, the origin of this coercive field being only due to the shape anisotropy. This field can theoretically be very large, on the order of

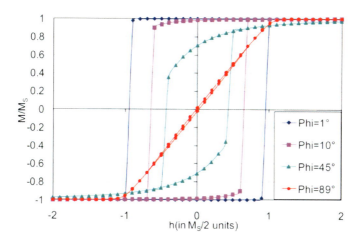

Figure 6.8 Hysteresis cycles of an ellipsoid in the Stoner–Wohlfarth model for different values of ϕ (angle between the long axis of the ellipsoid and the applied field; see Fig. 6.7). The magneto-crystalline anisotropy was assumed to be zero.

875 kA/m (= 1.1 T) in Fe and 710 kA/m (=0.88 T) in Co, provided the Stoner–Wohlfarth model applies. When the external magnetic field is applied perpendicular to the long axis ($\phi = 90°$), the hysteresis cycle is closed (red curve) and the saturation field is $H_c = M_s/2$. Note that for intermediate situations, the coercivity drops very quickly. For ϕ as small as $10°$ (purple curve), the coercivity drops down to $H_c = 0.65 \times M_s/2$. Thus if ones wants to exploit the shape anisotropy to produced materials with high coercivity, it is necessary to have a very good alignment of the wires (see Section 6.4).

6.3.3 Non-Uniform Magnetization Reversal Modes

The above description of anisotropic nano-objects within the Stoner–Wohlfarth model assumed that the particles remain single domain during the reversal process. This applies only for very small particles.

In the case of spheres it can be shown that a coherent rotation, i.e. the magnetic moments within the object remain all parallel during the magnetization reversal, is observed only for a radius smaller than

$$R_{\text{coh,sphere}} = \sqrt{24} l_{\text{ex}}$$

In the case of cylinders, this radius is further reduced to $R_{\text{coh,cyl}} = 3.65 l_{\text{ex}}$ [39].

For cylinder radius larger than $R_{\text{coh,cyl}}$, the magnetization reversal proceeds via non-fully collinear configurations of the magnetization. The reversal is referred to as incoherent reversal.

For even bigger radii, above the limit for the single domain (SD) being defined by $R_{\text{SD,cyl}} = 64 l_{\text{ex}}^2/\pi \delta$, the magnetization reversal proceeds via domain wall creation and propagation. Figure 6.9 summarizes the different situations in the case of Co nanowires.

In the region of radii $R_{\text{coh}} < R < R_{\text{SD}}$, various reversal modes may be observed. Figure 6.10 illustrates the curling and buckling reversal modes. In the curling mode, the magnetization is non-collinear across the section of the wire; in the buckling mode the magnetization is non-collinear along the wire axis.

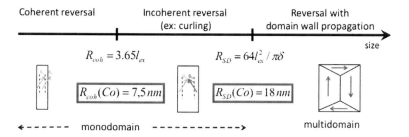

Figure 6.9 Magnetization reversal modes in Co nanowires. R_{SD} is the single-domain state radius below which the magnetization is single domain. R_{coh} is the radius below which the magnetization reversal is expected to be fully coherent.

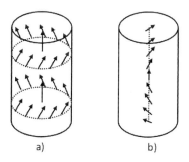

Figure 6.10 The (a) curling and (b) buckling nucleation modes in cylinders.

6.3.4 Description of the Magnetization Reversal Processes of Magnetic Nanoparticles within a Micromagnetic Modelling

In the previous sections, we have considered ideal objects such as ellipsoid or cylinders. The objects which are actually synthesized depart from these ideal shapes. In order to model as accurately as possible the properties of arbitrary shaped objects it is possible to perform numerical micromagnetic calculations. These micromagnetic computing codes are generally based on the solution of the Landau–Liftshitz–Gilbert–Langevin equation. Solving this equation can be performed either by a finite difference method (on a regular rectangular mesh) or by using finite elements (arbitrary mesh of the object). The first method is fast but more suited to films or rectangular objects. The second method is slower but is easier to use to describe spherical or cylindrical objects. Several finite difference codes are available on a freeware basis (OOMMF, MuMax, MicroMagnum), a few codes based on finite elements are also available (MagPar, Nmag). In the following, we will illustrate how micromagnetic simulations allow us to probe the details of the magnetic behaviour of nano-objects.

6.3.4.1 Effects of the shape of elongated magnetic particles on the coercive field

In this section we illustrate how the detailed shape of the magnetic nano-objects influences the coercive field. The first step is to model the objects with a mesh. Figure 6.11 illustrates how real nano-objects can be modelled very precisely. We underline the fact that we make a distinction between ellipsoids, cylinders with flat tips and cylinders with rounded tips.

Using micromagnetic simulations (NMag in this particular case), it is possible to calculate the hysteresis curves on these various objects. The coercivity of objects of various sizes and aspect ratio (length/diameter) is summarized in Fig. 6.12. To vary the aspect ratio, the length L of the objects is set at 100 nm and the diameter D is varied from 5 to 28 nm. These calculations give rise to various comments. In most cases, increasing the aspect ratio beyond 10 barely increases the coercive field. Significant differences are observed between the case of ellipsoidal particles and cylinders. The coercive field is reduced by about 20%. This enables quantifying the limits of the Stoner–Wohlfarth model. In the case of particles with less ideal shapes, a strong loss of coercive field is observed.

In order to understand where the loss of coercivity stems from, it is possible to observe the details of the reversal process. This is presented in Fig. 6.13. In the case of cylindrical objects, the sharp edges give rise to a "flower" state at the tips of the nanowires (Fig. 6.13a). The demagnetizing field is rather high but remains very localized at these tips (Fig. 6.13b). During reversal, the magnetization rotates in a C-shaped form (Fig. 6.13c). The magnetic potential isosurfaces are not flat any more at the tip as in the case of the capped nanowires and the demagnetizing field becomes very

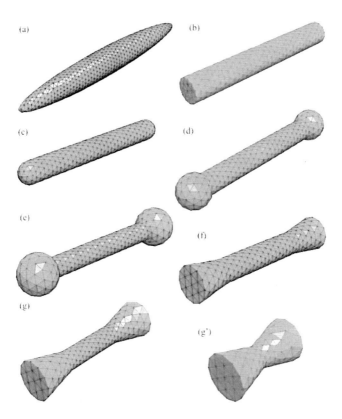

Figure 6.11 Different types of particles models and the corresponding meshes. (a) Ellipsoids, (b) cylinders, (c) capped cylinders, (d) dumbbells with small spherical ending, (dumbbell 1), (e) dumbbells with larger spherical endings ending (dumbbell 2), (f) cylinders with small cone endings (diabolo 1), (g) cylinders with larger cone endings (diabolo 2), and (g′) small diabolos. Adapted from [59].

high at one edge of the cylinder (Fig. 6.13d). This seed point is difficult to create because it costs locally quite a lot of energy. The coercivity of the wires remains very high.

6.3.4.2 Coherent and incoherent magnetization processes

As described in Fig. 6.9, the magnetization reversal rotation process is strongly dependent upon the diameter of the cylinder. Figure 6.14

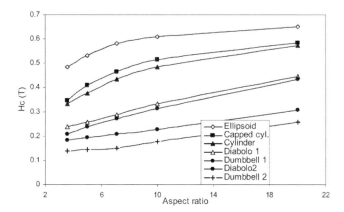

Figure 6.12 Variation of the coercivity of Co anisotropic particles as a function of their aspect ratio for different shapes of objects. The magnetic field is applied at 5.7° with respect to the object long axis. Adapted from [59].

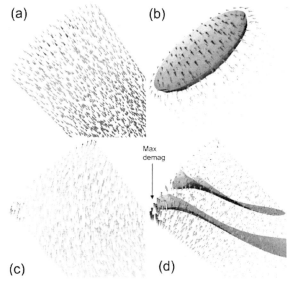

Figure 6.13 Magnetic state of a cylinder: (top) at remanence, magnetization and demagnetizing field; (bottom) before reversal, magnetization and demagnetizing field. (a) At remanence, the magnetization is almost perfectly collinear except at the edges of the wire; (b) the demagnetizing field is zero except at these edges; (c) before reversal, the magnetization rotates at the wire tip in a C-shaped form; (d) the demagnetizing field becomes very high at the edge of the tip. Adapted from [59].

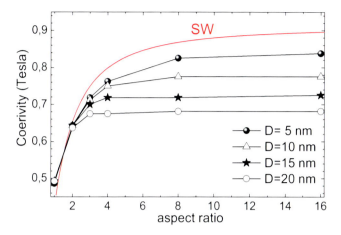

Figure 6.14 Evolution of the coercivity of Co cylinders as a function of their aspect ratio for different diameters D. The external field was at 22° with respect to the cylinder axis. The prediction of the Stoner–Wohlfarth (SW) model is plotted as a red continuous line (adapted from [62]). The bulk magneto-crystalline Co anisotropy was used for the calculation.

shows the evolution of the coercivity as a function of the wire diameter. These calculations confirm that for aspect ratios above 8–10, there is no more change in the reversal process and the coercivity has reached a saturation value. On the other hand for very small aspects ratios (AR < 3), there is little difference between the Stoner–Wohlfarth model and micromagnetic calculations. As the diameter of the wire increases, there is a continuous decrease of the coercive field. There is no abrupt change when the diameter becomes larger than the $D_{\text{coh}} = 15$ nm.

The details of the reversal process are illustrated in Fig. 6.15 for two diameters smaller ($D = 5$ nm) and larger ($D = 20$ nm) than D_{coh}. In the case of $D = 20$ nm $> D_{\text{coh}}$, the reversal process takes place via a buckling process (Fig. 6.10b), the magnetization becomes inhomogeneous along the wire axis. In the case of the small diameter $D = 5$ nm $< D_{\text{coh}}$ the reversal does not take place as a fully coherent rotation of the magnetization. A transverse domain wall is accommodated across the nanowire diameter and the reversal takes place as transverse domain wall propagation [60, 61]. In the other parts of the nanowire, the magnetization remains parallel to the wire

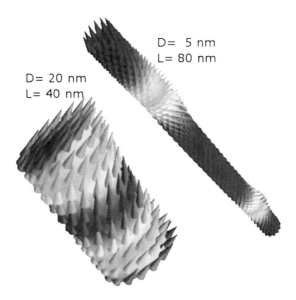

Figure 6.15 Magnetic states close to the coercivity of Co cylinders with different diameters and aspect ratios. The color variation corresponds to the magnetization component along the cylinder axis. The external field is applied with an angle $\varphi = 5°$ relative to the cylinder axis. Adapted from [62].

which is a very low energy state. This mechanism, which is different from the fully coherent rotation, is made possible by the fact that the tips of the wire create nucleation points for domain walls (Fig. 6.13).

6.3.5 Comparison of Experimental Data with Calculations

It is nowadays possible to produce magnetic objects close to model systems. In particular, Co cylinder with a crystalline structure can be grown with very low length and diameter dispersion by chemistry (polyol process or organometallic chemistry) (see Section 6.2). The nanorods show a very high coercivity due to the addition of shape and magneto-crystalline anisotropy (Fig. 6.16) [63]. On the other hand, the coercivity measured on nanorods grown by electrochemistry in alumina membranes is generally lower. This is usually accounted for by the fact that, in the latter case, the nanowires are not single crystals and that there are a lot of surface defects which are potential nucleation points for domain walls.

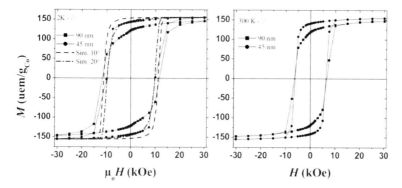

Figure 6.16 Magnetization curves for aligned Co nanorods parallel to the magnetic field, measured at 2 K (left) and 300 K (right). The curve labelled "Sim. 10°" and "Sim. 20°" are the hysteresis magnetization simulated for a system of interacting uniaxial nano-objects making an angle of 10° and 20° with respect to the magnetic field. Adapted with permission from [63].

Another point which should be mentioned is the effect of the temperature. All the above micromagnetic calculations assume a zero temperature, that is, the thermal fluctuations of the magnetization are not taken into account. As shown in Fig. 6.16, there is a large difference between measurements performed at 2 K and measurements performed at 300 K. A large part of this difference can be accounted for by the increase of the magnetocrystalline anisotropy of cobalt when the temperature is decreased. Another origin could be a lowering of the magnetic viscosity when the temperature is increased since viscosity measurements have shown that above 100 K magnetic fluctuations are important [64].

In the case of Co nanowires produced by the polyol process it was possible to produce wires with a wide range of diameters and lengths. The experimental results of the magnetometry measurements are summarized in Fig. 6.17. For aspect ratios below 10, one observes the same trend as what is numerically predicted. The coercivity doubles from small to large mean aspects ratios. For large aspects ratio ($AR_m > 10$), the experimental results are more scattered. This can be accounted for by the fact that crystalline defects are more likely to appear in such long wires. These defects

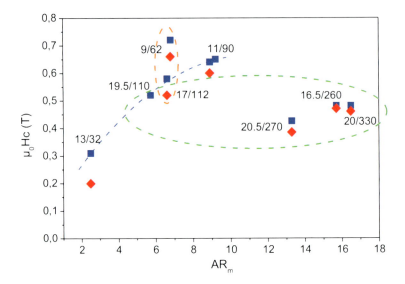

Figure 6.17 Coercivity of assemblies of Co nanorods vs. mean aspect ratio (AR_m) for different samples: aligned (blue squares) or randomly oriented (red diamonds). D_m/L_m are given for each sample (D_m and L_m: mean diameter and mean length, respectively). Adapted from [62].

will then act as nucleation point for the reversal and decrease the effective anisotropy.

6.4 Magnetic Properties of Assemblies of Nanoparticles

In the previous section, we have discussed the case of individual wires. However, in experimental measurements it is very difficult to measure single wires (except with techniques such as micro-SQUID). Thus all the experimental characterizations are performed on assemblies of particles.

6.4.1 Effect of the Particles' Orientation

In the case of randomly oriented nanoparticles (powder of nanoparticles), the canonical Stoner–Wohlfarth cycles of Fig. 6.8 need to be averaged over all possible orientations of the nanoparticles.

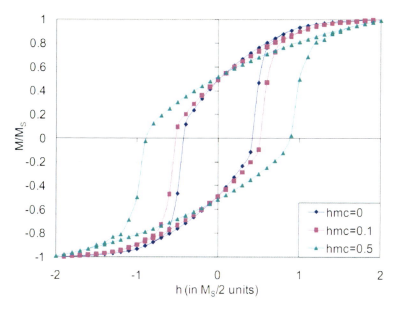

Figure 6.18 Hysteresis cycles of randomly oriented ellipsoids calculated in the Stoner–Wohlfarth model for different values of the magneto-crystalline anisotropy (h_{mc} along the long axis of the ellipsoid, h_{mc} in reduced units).

A rounded hysteresis cycle is obtained (dark blue diamonds in Fig. 6.18) with a coercive field $H_c = 0.43 \times M_s/2$ and a magnetization remanence of $M_r = 0.5$. This applies in the case of a random 3D orientation. In the case of a 2D random orientation which is typically obtained when powders are dried on a substrate for example, the value of the remanence is expected to be equal to $2/\pi$.

If the considered material has some magneto-crystalline anisotropy (magenta squares and cyan triangles in Fig. 6.18), the coercivity is increased but the remanent magnetization remains low, which is not desirable for applications such as the fabrication of permanent magnetic materials.

Experimental data confirm this behaviour. Indeed, Fig. 6.19 compares the case of aligned and disordered particles. Note that the remanence is above 0.5 and close to the value $2/\pi$, which suggests that some of the powders have a 2D orientation.

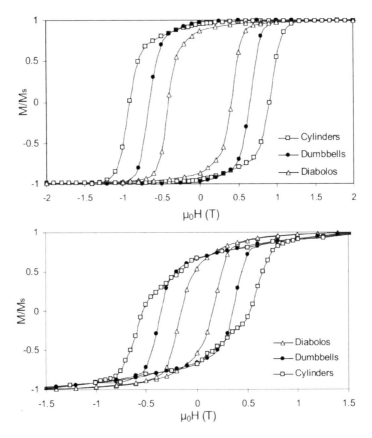

Figure 6.19 (Left) Hysteresis cycles at 150 K of different types of oriented nanowires: cylinders (Co nanorods), dumbbells (Co$_{80}$Ni$_{20}$ nanowires), and diabolos (short Co$_{50}$Ni$_{50}$ nanowires); (Right) Hysteresis cycles at 300 K of different powders of nanowires (non-oriented): cylinders (Co nanorods), dumbbells (Co$_{80}$Ni$_{20}$ nanowires), and diabolos (short Co$_{50}$Ni$_{50}$ nanowires). Adapted from [59].

6.4.2 Effect of the Macroscopic Demagnetizing Field

If one aims at fabricating materials with a high coercivity based on the shape anisotropy, it is clearly desirable to align the wires. This can be achieved in two ways: (i) fabricating magnetic nanowires in alumina templates (see Section 6.2), (ii) aligning a suspension of nanowires with a magnetic field and freeze or dry the solution so

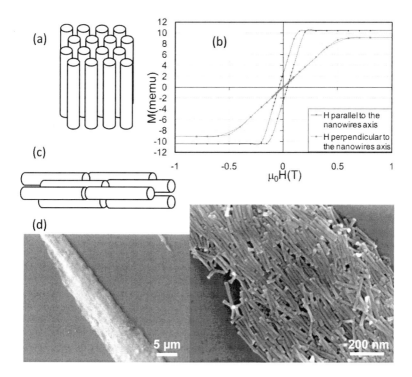

Figure 6.20 (a) Carpet of nanowires and corresponding hysteresis cycle measured along the wires axis (b) (the cycle is slanted due to the macroscopic shape anisotropy); scheme (c) and SEM images (d) of aggregates of nanorods obtained by aligning a suspension in a magnetic field.

as to trap the wires in a well-defined direction. These two ways of fabricating samples leads to very different magnetic behaviours due to very different macroscopic shape anisotropy.

Nanowires in alumina templates form a thin carpet of wires (Figs. 6.5a and 6.20a). In absence of magnetocrystalline anisotropy, the easy axis of individual wires is perpendicular to the film of wires. If the film was continuous, the demagnetizing field would be $H_d = N_z M_z$ with $N_z = 1$ (and $N_x = N_y = 0$), that is, the shape anisotropy induced by the demagnetizing field is in the film plane. In the case of a film of nanowires, the previous expression is modified as $H_d = P N_z M_z$, where P is the volume fraction of magnetic material in the film. The shape anisotropy of the film is reduced but remains in the film plane and is thus perpendicular to the shape anisotropy of the wires. The local and global shape anisotropies have directions at 90° with respect to each other. When the hysteresis curve is measured on such a carpet of nanowires, the hysteresis cycle is thus "slanted" due to the macroscopic shape anisotropy (Fig. 6.20b). The coercive field is not reduced but the remanence is significantly reduced.

On the other hand, in the case of elongated aggregates of nanowires described in Fig. 6.20c,d, when the magnetization curve in measured along the wires, the macroscopic demagnetizing field $H_d = N_z M_z$ with $N_z = 0$. Thus no macroscopic demagnetizing field is observed and the hysteresis cycle are expected to behave as shown in Fig. 6.8.

6.4.3 Effect of the Dipolar Interactions between Wires

Besides macroscopic demagnetizing effects, direct local dipolar interactions between wires can be important. Figure 6.21 represents the dipolar stray field at the tip of a nanowire ($L = 100$ nm, $r = 5$ nm, $\mu_0 M_s = 1$ T) calculated using Finite Element Method Magnetics (FEMM) [65]. It illustrates that the dipolar field is localized around the tip (in a volume with a typical size given by the radius of the nanowire). The stray fields are on the order of 0.1 T at a distance of 4 to 7 nm from the nanowire tip (see Fig. 6.21b). Thus, because the nanowires are separated by distances that can be as small as 2 to 4 nm, the dipolar field radiated by one wire on its neighbours can still be on the order of a fraction of 1 T.

These direct dipolar interactions leads to changes in the magnetization processes. In particular, for aligned wires it will lead to an effective tendency to anti-ferromagnetic coupling. In the case of carpets of nanowires this is illustrated in Fig. 6.22. In some configurations, at remanence, a long-range anti-ferromagnetic order is observed.

In the case of elongated aggregates of nanowires, similar effects arise. It is possible to quantify the magnitude of these interactions via Henkel plots (δM plots) or FORC plots. In order to construct the δM plots, the IRM (isothermal remanent magnetization) $I_R(H)$ and the DCD (direct current demagnetization) $I_D(H)$ curves are

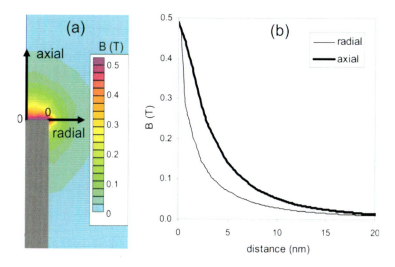

Figure 6.21 Dipolar field outside the tip of a magnetic nanowire ($L = 100$ nm, $r = 5$ nm, $\mu_0 M_s = 1$ T) (grey bar) calculated using FEMM [65]. (a) Axial mapping of the induction outside the wire. (b) Magnitude of the induction along the wire axis and the radial axis. The dipolar stray fields drop quickly but are still on the order of 0.1 T at a distance of 4 to 7 nm from the wire edges.

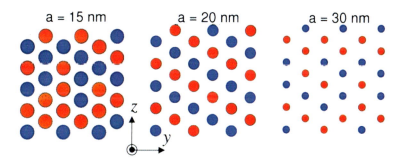

Figure 6.22 Array of 30 interacting nanowires. Remanent states after applying a saturating field along Oy for different inter-wire distances. Blue and red colors correspond to magnetization up and down inside the wires.

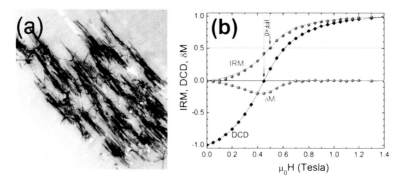

Figure 6.23 (a) TEM images of Co wires deposited under a magnetic field. (b) Construction of the δM plot by the IRM and DCD data for a sample dispersed in a PVP matrix (adapted from [66]).

measured [66]. δM plots are calculated by the following relation:

$$\delta M = 2I_R(H) - I_D(H) - 1$$

For non-interacting particles, $\delta M = 0$. Positive δM and negative δM are, respectively, interpreted as ferromagnetic and anti-ferromagnetic inter-particle interactions. The case of aggregates of nanowires is illustrated in Fig. 6.23. The Henkel plot shows negative δM values, which suggests that the local interactions between neighbouring wires are anti-ferromagnetic. The observed δM values are, however, quite small (0.05 to 0.2), which suggests that the dipolar interactions are limited. In order to get a more detailed insight of the nanowires interactions, FORC plots may be used [67].

6.5 Applications of Magnetic Nanowires

Magnetic nanowires and nanorods exhibit remarkable properties in terms of magnetization and coercivity. Cobalt NWs and NRs prepared by the chemical methods are particularly interesting because they exhibit very high coercivity resulting from the addition of shape and magnetocrystalline anisotropy. This makes them possible building blocks for the fabrication of permanent magnets. It is possible to numerically calculate the energy product of aligned powder of Co nanowires as a function of the magnetic volume

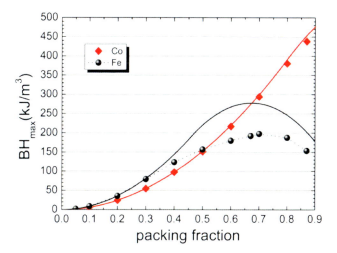

Figure 6.24 Energy product as a function of packing fraction for cobalt (red diamonds) and iron (black circles) nanowires. The continuous lines show the calculations under the assumption of square loops. Adapted from [68].

fraction of the assembly [68]. The results of such micro-magnetic calculations are shown in Fig. 6.24. For a realistic packing fraction of 60%, an energy product $(BH)_{max}$ of 220 kJ/m^3 (= 28 MGOe) is expected. This corresponds to the performances of low-grade NdFeB materials.

The advantage on permanent magnets based on nanowires would be that they are rare-earth free. The second advantage is that these materials can be used at rather high temperatures (up to 200 °C) while NdFeB materials need to be modified by the addition of very expensive *Dy* to be able to operate at temperatures higher than 100 °C.

It has recently been measured coercivity up to 1 T on cobalt nanowires prepared by a modified polyol process [69]. It was estimated from measurement on dilute systems that energy products as high as 350 kJ/m^3 could be expected for an ideal 100% density [69]. This would correspond to an energy product of 210 kJ/m^3 for a realistic 60% packing density. This is identical to the theoretical calculation of Fig. 6.24. The next step is to assess the actual performances in densely packed materials.

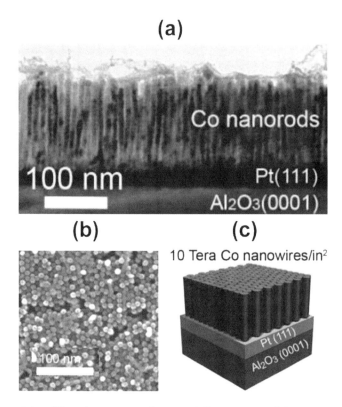

Figure 6.25 Ultra-dense array of epitaxial Co nanowires (a) cross section, (b) top view, (c) ideal representation as recording media. Adapted with permission from [1], copyright 2014 American Chemical Society.

A second possible application lies in the magnetic recording technology. For such applications nanowires grown in alumina templates may be considered as a bit pattern medium for perpendicular recording. A second fabrication process was demonstrated recently where nanowires with diameter of 5 nm can be grown epitaxially on crystalline substrates combining the growth of Pt thin layer by sputtering and the growth of cobalt nanowires by organometallic chemistry (Fig. 6.25) [1]. The small rod diameter, their strong perpendicular anisotropy and their organization into hexagonal arrays make them particularly interesting as magnetic recording media with expected capacities above 10 Terabits/in^2.

6.6 Conclusion and Outlook

Chemical synthesis now permits to produce highly anisotropic system with a very high crystalline quality and a good control of the particle shape (diameter, length). Due to their very good quality, these nano-objects are model systems to study the magnetism of anisotropic objects as well as studying the interactions between magnetic nanoparticles.

Micromagnetic modelling was very useful to describe the magnetization reversal in such objects as a function of their shape, diameter, aspect ratio and packing fraction in an assembly.

The excellent properties of cobalt nanowires in terms of magnetization ($M_s \sim 1.7$ T) and coercivity ($\mu_0 H \sim 1$ T) makes them interesting candidates for the fabrication of hard magnetic materials magnets. Consolidation experiments of magnetic nanowires into dense nanostructured anisotropic magnetic materials are expected to produce a new generation of permanent magnets.

Another outlook is to complexify the objects adding new functionalities. Starting from these nanowires, it is possible by chemistry to produce core–shell particles. For example, it is possible to add noble metal coating with optical properties in order to design a magneto-optical object with specific properties. We may also mention the use of anisotropic magnetic objects to target cancerous cell and destroy them via a mechanical mechanism.

References

1. Liakakos, N., Blon, T., Achkar, C., Vilar, V., Cormary, B., Tan, R. P., Benamara, O., Chaboussant, G., Ott, F., Warot-Fonrose, B., Snoeck, E., Chaudret, B., Soulantica, K., Respaud, M. (2014). Solution epitaxial growth of cobalt nanowires on crystalline substrates for data storage densities beyond 1 Tbit/in^2, *Nano Lett.*, **14**, pp. 3481–3486.

2. Maurer, T., Ott, F., Chaboussant, G., Soumare, Y., Piquemal, J.-Y., Viau, G. (2007). Magnetic nanowires as permanent magnet materials, *Appl. Phys. Lett.*, **91**, 172501.

3. Schrittwieser, S., Ludwig, F., Dieckhoff, J., Soulantica, K., Viau, G., Lacroix, L.-M., Mozo, Lentijo, S., Boubekri, R., Maynadié, J., Huetten, A., Brueckl,

H., Schotter, J. (2012). Modeling and development of a biosensor based on optical relaxation measurements of hybrid nanoparticles, *ACS Nano*, **6**, pp. 791–801.

4. Fiévet, F., Lagier, J.-P., Blin, B., Beaudouin, B., Figlarz, M. (1989). Homogeneous and heterogeneous nucleations in the polyol process for the preparation of micron and sub-micron size metal particles, *Solid State Ionics*, **32–33**, pp. 198–205.

5. Fiévet, F., Lagier, J.-P., Figlarz, M. (1989). Preparing monodisperse metal powders in micrometer and submicrometer sizes by the polyol process, *MRS Bull.*, **14**, pp. 29–34.

6. Viau, G., Fiévet-Vincent, F., Fiévet, F. (1996). Nucleation and growth of bimetallic CoNi and FeNi monodisperse particles prepared in polyols, *Solid State Ionics*, **84**, pp. 259–270.

7. Viau, G., Fiévet-Vincent, F., Fiévet, F. (1996). Monodisperse iron-based particles : Precipitation in liquid polyols, *J. Mater. Chem.*, **6**, pp. 1047–1053.

8. Toneguzzo, P., Viau, G., Acher, O., Fiévet-Vincent, F., Fiévet, F. (1998). Monodisperse ferromagnetic particles for microwave applications, *Adv. Mater.*, **10**, pp. 1032–1035.

9. Kodama, D., Shinoda, K., Sato, K., Konno, Y., Joseyphus, R. J., Motomiya, K., Takahashi, H., Matsumoto, T., Sato, Y., Tohji, K., Jeyadevan, B. (2006). Chemical synthesis of sub-micrometer- to nanometer-sized magnetic FeCo dice, *Adv. Mater.*, **18**, pp. 3154–3159.

10. Ducamp-Sanguesa, C., Herrera-Urbina, R., Figlarz, M. (1992). Synthesis and characterization of fine and monodisperse silver particles of uniform shape, *Solid State Chem.*, **100**, pp. 272–280.

11. Wiley, B., Sun, Y., Mayers, B., Xia, Y. (2005). Shape-controlled synthesis of metal nanostructures: The case of silver, *Chem. Eur. J.*, **11**, pp. 454–463.

12. Soumare, Y., Garcia, C., Maurer, T., Chaboussant, G., Ott, F., Fiévet, F., Piquemal, J.-Y., Viau, G. (2009). Kinetically controlled synthesis of cobalt nanorods with high magnetic coercivity, *Adv. Funct. Mater.*, **19**, pp. 1971–1977.

13. Aït Atmane, K., Michel, C., Piquemal, J.-Y., Sautet, P., Beaunier, P., Giraud, M., Sicard, M., Nowak, S., Losno, R., Viau, G. (2014). Control of anisotropic growth of cobalt nanorods in liquid phase: From experiment to theory ... and back, *Nanoscale*, **6**, pp. 2682–2692.

14. Ung, D., Viau, G., Ricolleau, C., Warmont, F., Gredin, P., Fiévet, F. (2005). CoNi nanowires synthesized by heterogeneous nucleation in liquid polyol, *Adv. Mater.*, **17**, pp. 338–344.

15. Soumare, Y., Piquemal, J.-Y., Maurer, T., Ott, F., Chaboussant, G., Falqui A., Viau, G. (2008). Oriented magnetic nanowires with high coercivity, *J. Mater. Chem.*, **18**, pp. 5696–5702.

16. Ung, D., Viau, G., Fiévet-Vincent, F., Herbst, F., Richard, V., Fiévet, F. (2005). Magnetic nanoparticles with hybrid shape, *Prog. Solid State Chem.*, **33**, pp. 137–145.

17. Viau, G., Garcia, C., Maurer, T., Chaboussant, G., Ott, F., Soumare, Y., Piquemal, J.-Y. (2009). Highly crystalline cobalt nanowires with high coercivity prepared by soft chemistry, *Phys. Status Solidi* A, **206**, pp. 663–666.

18. Cordente, N., Respaud, M., Senocq, F., Casanove, M.-J., Amiens, C., Chaudret, B. (2001). Synthesis and magnetic properties of nickel nanorods, *Nano Lett.*, **1**, pp. 565–568.

19. Lacroix, L.-M., Lachaize, S., Falqui, A., Respaud, M., Chaudret, B. (2009). Iron nanoparticle growth in organic superstructures, *J. Am. Chem. Soc.*, **131**, pp. 549–557.

20. Desvaux, C., Dumestre, F., Amiens, C., Respaud, M., Lecante, P., Snoeck, E., Fejes, P., Renaud, P., Chaudret, B. (2009). FeCo nanoparticles from an organometallic approach: Synthesis, organisation and physical properties, *J. Mater. Chem.*, **19**, pp. 3268–3275.

21. Dumestre, F., Chaudret, B., Amiens, C., Fromen, M.-C., Casanove, M.-J., Renaud, P., Zurcher, P. (2002). Shape control of thermodynamically stable cobalt nanorods through organometallic chemistry, *Angew. Chem. Int. Ed.*, **41**, pp. 4286–4289.

22. Dumestre, F., Chaudret, B., Amiens, C., Respaud, M., Fejes, P., Renaud, P., Zurcher, P. (2003). Unprecedented crystalline super-lattices of monodisperse cobalt nanorods, *Angew. Chem. Int. Ed.*, **42**, pp. 5213–5216.

23. Wetz, F., Soulantica, K., Respaud, M., Falqui, A., Chaudret, B. (2007). Synthesis and magnetic properties of Co nanorod superlattices, *Mater. Sci. Eng. C*, **27**, pp. 1162–1166.

24. Ciuculescu, D., Dumestre, F., Comesaña-Hermo, M., Chaudret, B., Spasova, M., Farle, M., Amiens, C. (2009). Single-crystalline Co nanowires: Synthesis, thermal stability, and carbon coating, *Chem. Mater.*, **21**, pp. 3987–3995.

25. Liakakos, N., Cormary, B., Li, X., Lecante, P., Respaud, M., Maron, L., Falqui, A., Genovese, A., Vendier, L., Koinis, S., Chaudret, B., Soulantica, K. (2012). The big impact of a small detail: Cobalt nanocrystal polymorphism as a result of precursor addition rate during Stock Solution preparation, *J. Am. Chem. Soc.*, **134**, pp. 17922–17931.

26. Liakakos, N., Gatel, C., Blon, T., Altantzis, T., Lentijo, S., Garcia, C., Lacroix, L.-M., Respaud, M., Bals, S., Van Tendeloo, G., Soulantica, K. (2014). Co–Fe nanodumbbells: Synthesis, structure, and magnetic properties,*Nano Lett.*, **14**, pp. 2747–2754.

27. Zadoina, L., Lonetti, B., Soulantica, K., Mingotaud, A.-F., Respaud, M., Chaudret, B., Mauzac, M. (2009). Liquid crystalline magnetic material, *J. Mat. Chem.*, **219**, pp. 8075–8078.

28. Riou, O., Lonetti, B., Davidson, P., Tan, R. P., Cormary, B., Mingotaud, A.-F., Di Cola, E., Respaud, M., Chaudret, B., Soulantica, K., Mauzac, M. (2014). Liquid crystalline polymer-co nanorod hybrids: Structural analysis and response to a magnetic field, *J. Phys. Chem. B*, **118**, pp. 3218–3225.

29. Puntes, V., Zanchet, D., Erdonmez, C. K., Alivisatos, A. P. (2002). Synthesis of hcp-Co nanodisks, *J. Am. Chem. Soc.*, **124**, pp. 12874–12880.

30. Park, S.-J., Kim, S., Lee, S., Khim, Z. G., Char, K., Hyeon, T. (2000). Synthesis and magnetic studies of uniform iron nanorods and nanospheres, *J. Am. Chem. Soc.*, **122**, pp. 8581–8582.

31. Wang, C., Hou, Y., Kim, J., Sun, S. (2007). A general strategy for synthesizing FePt nanowires and nanorods, *Angew. Chem. Int. Ed.*, **46**, pp. 6333–6335.

32. Aranda, P., García, J. M. (2002). Porous membranes for the preparation of magnetic nanostructures, *J. Magn. Magn. Mater.*, **249**, pp. 214–219.

33. Sanz, R., Hernández-Vélez, M., Pirota, K. R., Baldonedo, J. L., Vázquez, M. (2007). Fabrication and magnetic functionatization of cylindrical porous anodic alumina, *Small*, **3**, pp. 434–437.

34. Schmid, G., (2002). Materials in nanoporous alumina,*J. Mater. Chem.*, **12**, pp. 1231–1238.

35. Martin, C. R. (1996). Membrane-based synthesis of nanomaterials, *Chem. Mater.*, **8**, pp. 1739–1746.

36. Huczko, A. (2000). Template-based synthesis of nanomaterials, *Appl. Phys. A*, **70**, pp. 365–376.

37. Sellmyer, D. J., Zheng, M., Skomski, R. (2001). Magnetism of Fe, Co and Ni nanowires in self-assembled arrays, *J. Phys. Condes. Mater.*, **13**, pp. R433–R460.

38. Kawai, S., Ishiguro, I. (1976). Recording characteristics of anodic oxide-films on aluminium containing electrodeposited ferromagnetic metals and alloys,*J. Electrochem. Soc.*, **123**, pp. 1047–1051.

39. Skomski, R., Zeng, H., Zheng, M., Sellmyer, D. J. (2000). Magnetic localization in transition-metal nanowires, *Phys. Rev. B*, **62**, pp. 3900–3904.

40. Paulus, P. M., Luis, F., Kroll, M., Schmid, G., de Jongh, L. J. (2001). Low-temperature study of the magnetization reversal and magnetic anisotropy of Fe, Ni, and Co nanowires, *J. Magn. Magn. Mater.*, **224**, pp. 180–196.

41. Bran, C., Ivanov, Yu. P., Trabada, D. G., Tomkowicz, J., del Real, R. P., Chubykalo-Fesenko, O., Vazquez, M. (2013). Structural dependence of magnetic properties in Co-based nanowires: Experiments and micromagnetic simulations, *IEEE Trans. Magn.*, **49**, pp. 4491–4497.

42. Bran, C., Ivanov, Yu. P., García, J., del Real, R. P., Prida, V. M., Chubykalo-Fesenko, O., Vazquez, M. (2013). Tuning the magnetization reversal process of FeCoCu nanowire arrays by thermal annealing, *J. Appl. Phys.* **114**, 043908.

43. Kröll, M., Blau, W. J., Grandjean, D., Benfield, R. E., Luis, F., Paulus, P. M., de Jongh, L. J. (2002). Magnetic properties of ferromagnetic nanowires embedded in nanoporous alumina membranes, *J. Magn. Magn. Mater.*, **249**, pp. 241–245.

44. Huang, X. H., Li, L., Luo, X., Zhu, X. G., Li, G. H. (2008). Orientation-controlled synthesis and ferromagnetism of single crystalline Co nanowire arrays, *J. Phys. Chem. C*, **112**, pp. 1468–1472.

45. Caffarena, V. R., Guimarães, A. P., Folly, W. S. D., Silva, E. M., Capitaneo, J. L. (2008). Magnetic behavior of electrodeposited cobalt nanowires using different electrolytic bath acidities, *Mater. Chem. Phys.*, **107**, pp. 297–304.

46. Darques, M., Encinas-Oropesa, A., Villa, L., Piraux, L. (2004). Controlled changes in the microstructure and magnetic anisotropy in arrays of electrodeposited Co nanowires induced by the solution pH, *J. Phys. D: Appl. Phys.*, **37**, pp. 1411–1416.

47. Darques, M., Piraux, L., Encinas, A., Bayle-Guillemaud, P., Popa, A., Ebels, U. (2005). Electrochemical control and selection of the structural and magnetic properties of cobalt nanowires, *Appl. Phys. Lett.*, **86**, 072508.

48. Hernández-Vélez, M. (2006). Nanowires and 1D arrays fabrication: An overview, *Thin Solid Films*, **495**, pp. 51–63.

49. Chaure, N. B., Stamenov, P., Rhen, F. M. F., Coey, J. M. D. (2005). Oriented cobalt nanowires prepared by electrodeposition in a porous membrane, *J. Magn. Magn. Mater.*, **290–291**, pp. 1210–1213.

50. Mohaddes-Ardabili, L., Zheng, H., Ogale, S. B., Hannoyer, B., Tian, W., Wang, J., Lofland, S. E., Shinde, S. R., Zhao, T., Jia, Y., Salamanca-Riba, L., Schlom, D. G., Wuttig, M., Ramesh, R. (2004). Self-assembled single-

crystal ferromagnetic iron nanowires formed by decomposition. *Nat. Mater.*, **3**, pp. 533–538.

51. Mohaddes-Ardabili, L., Zheng, H., Zhan, Q., Yang, S. Y., Ramesh, R., Salamanca-Riba, L., Wuttig, M., Ogale, S. B., Pan, X. (2005). Size and shape evolution of embedded single-crystal alpha-Fe nanowires, *Appl. Phys. Lett.*, **87**, 203110.

52. Zheng, H., Wang, J., Mohaddes-Ardabili, L., Wuttig, M., Salamanca-Riba, L., Schlom, D. G., Ramesh, R. (2004). Three-dimensional heteroepitaxy in self-assembled $BaTiO_3$-$CoFe_2O_4$ nanostructures, *Appl. Phys. Lett.*, **85**, pp. 2035–2037.

53. Aimon, N. M., Kim, D. H., Choi, H. K., Ross, C. A. (2012). Deposition of epitaxial $BiFeO_3$/$CoFe_2O_4$ nanocomposites on (001) $SrTiO_3$ by combinatorial pulsed laser deposition, *Appl. Phys. Lett.*, **100**, 092901.

54. Schio, P., Vidal, F., Zheng, Y., Milano, J., Fonda, E., Demaille, D., Vodungbo, B., Varalda, J., de Oliveira, A. J. A., Etgens, V. H. (2010). Magnetic response of cobalt nanowires with diameter below 5 nm, *Phys. Rev. B*, **82**, 094436.

55. Vidal, F., Zheng, Y., Schio, P., Bonilla, F. J., Barturen, M., Milano, J., Demaille, D., Fonda, E., de Oliveira, A. J. A., Etgens, V. H. (2012). Mechanism of localization of the magnetization reversal in 3 nm wide Co nanowires, *Phys. Rev. Lett.*, **109**, 117205.

56. Bonilla, F. J., Novikova, A., Vidal, F., Zheng, Y., Fonda, E., Demaille, D., Schuler, V., Coati, A., Vlad, A., Garreau, Y., Sauvage Simkin, M., Dumont, Y., Hidki, S., Etgens, V. (2013). Combinatorial growth and anisotropy control of self-assembled epitaxial ultrathin alloy nanowires, *ACS Nano*, **7**, pp. 4022–4029.

57. Shin, J., Goyal, A., Cantoni, C., Sinclair, J. W., Thompson, J. R. (2012). Self-assembled ferromagnetic cobalt/yttria-stabilized zirconia nanocomposites for ultrahigh density storage applications, *Nanotechnology*, **23**, 155602.

58. Skomski, R., Coey, J. M. D. (1999). *Permanent Magnetism*, Institute of Physics Publishing, Bristol and Philadelphia.

59. Ott, F., Maurer, T., Chaboussant, G., Soumare, Y., Piquemal J.-Y., Viau, G. (2009). Effects of the shape of elongated magnetic particles on the coercive field, *J. Appl. Phys.*, **105**, 013915.

60. Vivas, L. G., Escrig, J., Trabada, D. G., Badini-Confalonieri, G. A., Vazquez, M. (2012). Magnetic anisotropy in ordered textured Co nanowires, *Appl. Phys. Lett.*, **100**, 252405.

61. Vivas, L. G., Vazquez, M., Escrig, J., Allende, S., Altbir, D., Leitao, D. C., Araujo, J. P. (2012). Magnetic anisotropy in CoNi nanowire arrays: Analytical calculations and experiments, *Phys. Rev. B*, **85**, 035439.

62. Pousthomis, M., Anagnostopoulou, E., Panagiotopoulos, I., Boubekri, R., Fang, W., Ott, F., Aït Atmane, K., Piquemal, J.-Y., Lacroix, L.-M., Viau, G. (2015). Localized magnetization reversal processes in cobalt nanorods with different aspect ratios, *Nano Res.*, **8**, pp. 2231–2241.

63. Soulantica, K., Wetz, F., Maynadié, J., Falqui, A., Tan, R. P., Blon, T., Chaudret, B., Respaud, M. (2009). Magnetism of single-crystalline Co nanorods, *Appl. Phys. Lett.*, **95**, 152504.

64. Maurer, T., Zighem, F., Ott, F., Chaboussant, G., André, G., Soumare, Y., Piquemal, J.-Y., Viau, G., Gatel, C. (2009). Exchange bias in Co/CoO core-shell nanowires: Role of the antiferromagnetic superparamagnetic fluctuations, *Phys. Rev. B*, **80**, 064427.

65. Meeker, D. C., Finite element method magnetics, http://www.femm.info.

66. Panagiotopoulos, I., Fang, W., Ait-Atmane, K., Piquemal, J.-Y., Viau, G., Dalmas, F., Boué, F., Ott, F. (2013). Low dipolar interactions in dense aggregates of aligned magnetic nanowires, *J. Appl. Phys.*, **114**, 233909.

67. Gilbert, D. A., Zimanyi, G. T., Dumas, R. K., Winklhofer, M., Gomez, A., Eibagi, N., Vicent, J. L., Liu, K. (2014). Quantitative decoding of interactions in tunable nanomagnet arrays using first order reversal curves, *Sci. Rep.*, **4**, 4204.

68. Panagiotopoulos, I., Fang, W., Ott, F., Boué, F., Aït-Atmane, K., Piquemal, J.-Y., Viau, G. (2013). Packing fraction dependence of the coercivity and the energy product in nanowire based permanent magnets, *J. Appl. Phys.*, **114**, 143902.

69. Gandha, K., Elkins, K., Poudyal, N., Liu, X., Liu, J. P. (2014). High energy product developed from cobalt nanowires, *Sci. Rep.*, **4**, 5345.

Chapter 7

Magnetoferritin Nanoparticles as a Promising Building Blocks for Three-Dimensional Magnonic Crystals

Sławomir Mamica

Faculty of Physics, Adam Mickiewicz University in Poznan,
ul. Umultowska 85, Poznan, Poland
mamica@amu.edu.pl

7.1 Introduction

Magnetic nanoparticles (NPs) are one of the hottest topics nowadays because of their unusual physical properties as well as promising applications in a wide variety of fields that range from medicine to nanoelectronics (Reddy et al., 2012; Singamaneni et al., 2011). The very special examples of such structures are biomimetic magnetic nanoparticles, i.e., NPs grown in a biomineralization process inside biological protein cages (Flenniken et al., 2009; Henry and Debarbieux, 2012; Klem et al., 2005; Yamashita et al., 2010). They are special not only because of the peculiar way of growing but mostly due to some extra properties inherited from the biological material. Combining these properties with magnetic

---

Magnetic Structures of 2D and 3D Nanoparticles: Properties and Applications
Edited by Jean-Claude Levy
Copyright © 2016 Pan Stanford Publishing Pte. Ltd.
ISBN 978-981-4613-67-5 (Hardcover), 978-981-4613-68-2 (eBook)
www.panstanford.com

features gives possibility to apply biomimetic NPs in different branches of medicine and technology. In medicine, e.g., they can be used for targeting and visualizing tumor tissues (Fan et al., 2012), as resonance imaging contrast agents (Sana et al., 2010), in hyperthermia (Fantechi et al., 2014; Martinez-Boubeta et al., 2013), or as drug delivery vessels (Uchida et al., 2006). In other fields their applications include the decontamination of radioactive waste streams (Urban et al., 2012) as well as the position-controlled growth of carbon nanotubes (Hanasaki et al., 2011; Kumagai et al., 2010). Finally, in electronic applications they can be used in the fabrication of new types of battery electrodes (Nam et al., 2008; Yang et al., 2013) or in nanodot floating gate memory (Sarkar et al., 2007). Due to their tendency to self-organization, biomimetic NPs prove to be useful as building blocks for fabricating two- (2D) and three-dimensional (3D) highly ordered structures (Feng and Damodaran, 2000; Majetich et al., 2011; Payne et al., 2013; Rong et al., 2011).

In this chapter, we point out another possible application of biomimetic magnetic NPs, namely fabrication of magnonic crystals (MCs), a magnetic counterpart of photonic crystals, in which spin waves propagate instead of electromagnetic waves (Nikitov et al., 2001; Vasseur et al., 1996) (see also Chapter 8 [by Klos and Krawczyk]). MCs traditionally consist of two magnetic materials,[a] one (inclusions) disposed periodically in another (a matrix). Similar, to other periodic composites, such as semiconductor superlattices (Kłos and Krawczyk, 2010), photonic (Joannopoulos et al., 2008) or phononic crystals (Halevi, 2008), by adjusting of the shape and periodic arrangement of the inclusions as well as constituent materials, the dispersion of spin waves in MCs can be tailored in a way unavailable in natural materials (Kłos et al., 2012; Krawczyk et al., 2002; Lenk et al., 2011; Mamica et al., 2012b; Serga et al., 2010). This entails the modeling of the velocity and the direction and effective damping of the spin waves (Neusser et al., 2010; Romero-Vivas et al., 2012). However, the fundamental characteristic of periodic structures is the occurrence of a gap in

---

[a]Such MCs are called bicomponent MCs, and in first (theoretical) papers just such cases are studied. Nowadays also single-component MCs are very popular, such as antidot lattices (Neusser et al., 2011) or systems of coupled magnetic dots (Galkin et al., 2005). However, in this chapter only bicomponent MCs are considered.

their frequency spectrum, i.e., a certain frequency range in which the wave propagation is prohibited, regardless of the direction of propagation. In MCs this gap is referred to as magnonic bandgap and the possibility of its tuning makes MCs the basic materials for the new branch of physics referred to as magnonics (Krawczyk and Grundler, 2014; Kruglyak et al., 2010).

One of the inherent features of ferromagnetic materials is the concurrence of the exchange and dipolar interactions. In Chapter 3 we study the origin and effects of the competition between these two kinds of interaction in 2D nanodots. Several features are highly influenced by this competition: the stability of the magnetic (vortex) configuration in dots (Mamica et al., 2011) and rings (Mamica, 2013b; Mamica et al., 2012c), the spin-wave excitations spectrum (Mamica, 2013c; Mamica et al., 2014), and the spontaneous reversal of the vortex core (Depondt et al., 2013)). However, the coexistence of the short- and long-range interactions is of much impact also in systems completely different from the spatially confined 2D nanodots. In uniform thin films, it is responsible for several dynamical effects, including surface and subsurface localization (Mamica et al., 1998, 2000; Puszkarski et al., 1998). In patterned thin films with perpendicular anisotropy it results in the splitting of the spectrum into subbands (Krawczyk et al., 2011; Pal et al., 2012). The dipolar interactions are also responsible for the non-reciprocity effect (Mruczkiewicz et al., 2013) and negative group velocity (Krawczyk et al., 2013) in the spin wave spectrum of thin films. In 3D MCs Krawczyk and Puszkarski have shown the existence of a magnonic bandgap with a maximum of the width in its lattice-constant dependence for 10–20 nm (Krawczyk and Puszkarski, 2008). In our papers (Mamica, 2013a; Mamica et al., 2012a), we explain this behavior by addressing it just to the concurrence of the exchange and dipolar interactions. We also show that inclusions of 8 nm in diameter in the fcc lattice are most efficient for the magnonic gap opening and the maximum of the gap width occurs for the lattice constant about 13 nm. In our theoretical study, we obtain a few hundred GHz broad gap in the subterahertz frequency range. The production of such MCs would be an enormous step forward in magnonics, but the experimental realization of 3D MCs with the periodicity in the range of an over a dozen nanometer is rather

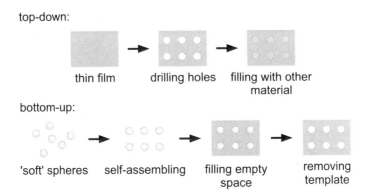

Figure 7.1 Philosophy of nanocomposite fabrication: top-down vs. bottom-up.

challenging and such structures remain the subject of theoretical studies.

There are two kinds of methods of MCs (and similar structures) fabrication: top-down and bottom-up (Kruglyak et al., 2012) (see Fig. 7.1). Top-down methods are very common in the case of 2D MCs. First of all a magnetic thin film is prepared in which a regular lattice of holes is drilled. Finally, the holes are filled with another magnetic material. The idea of the bottom-up technique is as follows. Let us take self-assembling spheres made from a material ease to remove (e.g., polystyrene). Build 3D structure and fill empty spaces with the desired material. Then remove the template. This approach is used in the fabrication of 3D photonic crystals with so-called inverse opal structure (Kekesi et al., 2013). Unfortunately, both methods are rather useless for 3D MCs with nanometer periodicity and the main problem is the size of inclusions. In this context self-assembled biomimetic magnetic NPs appears very promising as a magnetic template for bottom-up technology for producing of nanoscale 3D MCs. Especially, as we will see later, magnetoferritin crystals have the crystallographic structure and the lattice constant almost optimized for the occurrence of a magnonic bandgap (Mamica et al., 2012a).

In the next section, we introduce very popular example of the biomimetic magnetic NP called magnetoferritin (mFT). We also shortly characterize 3D mFT crystals obtained by protein crystallization technique. In Section 7.3, we describe a 3D version of the plane wave method (PWM), a powerful theoretical tool for studying MCs. The method is derived in Chapter 8 of this book together with applications to thin film 2D MCs. In Section 7.4 we use PWM to examine the spin wave spectra in 3D MCs based on mFT crystals. Particular attention is paid to the magnonic bandgap and its evolution with the lattice constant and material parameters. On the basis of the spin wave profiles, we explain the behavior of the gap width. Finally, in Section 7.5, we sum up the most important results and conclusions.

7.2 Magnetoferritin Nanoparticles and Their 3D Assemblies

The biomimetic way of NPs fabrication, i.e., the use of cage-like proteins as reaction chambers, has a number of advantages. One of them is an extremely high level of homogeneity of the size distribution and the properties of the biological material.[a] This can be used to perform highly uniform NPs in terms of size and shape. Another advantage is diversity of the size and properties of cage-like proteins. The internal diameter of the semi-spherical protein cages ranges from 4.5 nm for smallest Dps (DNA protection during starvation, i.e., DNA-binding protein from starved cells) (Douglas and Young, 1999) to 90 nm for herpes virus (Tatman et al., 1994) (see Table 7.1 for more examples). Diversity of protein shells in functionality and chemical and thermal stabilities gives possibility for materials synthesis under a variety of conditions (Flenniken et al., 2009). From the point of view of our study, the self-assembly tendency of biomimetic NPs is very important, which can be utilized for the fabrication of highly ordered 3D structures. And there is one more very useful property: the existence of three distinct surfaces—internal, external, and intermediate. The internal surface of the cage is responsible for biomineralization

---

[a]Concerning particular species. Protein cages of the same type may differ from species to species.

Table 7.1 A few examples of protein cages

Protein cage	Inner/outer diameter [nm]
Dps (from *L. innocua*) (Klem et al., 2005)	6/9
HspG41C (Abedin et al., 2009)	6.5/12
Apoferritin (from Horse spleen) (Klem et al., 2005)	8/12
Lumazine synthase (from *A. aeolicus*) (Zhang et al., 2001)	8/15.4
Cowpea chlorotic mottle virus (CCMV) (Klem et al., 2005)	24/28
Cauliflower mosaic virus (Hoh et al., 2010)	25/52
HSV-1 (herpes simplex virus type 1) (Tatman et al., 1994)	90/125

process. It determines the size, shape, and internal structure of NPs grown inside. The external surface defines the tendency to self-organization: the structure and lattice constant of the NPs crystal (or specify the cell targeting). The "intermediate surface" is the interface between the proteins the cage consists of; usually it has a form of channels specialized in delivering of particular substances inside the cage. These three surfaces could be functionalized independently via genetic or chemical modification, which gives a large number of possibilities in designing the final structure (Sapsford et al., 2013). This means, e.g., the characteristics of the external surface can be modified without affecting the internal one, which allows one to control the self-assembly process without modifying the NPs obtained inside the protein cage.

One of the most commonly studied biomimetic magnetic NPs are magnetoferritins. This very numerous superfamily is based on a ferritin, a protein used in living organisms to store an iron in a non-toxic form (Andrews, 2010; Harrison and Arosio, 1996). Natural ferritin is composed of the protein shell and ferrihydrite core (Fig. 7.2). The shell, called apoferritin, consists of 24 protein subunits forming a spherical cage that can store approximately 4500 Fe atoms (Liu and Theil, 2005).[a] Different composition of subunits is observed between animal ferritins and the others (plant and bacterial), but the size of apoferritin is similar: The external diameter is approximately 12 nm and the internal one 8 nm (Theil et al., 2013).

---

[a]We take into account only "normal" ferritin, in contrast to so-called mini-ferritin, a Dps protein built up of 12 subunits (Theil et al., 2006).

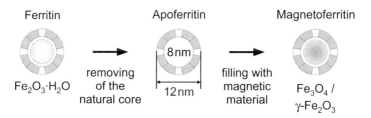

Figure 7.2 Mimicking the nature: a strategy for fabrication of magnetic nanoparticles.

The first step in the fabrication of biomimetic NPs is the removal of the natural core and then the filling of the empty apoferritin with the accumulated material (Meldrum et al., 1991) (Fig. 7.2). The list of possible materials stored in apoferritin is quite impressive; for a survey see, e.g., Yoshimura (2006), Flenniken et al. (2009), Uchida et al. (2010), Yamashita et al. (2010). Among magnetic materials one can find Mn, Co, CoPt, Ni, or magnetite/maghemite (Fe_3O_4/γ-Fe_2O_3), and if the obtained core has a significant total magnetic moment, the nanoparticle is referred to as magnetoferritin (Meldrum et al., 1992). Next step is three-stage purification of magnetoferritins, i.e., removing of impurities, empty or partially filled ferritins, and finally dimers and higher oligomers (Okuda et al., 2012). This process allows the fabrication of mFT NPs highly monodisperse in terms of size as well as magnetic parameters with apoferritin completely filled with the magnetic material. For example, magnetite/maghemite core is reported to have a diameter of approximately 8 nm and a total magnetic moment of the order of $10^4 \mu_B$ (Kasyutich et al., 2008). This gives the saturation magnetization $M_S = 346$ kA/m and exchange constant $A = 1.3 \times 10^{-11}$ J/m, which agrees well with the experimental data given in Koralewski et al. (2012).

There are different techniques for the fabrication of 3D NPs crystals. For example, functionalized NPs can form well-ordered 3D structures via electrostatic interactions (Sun et al., 2000). Such colloidal crystals made of sub-10 nm Fe oxide NPs can be as large as 25 μm (Zeng et al., 2006), what is, in fact, not too much. Another technique is based on DNA conjugation used to assemble Au NPs (Macfarlane et al., 2011). This approach gives also rather small structures; moreover, to our knowledge it has not been applied to magnetic NPs yet. On the other hand, the protein crystallization technique, i.e., approach used to crystallize mFT NPs, allows the production of highly ordered crystals with all three dimensions in the range of a few 100 μm (Kasyutich et al., 2008). The latest papers report crystals as big as 0.4 mm made of one of NPs types: empty apoferritin, magnetite/maghemite mFT NPs, or pure magnetite mFT NPs (Okuda et al., 2012).

The mFT NPs crystals obtained with use of protein crystallization have high-quality fcc structure and the lattice constant about 18.5 nm (Kasyutich et al., 2009). An interesting effect is a substantial reduction of the lattice constant as a result of dehydration: The lattice constant decreases to ca. 14 nm when the crystals are taken off the mother liquor and dried (Kasyutich et al., 2009). This could be the way of controlling the lattice constant within the range of ten-odd nanometers, especially in combination with the possibility of functionalization of the apoferritin external surface. As we will see later, in this range even slight changes in the lattice constant modifies drastically the spin-wave spectrum of MCs.

7.3 Theoretical Approach: 3D PWM

The considered system is a crystal consisting of mFT NPs (scatterers or inclusions, material A) arranged periodically in a ferromagnetic matrix (material B). Each constituent material is characterized by three magnetic material parameters: the saturation magnetization M_S, the exchange constant A, and the exchange length λ_{ex}. Since $\lambda_{ex} = \sqrt{2A/\mu_0 M_S^2}$, only two of these parameters are independent. For each parameter a contrast is defined as the ratio of its value in mFT to its value in the matrix. The structure of the MC is characterized by the type of the 3D lattice, the lattice constant, and the diameter of the scatterers. In the case considered the lattice type is limited to fcc, in which mFT NPs crystallize. Also, the diameter of the scatterers is constant, fixed at 8 nm, i.e., the diameter of the core of fully loaded mFT. In such an MC the minimal lattice constant, corresponding to touching mFT NPs, is 11.314 nm (which means apoferritins are fully removed). A 0.1 T external magnetic

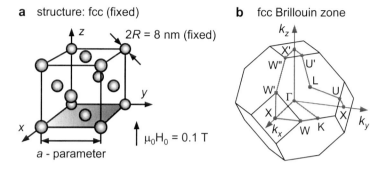

Figure 7.3 (a) Schematic depiction of the MC structure used in this study. (b) First Brillouin zone for fcc structure with high symmetry path, i.e., the line connecting high-symmetry points. The magnetic field causes the non-equivalency of some points, e.g., W, W′, and W″.

field oriented along one of the crystal axes is applied to the system (Fig. 7.3a); the field is strong enough to saturate the system, which allows the use of linear approximation. In Fig. 7.3b we show the first Brillouin zone for the fcc structure along with high symmetry points marked with dots. The set of high symmetry points for the fcc crystal contains Γ, X, W, K, U, and L. However, the magnetic field breaks the symmetry of the fcc lattice and cause some points to be non-equivalent, e.g., propagation directions corresponding to points W, W′, and W″ are oriented at different angles with respect to the magnetic field and thus they are not equivalent anymore.

For the description of the magnetization dynamics, we use the damping-free Landau–Lifshitz equation, with the effective field that includes terms related to the external field, the exchange interaction and the magnetostatic interaction. The latter, besides the dipolar interaction, includes higher multipole interaction, which, however, is of much lesser importance in the systems studied here. The detailed form of the magnetostatic field, taking account of the non-uniformity of the demagnetizing field, is given in Mamica et al. (2012a), and different types of the exchange component are discussed in Krawczyk et al. (2012) (see also detailed derivation provided in Chapter 8).

The idea of the PWM is to expand the dynamic functions (the dynamic component of the magnetization and the magnetostatic potential) into a series of plane waves with the use of Bloch's theorem and apply the Fourier transformation to the periodically distributed magnetic parameters. The final outcome is an eigenvalue problem, the numerical diagonalization of which yields the spin-wave frequencies and the Fourier coefficients for the periodic factor of the Bloch function describing the dynamic components of the magnetization (the elements of the matrix to diagonalize are given in (Mamica et al., 2012a)). The number of plane waves (reciprocal lattice vectors) used in the expansion of the dynamic functions conditions the convergence of the results and the computation time. Here the use of 1241 plane waves ensured convergent results concerning the width and center of the considered magnonic bandgap with an error of less than 2%.

In MCs with a substantial magnetic parameter contrast, in particular that of saturation magnetization, the amplitude of magnetization precession in modes with the lowest frequencies tends to concentrate in one of the constituent materials. As a measure of the degree of concentration we introduce the concentration coefficient, defined as follows for material A:

$$C_A = \frac{\widetilde{m}_A}{\widetilde{m}_A + \widetilde{m}_B},\qquad(7.1)$$

where $\widetilde{m}_X = \frac{1}{V_X}\int_{V_X}|\vec{m}|^2\,dv$ is the mean value of the square of the magnetization precession amplitude in the volume V_X, with X denoting the scattering centers ($X = A$) or the matrix ($X = B$). A concentration coefficient above 0.5 means that the concentration of the excitation is higher in the nanoparticles than in the matrix.

7.4 Spin-Wave Spectra of Magnetoferritin-Based Magnonic Crystals

Using the PWM described earlier in the text, we will determine the spin-wave spectrum of MC based on mFT NPs. Figure 7.4 presents three sample spin-wave spectra plotted along the line connecting high-symmetry points in the first Brillouin zone (BZ). The spectra have been obtained for mFT NPs arranged in an fcc lattice and embedded in a ferromagnetic matrix. The matrix

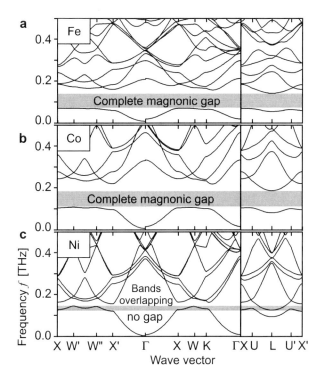

Figure 7.4 Spin-wave spectra (up to 0.5 THz) of MC with mFT NPs embedded in an (a) iron, (b) cobalt, and (c) nickel matrix and arranged in an fcc lattice with lattice constant 18.5 nm (mFT crystal as prepared). The spectra are plotted along the high-symmetry path in the first Brillouin zone shown in Fig. 7.3b. The shaded area in (a) and (b) represents the complete magnonic bandgap and in (c) bands' overlapping.

materials considered are iron, cobalt, and nickel, and the material parameters are specified in Table 7.2.[a] The lattice constant is 18.5 nm (mFT crystal as prepared). In first two cases, there is a wide complete magnonic bandgap between the first and second bands of almost the same width (about 100 GHz); however, the gap is shifted by ca. 50 GHz toward higher frequencies for Co (panel b). For Ni matrix the complete magnonic gap failed to open. As it was shown in

---

[a]Parameter values for mFT are calculated on the basis of Kasyutich et al. (2008).

Table 7.2 Values of material parameters: spontaneous magnetization M_S and exchange stiffness constant A, in the materials considered in this study

Material	Fe	Co	Ni	mFT
M_S [kA/m]	1752	1390	480	346
Contrast: $M_S(mFT)/M_S(matrix)$	0.20	0.25	0.72	
A [J/m] 10^{-11}	2.1	2.8	0.86	1.0
Contrast: $A(mFT)/A(matrix)$	0.48	0.36	1.16	

Krawczyk and Puszkarski (2008) the condition for the gap opening is sufficiently strong saturation magnetization contrast.[a]

The existence of a threshold magnetic parameter contrast below which the complete bandgap does not open is explained in Mamica et al. (2012a). It stems from the fact that in a spectrum for 3D structure an eventual complete bandgap is so-called *indirect* gap, with the bottom and top related to different propagation directions, in contrast to the *directional* gap, i.e., the gap for particular direction of the propagation (Fig. 7.5). One-dimensional systems have no threshold contrast: Any contrast will cause the bandgap to open at the boundary of the BZ (Krawczyk et al., 2001). A similar effect is observed in 3D systems for a specific propagation direction, in which case the bandgap is qualified as directional. Roughly speaking, the width of the directional bandgap depends on the contrast of magnetic parameters, and its central frequency on the value of the wave vector at the boundary of the BZ. In periodic 3D structures the wave vector at the boundary of the BZ has in general different values for different directions of propagation, hence the overlapping of neighboring bands if the directional bandgap is too narrow. This very effect underlies the lack of complete magnonic bandgap for the Ni matrix (Fig. 7.4c). Also, consequently, the threshold contrast can be expected to be the lowest in structures with a BZ closest to a sphere. This explains another result presented in Krawczyk and Puszkarski (2008), where the threshold contrast is found to be the lowest for the fcc lattice and the largest for the sc structure. The difference in the length of the shortest and longest wave-vector from

---

[a] Please notice the contrast below 1 is stronger when it has lower value. Thus, in Table 7.2 the strongest contrast of M_S stands for Fe.

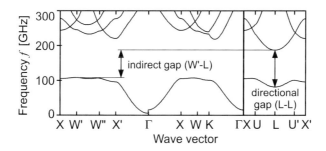

Figure 7.5 Directional bandgap vs. indirect one.

BZ border (the distance of the extreme BZ boundary points from the zone center) in the fcc lattice is the smallest (as small as 0.504 π/a). For the bcc structure the difference is 0.586 π/a, and for the sc is the highest (0.732 π/a). This line of reasoning applies to moderate contrasts, which can be regarded as a perturbation. If the contrast is too large, additional effects, such as repulsion of modes, will become sensible in the spin-wave spectrum and may cause the complete magnonic bandgap to shrink. On the other hand, the argumentation is general and works for other composites, e.g., photonic crystals (Joannopoulos et al., 2008).

Yet another effect to be observed in the spectra shown in Fig. 7.4 is a stepwise change in the frequency of the lowest mode at the Γ point, i.e., upon the change of the propagation direction from parallel (X′ Γ) to perpendicular (Γ–X) to the applied field. This frequency step results from taking into account the magnetostatic interaction and occurs in homogeneous media, too (Stancil and Prabhakar, 2009).

Influence of the lattice constant

As we have already mentioned, the lattice constant a of mFT crystal is reduced to 14 nm after drying. Along with the possibility of functionalization of the external surface of mFT NPs, it is a promising feature in terms of controlling of the lattice constant of the MC. Figure 7.6 shows the width and central frequency of the magnonic bandgap plotted versus a for two matrix materials: Fe and Co. In both cases the maximal width of the bandgap falls for

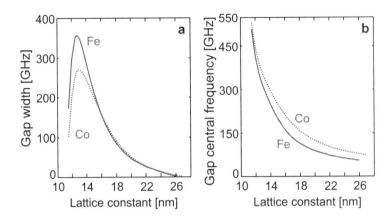

Figure 7.6 (a) The width and (b) the central frequency of the complete magnonic bandgap versus lattice constant for mFT-based MC with an iron and cobalt matrix.

$a \approx 13$ nm, the precise value depending on the matrix material. (We have considered also nickel matrix, for which, however, the bandgap failed to open at all due to weak saturation magnetization contrast.) For small a (up to 15 nm) the bandgap is wider for the Fe matrix, while for $a > 15$ nm for both matrices the gap width is almost the same. For Co matrix the gap is shifted toward higher frequencies in all range of a. Thus, the combination of the lattice constant and matrix material gives possibility to adjust the magnonic gap in quite a broad range of frequencies.

The detailed elucidation of the origin of the maximum of the bandgap width in its dependence on the lattice constant is provided in Mamica et al. (2012a). Thus, here we briefly recollect main points of that explanation. In Fig. 7.7 we present the top of the first band (bottom of the magnonic bandgap) and the bottom of the second band (top of the bandgap) plotted versus a for the Fe matrix. The bottom of the bandgap falls at the point W' and its top at L (compare Fig. 7.4). Both points lie at the boundary of the BZ. As the lattice constant increases, the wave vectors at the boundary of the BZ shorten and, consequently, the corresponding spin-wave frequencies decrease (Stancil and Prabhakar, 2009). Also, in the long-wave limit, i.e., for short wave vectors, the impact of the exchange interaction lessens. Thus, typically, the rate of the decrease

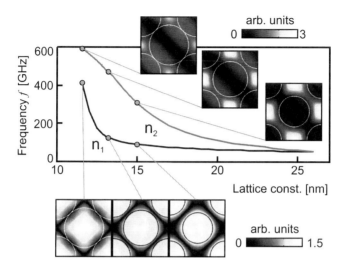

Figure 7.7 The bottom of the gap (top of the first band, n_1) and the top of the gap (bottom of the second band, n_2) versus the lattice constant for Fe matrix. The plot is accompanied by the spin wave profiles in a plane perpendicular to the external field and passing through the centers of mFT NPs, the contours of which are represented by circles (shaded plane in Fig. 7.3a).

in frequency with increasing lattice constant is high for small lattice constants and moderate for larger values of this parameter. Both curves shown in Fig. 7.7 present additional effects, though. In the lattice constant range from 11.5 nm to 13 nm the top of the first band (n_1) drops steeply. Also, the frequency of the bottom of the second band (n_2) decreases, but at a much lower rate. As a result, the bandgap widens and its central frequency decreases. For $a \approx 13$ nm the slope of both curves changes substantially. First of all, the rate of fall of the bottom of the bandgap becomes much lower than that of its top with further increase in the lattice constant; as a result, the bandgap shrinks.

This is a consequence of the change in the distribution of the dynamic component of the magnetization (the spin-wave profile). As we can see in Fig. 7.7, the lowest mode is strongly concentrated in the mFT NPs, their saturation magnetization being much lower than that of the matrix. The concentration coefficient (7.1) is

$C_A = 0.69$ for $a = 11.5$ nm, 0.70 for 13 nm, and 0.73 for 15 nm. In the range of small a, for touching mFT NPs, the profile of the lowest mode has a bulk character. This results from the interaction between neighboring mFT NPs. As the lattice constant grows to ca. 13 nm, the profile very quickly becomes nearly uniform in the mFT NPs and rapidly evanescent in the matrix. This is due to the fading interaction between the increasingly isolated excitations. In combination with the above-discussed decline in the impact of the exchange interaction with shrinking BZ, this causes the top of the lowest band to drop steeply. Further increase in the lattice constant does not modify the profile and only increases the spacing between the strongly isolated excitations. Thus, for $a > 13$ nm the bottom of the bandgap depends on the magnetostatic interaction, which results in the weak lattice-constant dependence.

The second mode is concentrated mainly in the matrix, though in the range of small lattice constants the amplitude of magnetization precession penetrates into the scatterers, too ($C_A = 0.38$ for $a = 11.5$ nm). As a increases, the concentration of the second mode in the matrix grows, mostly in small regions between nanoparticles ($C_A = 0.23$ for $a = 13$ nm). Increase in the concentration of the second mode in this relatively small volume should result in increased frequency of this mode; however, combined with the increase in the lattice constant, it only compensates to some extent for the decrease in frequency. As a grows above 13 nm, n_2 continues to leave the mFT NPs ($C_A = 0.08$ for $a = 15$ nm), but the increase in its matrix concentration cannot compensate for the increase in the volume of the matrix material.[a] As a result, for $a > 13$ nm the frequency of the second mode decreases at a higher rate than that of the first mode, since the volume of the mFT NPs, in which the first mode is concentrated, does not change. As a consequence, the bandgap closes for larger lattice constants.

Influence of the mFT magnetic moment

The saturation magnetization contrast can be modified by changing not only the matrix but also the core material in the mFT NPs.

---

[a]The change in the volume of the matrix material has similar effect as the change in the size of the potential well.

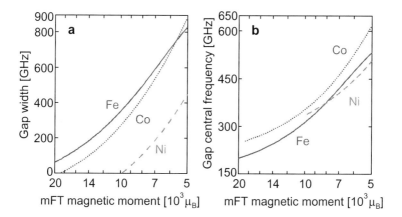

Figure 7.8 (a) The width and (b) the central frequency of the complete magnonic bandgap versus the magnetic moment of the mFT NPs for mFT-based MC with an iron, cobalt, and nickel matrix.

Considering the limitations of the magnetic materials that could be used in potential methods for the introduction of a ferromagnetic matrix into an mFT crystal, the possibility of controlling the magnetic characteristics of mFT NPs is of key importance for tailoring the magnonic spectrum. Moreover, variety of materials used to fill apoferritin gives the possibility of adjusting the saturation magnetization of the core in a wide range. In Fig. 7.8 we show the width and central frequency of the magnonic gap vs. the total magnetic moment of mFT NPs the MC is based on. We use three matrix materials—Fe, Co, and Ni—and the lattice constant is fixed to 13 nm (maximum of the gap width). Typically, an mFT NP fully loaded with a magnetite/maghemite has magnetic moment $10^4 \mu_B$ (Kasyutich et al., 2008). The dependences in Fig. 7.8 range from the magnetic moment twice the typical for mFT, which is gradually reduced to half of this value. The gap widens quickly with increasing M_S contrast (decreasing magnetic moment of the mFT NPs), at the same time the central frequency of the gap is shifted to higher values. Moreover, starting at $9.9 \times 10^3 \mu_B$ (only 1% less than in typical mFT NPs) the complete magnonic gap opens even for the Ni matrix.

7.5 Final Remarks

In the light of the results presented in this chapter, mFT crystals produced by protein crystallization have lattice constants in the range in which the exchange and dipolar interactions are of equal impact on the spin-wave spectrum, which provides special conditions for the magnonic bandgap tuning by reducing the lattice constant and adjustment of the saturation magnetization contrast. Especially, dried mFT crystals are almost optimized for the widest gap. Therefore, they appear to be very promising materials for the fabrication of 3D MCs with a high-quality structural order and a wide magnonic bandgap in the range of a few hundred GHz. On the other hand, the introduction of a ferromagnetic matrix into an mFT crystal is definitely a difficult task. One of the techniques that could be used to solve this problem is atomic layer deposition (Huber et al., 2011; Ritala and Leskela, 1999). This method involves significant limitations in terms of the deposited magnetic materials. Therefore, the possibility of precise adjustment of the saturation magnetization of the mFT core is a major advantage of these biomimetic nanoparticles.

There are also some crude approximations in the model. First of all, we do not take into account the protein shell around mFT NPs. Such shell could introduce additional anisotropy at the interface inclusion/matrix, which will change the pinning of the magnetization and thus influence the spin wave spectrum. Another approximation is the assumption of the uniform magnetization inside the mFT core. Usually the core starts to crystallize in a few nucleation points, which results in a non-uniform internal structure. Also, some properties of mFT NPs indicates core–shell structure of the magnetic nanoparticle (Kostiainen et al., 2011). However, with the magnonic gap as wide as almost 1 THz in our idealized model, we expect it will survive even in real magnetoferritin-based magnonic crystals.

Acknowledgments

The author acknowledges the support from the National Science Centre (NCN) of Poland, Project DEC-2-12/07/E/ST3/00538.

References

Abedin, M. J., Liepold, L., Suci, P., Young, M., and Douglas, T. (2009). Synthesis of a cross-linked branched polymer network in the interior of a protein cage, *J. Am. Chem. Soc.* **131**, pp. 4346–4354.

Andrews, S. C. (2010). The ferritin-like superfamily: Evolution of the biological iron storeman from a rubrerythrin-like ancestor, *Biochim. Biophys. Acta* **1800**, pp. 691–705.

Depondt, P., Levy, J.-C. S., and Mamica, S. (2013). Vortex polarization dynamics in a square magnetic nanodot, *J. Phys.: Condens. Matter* **25**, p. 466001.

Douglas, T., and Young, M. (1999). Virus particles as templates for materials synthesis, *Adv. Mater.* **11**, p. 679.

Fan, K., Cao, C., Pan, Y., Lu, D., Yang, D., Feng, J., Song, L., Liang, M., and Yan, X. (2012). Magnetoferritin nanoparticles for targeting and visualizing tumour tissues, *Nat. Nanotechnol.* **7**, p. 459.

Fantechi, E., Innocenti, C., Zanardelli, M., Fittipaldi, M., Falvo, E., Carbo, M., Shullani, V., Mannelli, L. D. C., Ghelardini, C., Ferretti, A. M., Ponti, A., Sangregorio, C., and Ceci, P. (2014). A smart platform for hyperthermia application in cancer treatment: Cobalt-doped ferrite nanoparticles mineralized in human ferritin cages, *ACS Nano* **8**, p. 4705.

Feng, L., and Damodaran, S. (2000). Two dimensional array of magnetoferritin on quartz surface, *Thin Solid Films* **365**, pp. 99–103.

Flenniken, M. L., Uchida, M., Liepold, L. O., Kang, S., Young, M. J., and Douglas, T. (2009). A library of protein cage architectures as nanomaterials, *Curr. Top. Microbiol. Immunol.* **327**, p. 71.

Galkin, A. Y., Ivanov, B. A., and Zaspel, C. E. (2005). Collective magnon modes for magnetic dot arrays, *J. Magn. Magn. Mater.* **286**, pp. 351–355.

Halevi, P. (2008). *Photonic and Phononic Crystals* (John Wiley and Sons, Inc.).

Hanasaki, I., Tanaka, T., Isono, Y., Zheng, B., Uraoka, Y., and Yamashita, I. (2011). Location and density control of carbon nanotubes synthesized using ferritin molecules, *Jpn. J. Appl. Phys.* **50**, p. 075102.

Harrison, P. M., and Arosio, P. (1996). The ferritins: Molecular properties, iron storage function and cellular regulation, *Biochim. Biophys. Acta* **1275**, pp. 161–203.

Henry, M., and Debarbieux, L. (2012). Tools from viruses: Bacteriophage successes and beyond, *Virology* **434**, pp. 151–161.

Hoh, F., Uzest, M., Drucker, M., Plisson-Chastang, C., Bron, P., Blanc, S., and Dumas, C. (2010). Structural insights into the molecular mechanisms of cauliflower mosaic virus transmission by its insect vector, *J. Virol.* **84**, pp. 4706–4713.

Huber, R., Schwarze, T., Berberich, P., Rapp, T., and Grundler, D. (2011). Atomic layer deposition for the fabrication of magnonic metamaterials, in *Metamaterials'2011: The 5th International Congress on Advanced Electromagnetic Materials in Microwaves and Optics, Metamorphose-VI*, ISBN 978-952-67611-0-7, p. 588.

Joannopoulos, J. D., Meade, R. D., and Winn, J. N. (2008). *Photonic Crystals: Molding the Flow of Light* (Princeton University Press).

Kasyutich, O., Sarua, A., and Schwarzacher, W. (2008). Bioengineered magnetic crystals, *J. Phys. D* **41**, p. 134022.

Kasyutich, O., Tatchev, D., Hoell, A., Ogrin, F., Dewhurst, C., and Schwarzacher, W. (2009). Small angle X-ray and neutron scattering study of disordered and three dimensionalordered magnetic protein arrays, *J. Appl. Phys.* **105**, p. 07B528.

Kekesi, R., Royer, F., Jamon, D., Mignon, M. F. B., Abou-Diwan, E., Chatelon, J. P., Neveu, S., and Tombacz, E. (2013). 3D magneto-photonic crystal made with cobalt ferrite nanoparticles silica composite structured as inverse opal, *Opt. Mater. Express* **3**, pp. 935–947.

Klem, M. T., Young, M., and Douglas, T. (2005). Biomimetic magnetic nanoparticles, *Mater. Today* **8**, p. 28.

Kłos, J., and Krawczyk, M. (2010). Electronic and hole spectra of layered systems of cylindrical rod arrays: Solar cell application, *J. Appl. Phys.* **107**, p. 043706.

Kłos, J., Sokolovskyy, M. L., Mamica, S., and Krawczyk, M. (2012). The impact of the lattice symmetry and the inclusion shape on the spectrum of 2D magnonic crystals, *J. Appl. Phys.* **111**, p. 123910.

Koralewski, M., Kłos, J. W., Baranowski, M., Mitroova, Z., Kopcansky, P., Melnikova, L., Okuda, M., and Schwarzacher, W. (2012). The faraday effect of natural and artificial ferritins, *Nanotechnology* **23**, p. 355704.

Kostiainen, M. A., Ceci, P., Fornara, M., Hiekkataipale, P., Kasyutich, O., Nolte, R. J. M., Cornelissen, J. J. L. M., Desautels, R. D., and van Lierop, J. (2011). Hierarchical self-assembly and optical disassembly for controlled switching of magnetoferritin nanoparticle magnetism, *ACS Nano* **5**, pp. 6394–6402.

Krawczyk, M., and Grundler, D. (2014). Review and prospects of magnonic crystals and devices with reprogrammable band structure, *J. Phys.: Condens. Matter* **26**, p. 123202.

Krawczyk, M., Lévy, J.-C., Mercier, D., and Puszkarski, H. (2001). Forbidden frequency gaps in magnonic spectra of ferromagnetic layered composites, *Phys. Lett. A* **282**, p. 186.

Krawczyk, M., Mamica, S., Kłos, J. W., Romero-Vivas, J., Mruczkiewicz, M. and Barman, A. (2011). Calculation of spin wave spectra in magnetic nanograins and patterned multilayers with perpendicular anisotropy, *J. Appl. Phys.* **109**, p. 113903.

Krawczyk, M., Mamica, S., Mruczkiewicz, M., Kłos, J. W., Tacchi, S., Madami, M., Gubbiotti, G., Duerr, G., and Grundler, D. (2013). Magnonic band structures in two-dimensional bi-component magnonic crystals with in-plane magnetization, *J. Phys. D* **46**, p. 495003.

Krawczyk, M., and Puszkarski, H. (2008). Plane-wave theory of three-dimensional magnonic crystals, *Phys. Rev. B* **77**, p. 054437.

Krawczyk, M., Puszkarski, H., Lévy, J.-C. S., Mamica, S., and Mercier, D. (2002). Theoretical study of spin wave resonance filling fraction effect in composite ferromagnetic [A|B|A] trilayer, *J. Magn. Magn. Mater.* **246**, 1–2, pp. 93–100.

Krawczyk, M., Sokolovskyy, M. L., Kłos, J. W., and Mamica, S. (2012). On the formulation of the exchange field in the Landau-Lifshitz equation for spin-wave calculation in magnonic crystals, *Adv. Cond. Mat. Phys.* **2012**, p. 764783.

Kruglyak, V. V., Demokritov, S. O., and Grundler, D. (2010). Magnonics, *J. Phys. D* **43**, p. 264001.

Kruglyak, V. V., Dvornik, M., Mikhaylovskiy, R. V., Dmytriiev, O., Gubbiotti, G., Tacchi, S., Madami, M., Carlotti, G., Montoncello, F., Giovannini, L., Zivieri, R., Kłos, J. W., Sokolovskyy, M. L., Mamica, S., Krawczyk, M., Okuda, M., Eloi, J. C., Jones, S. W., Schwarzacher, W., Schwarze, T., Brandl, F., Grundler, D., Berkov, D. V., Semenova, E., and Gorn, N. (2012). Magnonic metamaterials, in X.-Y. Jiang (ed.), *Metamaterial* (InTech), Available from: http://www.intechopen.com/books/metamaterial/magnonic-metamaterials.

Kumagai, S., Ono, T., Yoshii, S., Kadotani, A., Tsukamoto, R., Nishio, K., Okuda, M., and Yamashita, I. (2010). Position-controlled vertical growths of individual carbon nanotubes using a cage-shaped protein, *Appl. Phys. Express* **3**, p. 015101.

Lenk, B., Ulrichs, H., Garbs, F., and Munzenberg, M. (2011). The building blocks of magnonics, *Phys. Rep.* **507**, p. 107.

Liu, X., and Theil, E. C. (2005). Ferritins: Dynamic management of biological iron and oxygen chemistry, *Acc. Chem. Res.* **38**, pp. 167–175.

Macfarlane, R. J., Lee, B., Harris, N., Schatz, G.-C., and Mirkin, C.-A. (2011). Nanoparticle superlattice engineering with DNA, *Science* **334**, p. 204.

Majetich, S. A., Wen, T., and Booth, R. A. (2011). Functional magnetic nanoparticle assemblies: Formation, collective behavior, and future directions, *ACS Nano* **5**, pp. 6081–6084.

Mamica, S. (2013a). Tailoring of the partial magnonic gap in three-dimensional magnetoferritin-based magnonic crystals, *J. Appl. Phys.* **114**, p. 043912.

Mamica, S. (2013b). Stabilization of the in-plane vortex state in two-dimensional circular nanorings, *J. Appl. Phys.* **113**, p. 093901.

Mamica, S. (2013c). Spin-wave spectra and stability of the in-plane vortex state in two-dimensional magnetic nanorings, *J. Appl. Phys.* **114**, p. 233906.

Mamica, S., Józefowicz, R., and Puszkarski, H. (1998). The role of oblique-to-surface disposition of neighbours in the emergence of surface spin waves in magnetic films, *Acta Phys. Pol. A* **94**, pp. 79–91.

Mamica, S., Krawczyk, M., and Kłos, J. W. (2012b). Spin-wave band structure in 2D magnonic crystals with elliptically shaped scattering centres, *Adv. Cond. Mat. Phys.* **2012**, p. 161387.

Mamica, S., Krawczyk, M., Sokolovskyy, M. L., and Romero-Vivas, J. (2012a). Large magnonic band gaps and spectra evolution in three-dimensional magnonic crystals based on magnetoferritin nanoparticles, *Phys. Rev. B* **86**, p. 144402.

Mamica, S., Lévy, J. C. S., Depondt, P., and Krawczyk, M. (2011). The effect of the single-spin defect on the stability of the in-plane vortex state in 2D magnetic nanodots, *J. Nanopart. Res.* **13**, pp. 6075–6083.

Mamica, S., Levy, J.-C. S., and Krawczyk, M. (2014). Effects of the competition between the exchange and dipolar interactions in the spin-wave spectrum of two-dimensional circularly magnetized nanodots, *J. Phys. D* **47**, p. 015003.

Mamica, S., Lévy, J.-C. S., Krawczyk, M., and Depondt, P. (2012c). Stability of the landau state in square two-dimensional magnetic nanorings, *J. Appl. Phys.* **112**, p. 043901.

Mamica, S., Puszkarski, H., and Lévy, J.-C. S. (2000). The role of next-nearest neighbours for the existence conditions of subsurface spin waves in magnetic films, *phys. stat. sol. (b)* **218**, pp. 561–569.

Martinez-Boubeta, C., Simeonidis, K., Makridis, A., Angelakeris, M., Iglesias, O., Guardia, P., Cabot, A., Yedra, L., Estradé, S., Peiró, F., Saghi, Z., Midgley,

P. A., Conde-Leborán, I., Serantes, D., and Baldomir, D. (2013). Learning from nature to improve the heat generation of iron-oxide nanoparticles for magnetic hyperthermia applications, *Sci. Rep.* **3**, p. 1652.

Meldrum, F. C., Heywood, B. R., and Mann, S. (1992). Magnetoferritin: in vitro synthesis of a novel magnetic protein, *Science* **257**, pp. 522–523.

Meldrum, F. C., Wade, V. J., Nimmo, D. L., Heywood, B. R., and Mann, S. (1991). Synthesis of inorganic nanophase materials in supramolecular protein cages, *Nature (London)* **349**, pp. 684–687.

Mruczkiewicz, M., Krawczyk, M., Gubbiotti, G., Tacchi, S., Filimonov, Y. A., Kalyabin, D. V., Lisenkov, I. V., and Nikitov, S. A. (2013). Nonreciprocity of spin waves in metallized magnonic crystal, *New J. Phys.* **15**, p. 113023.

Nam, K. T., Wartena, R., Yoo, P. J., Liau, F. W., Lee, Y. J., Chiang, Y. M., Hammond, P. T., and Belcher, A. M. (2008). Stamped microbattery electrodes based on self-assembled M13 viruses, *Proc. Natl. Acad. Sci. U.S.A.* **105**, p. 17227.

Neusser, S., Bauer, H. G., Duerr, G., Huber, R., Mamica, S., Woltersdorf, G., Krawczyk, M., Back, C. H., and Grundler, D. (2011). Tunable metamaterial response of a Ni80Fe20 antidot lattice for spin waves, *Phys. Rev. B* **84**, p. 184411.

Neusser, S., Duerr, G., Bauer, H. G., Tacchi, S., Madami, M., Woltersdorf, G., Gubbiotti, M., Back, C. H., and Grundler, D. (2010). Anisotropic propagation and damping of spin waves in a nanopatterned antidot lattice, *Phys. Rev. Lett.* **105**, p. 067208.

Nikitov, S. A., Tailhades, P., and Tsai, C. S. (2001). Spin waves in periodic magnetic structures - magnonic crystals, *J. Magn. Magn. Mater.* **236**, p. 320.

Okuda, M., Eloi, J.-C., Jones, S. E. W., Sarua, A., Richardson, R. M., and Schwarzacher, W. (2012). Fe3O4 nanoparticles: Protein-mediated crystalline magnetic superstructures, *Nanotechnology* **23**, p. 415601.

Pal, S., Rana, B., Saha, S., Mandal, R., Hellwig, O., Romero-Vivas, J., Mamica, S., Kłos, J. W., Mruczkiewicz, M., Sokolovskyy, M. L., Krawczyk, M., and Barman, A. (2012). Time-resolved measurement of spin-wave spectra in CoO capped [Co(t)/Pt(7 Å)](n-1) Co(t) multilayer systems, *J. Appl. Phys.* **111**, p. 07C507.

Payne, G. F., Kim, E., Cheng, Y., Wu, H.-C., Ghodssi, R., Rubloff, G. W., Raghavan, S. R., Culverag, J. N., and Bentley, W. E. (2013). Accessing biology's toolbox for the mesoscale biofabrication of soft matter, *Soft Matter* **9**, p. 6019.

Puszkarski, H., Lévy, J.-C. S., and Mamica, S. (1998). Does the generation of surface spin-waves hinge critically on the range of neighbour interaction? *Phys. Lett. A* **246**, pp. 347–352.

Reddy, L. H., Arias, J. L., Nicolas, J., and Couvreur, P. (2012). Magnetic nanoparticles: design and characterization, toxicity and biocompatibility, pharmaceutical and biomedical applications, *Chem. Rev.* **112**, pp. 5818–5878.

Ritala, M., and Leskela, M. (1999). Atomic layer epitaxy: A valuable tool for nanotechnology? *Nanotechnology* **10**, p. 19.

Romero-Vivas, J., Mamica, S., Krawczyk, M., and Kruglyak, V. V. (2012). Investigation of spin wave damping in three-dimensional magnonic crystals using the plane wave method, *Phys. Rev. B* **86**, p. 144417.

Rong, J., Niu, Z., Lee, L. A., and Wang, Q. (2011). Self-assembly of viral particles, *Curr. Opin. Colloid Interface Sci.* **16**, pp. 441–450.

Sana, B., Johnson, E., Sheah, K., Poh, C. L., and Lim, S. (2010). Iron-based ferritin nanocore as a contrast agent, *BioInterphases* **5**, p. FA48.

Sapsford, K. E., Algar, W. R., Berti, L., Gemmill, K. B., Casey, B. J., Oh, E., Stewart, M. H., and Medintz, I. L. (2013). Functionalizing nanoparticles with biological molecules: Developing chemistries that facilitate nanotechnology, *Chem. Rev.* **113**, pp. 1904–2074.

Sarkar, J., Tang, S., Shahrjerdi, D., and Banerjee, S. K. (2007). Vertical flash memory with protein-mediated assembly of nanocrystal floating gate, *Appl. Phys. Lett.* **90**, p. 103512.

Serga, A. A., Chumak, A. V., and Hillebrands, B. (2010). YIG magnonics, *J. Phys. D* **43**, p. 264002.

Singamaneni, S., Bliznyuk, V. N., Binekc, C., and Tsymbal, E. Y. (2011). Magnetic nanoparticles: Recent advances in synthesis, self-assembly and applications, *J. Mater. Chem.* **21**, p. 16819.

Stancil, D. D., and Prabhakar, A. (2009). *Spin Waves. Theory and Applications* (Springer).

Sun, S. H., Murray, C. B., Weller, D., Folks, L., and Moser, A. (2000). Monodisperse FePt nanoparticles and ferromagnetic FePt nanocrystal superlattices, *Science* **287**, p. 1989.

Tatman, J. D., Preston, V. G., Nicholson, P., Elliott, R. M., and Rixon, F. J. (1994). Assembly of herpes simplex virus type 1 capsids using a panel of recombinant baculoviruses, *J. General Virol.* **75**, pp. 1101–111.

Theil, E. C., Beheraa, R. K., and Toshaa, T. (2013). Ferritins for chemistry and for life, *Coord. Chem. Rev.* **257**, pp. 579–586.

Theil, E. C., Matzapetakis, M., and Liu, X. (2006). Ferritins: Iron/oxygen biominerals in protein nanocages, *J. Biol. Inorg. Chem.* **11**, pp. 803–810.

Uchida, M., Flenniken, M. L., Allen, M., Willits, D. A., Crowley, B. E., Brumfield, S., Willis, A. F., Jackiw, L., Jutila, M., Young, M. J., and Douglas, T. (2006). Targeting of cancer cells with ferrimagnetic ferritin cage nanoparticles, *J. Am. Chem. Soc.* **128**, p. 16626.

Uchida, M., Kang, S., Reichhardt, C., Harlen, K., and Douglas, T. (2010). The ferritin superfamily: Supramolecular templates for materials synthesis, *Biochim. Biophys. Acta* **1800**, pp. 834–845.

Urban, I., Ratcliffe, N. M., Duffield, J. R., Elder, G. R., and Patton, D. (2012). Functionalized paramagnetic nanoparticles for waste water treatment, *Chem. Commun.* **46**, p. 4583.

Vasseur, J. O., Dobrzynski, L., Djafari-Rouhani, B., and Puszkarski, H. (1996). Magnon band structure of periodic composites, *Phys. Rev. B* **54**, p. 1043.

Yamashita, I., Iwahori, K., and Kumagai, S. (2010). Ferritin in the field of nanodevices, *Biochim. Biophys. Acta* **1800**, p. 846.

Yang, D., Zhou, Y., Rui, X., Zhu, J., Lu, Z., Fong, E., and Yan, Q. (2013). Fe3O4 nanoparticle chains with n-doped carbon coating: Magnetotactic bacteria assisted synthesis and high-rate lithium storage, *RSC Adv.* **3**, pp. 14960–14962.

Yoshimura, H. (2006). Protein-assisted nanoparticle synthesis, *Colloids Surfaces A* **282-283**, pp. 464–470.

Zeng, H., Black, C. T., Sandstrom, R. L., Rice, P. M., Murray, C. B., and Sun, S. H. (2006). Magnetotransport of magnetite nanoparticle arrays, *Phys. Rev. B* **73**, p. 020402(R).

Zhang, X., Meining, W., Fischer, M., Bacher, A., and Ladenstein, R. (2001). X-ray structure analysis and crystallographic refinement of lumazine synthase from the hyperthermophile aquifex aeolicus at 1.6 Å resolution: Determinants of thermostability revealed from structural comparisons, *J. Mol. Biol.* **306**, pp. 1099–1114.

Chapter 8

Magnonic Crystals: From Simple Models toward Applications

Jarosław W. Kłos and Maciej Krawczyk

Faculty of Physics, Adam Mickiewicz University in Poznań,
Umultowska 85, 61-641 Poznan, Poland
klos@amu.edu.pl, krawczyk@amu.edu.pl

8.1 Introduction

Influence of periodic modulation of the material on the wave propagation was considered by Lord Rayleigh already in 1887 for vibrations (Strutt, 1887). After that, in the twenties of the last century the theory of electrons in crystal lattice was developed, and also at that time, the most important theorem for waves in periodic media—the Bloch theorem—was introduced to describe propagation of electrons in a periodic potential (Bloch, 1929).[a] Although the theory of differential equation with coefficients being periodic functions of independent variable was developed by Floquet and Hill at the end of the 19th century (Floquet, 1883; Hill,

---

[a]It is worth to mention that F. Bloch has developed also the theory of spin waves (Bloch, 1930, 1932).

Magnetic Structures of 2D and 3D Nanoparticles: Properties and Applications
Edited by Jean-Claude Levy
Copyright © 2016 Pan Stanford Publishing Pte. Ltd.
ISBN 978-981-4613-67-5 (Hardcover), 978-981-4613-68-2 (eBook)
www.panstanford.com

1886), the extension of the theory to three-dimensional space and its application to solid-state physics calculations remained a vital step in physics. The concept of pass and stop bands (band gaps) for waves propagating in periodic media were developed in the following years and various kinds of filters and other devices for electromagnetic, mechanical or transmission lines were designed (Brillouin, 1953; Elachi, 1976; Yeh, 1977). The periodicity was introduced also into ferromagnetic materials for tailoring the transmission of magnetostatic waves in late 1970s (Parekh and Tuan, 1977; Sykes et al., 1976). Periodic distribution of the saturation magnetization by ion implantation (Hartemann, 1987), regular lattice of etched grooves, periodic modulation by metallic stripes or dots on top of a ferromagnetic film (Owens et al., 1977, 1978) and periodic perturbation of effective magnetic field were among the potential ways of creating periodicity in ferromagnetic materials considered that time. However, the discovery of photonic crystals in 1987 by E. Yablonovitch and S. John (John, 1987; Yablonovitch, 1987) renewed interest in periodic structures, with many new ideas and abundant new physics pushing it into unexplored directions. The ideas developed in the field of photonic crystals were applied to other types of waves, among others also to spin waves (SWs), were *magnonic crystals* (MC) are considered as spin wave counterpart of photonic crystals (Gulyaev and Nikitov, 2001; Krawczyk and Puszkarski, 1998; Puszkarski and Krawczyk, 2003; Vasseur et al., 1996). MCs are magnetic materials with periodic distribution of the constituent materials or some magnetic parameters (e.g., saturation magnetization or magnetocrystalline anisotropy), or modulated other parameters of importance for the propagation of SWs (such as external magnetic field, film thickness, stress or surroundings of the homogeneous ferromagnetic film).

In any periodic media the eigensolutions of the wave equation in linear regime fulfill the Bloch theorem and, regardless of the type of excitation, form the band structure in the frequency–wave vector space. Thus the qualitative understanding of the magnonic (i.e., spin wave) band formation can be based on the classical textbooks for the solid-state physics and solutions of the scalar Schrödinger equation in the periodic potential (Ashcroft and Mermin, 1976) or wave equation for electromagnetic wave propagation (Joannopoulos et al.,

2008). Nevertheless, the detailed shape of the band structure can be derived only from numerical calculations. This is most common way for study magnonic bands sensitivity to the structure and material parameters or to an external field. Moreover, spin waves have complex dispersion relation already in thin homogeneous films. This dispersion depends on the wavenumber, relative contribution of the short-range exchange and long-range dipole interactions, the external shape of the sample, the magnitude and orientation of the external static magnetic field with respect to the propagation direction and the magnetocrystalline anisotropy. This means that the spin wave band structure of magnonic crystals will be influenced by many additional factors apart from those they have in common with other types of artificial crystals. This makes MCs an intriguing topic for scientific studies and constitutes MCs the main subject of research in the field of *magnonics* (Krawczyk and Grundler, 2014).

Magnonics is a young field of nanoscience and technology dedicated to exploration of the coherent magnetization dynamics in nanoscale (Demokritov and Slavin, 2013; Kruglyak et al., 2010; Lenk et al., 2011). One of its main objectives is to exploit the spin wave dynamics for technological applications, including use of spin waves for carrying and manipulating information (Krawczyk and Grundler, 2014). The advantages of magnonics over electronics and photonics include low energy consumption and fast operation rates compared with electronic devices (Cherepov et al., 2014), and possible integration with standard CMOS technology, at levels impossible to achieve with electromagnetic waves (Khitun et al., 2008). Besides that, processing with spin waves allows easy tunability by the external magnetic field (Gurevich and Melkov, 1996), low energy costs tunability by electric field or stress when combined with magnetoelectric or magnetostrictive materials (Dreher et al., 2012; Zighem et al., 2013). It can be influenced also by electric current (Brataas et al., 2012; Edwards et al., 2012) and allows for magnetic momentum transfer without charge transport, thus magnon spintronics (Pirro et al., 2014; Yu et al., 2013) gives a chance for competing with electronics and spintronics, which is based on charge transport. Moreover, magnetic structures offer possibility for playing with the magnetic configuration. It means that the same element can have various spin wave dynamics in dependence on the static magnetization configuration, similar to changes of the resistivity in GMR and TMR structures for electric current. The operational functionality of magnonic device or its subunit can be reprogrammable effectively (Ding et al., 2012; Huber et al., 2013; Topp et al., 2011) and used for instance to prototype magnonic transistors (Krawczyk and Grundler, 2014). This makes magnonics and especially MCs of thin-film geometry interesting for a number of other fields of nanoscience and nanotechnology because of complementarity and compatibility to semiconductor technology, spintronics, and microwave devices (Krawczyk and Grundler, 2014; Kruglyak et al., 2010; Neusser and Grundler, 2009).

For technological applications of magnonics, especially in integrated devices and magnon spintronics, the MCs of thin-film geometry are expected to play a crucial role (Demokritov and Slavin, 2013; Gubbiotti et al., 2010; Kruglyak et al., 2010; Lenk et al., 2011; Neusser and Grundler, 2009). However, study of spin wave dynamics in planar magnonic films is also very interesting from the scientific point of view. This is because the dispersion relation of spin waves in thin films is strongly anisotropic, easily modulated by external factors, which combines the positive and negative group velocities. Such metamaterial properties in photonics and other fields are only hardly accessible with sophisticated structuralization (Kadic et al., 2013; Soukoulis and Wegener, 2011). In this sense, magnonics goes beyond photonics or plasmonics, because it allows for experimental investigation of the band structure formation in periodically structuralized materials with positive and negative group velocities; study of a transmission and reflection of waves from negative/positive refractive index media; consideration of the structures with sharp, graded, or periodic interfaces; and utilization of nonreciprocal properties in wave propagation due to permanent violation of the reciprocity principle or to study influence of nonlinearity—an inherent element of the magnetization dynamics.

For these reasons, we will concentrate here on thin-film magnonic crystals with periodicity along one direction (one-dimensional (1D) MCs) and along two directions (two-dimensional (2D) MCs). We will elucidate the influence of magnetostatic and exchange interactions on the magnonic band formation and distinguish the contributions having the source in these two types of interactions. However, first we show the main features of spin wave dispersion relation in homogeneous thin films in Section 8.2. Then we introduce selected excerpts from the theory of spin waves especially important for spin wave propagation in magnonic crystals Section 8.3.1 and present the plane wave method (PWM)suitable for calculations of the magnonic band structure in Section 8.3.3. After these introductory description, we explain basic properties of spin waves propagating in periodic structures and point out the most interested features from the scientific and technological point of view in Section 8.4.1. In Sections 8.4.2 and 8.4.3 the formation of the magnonic band structure and the opening of the magnonic band gap will be described with simple model based on analysis of the dispersion relation of the homogeneous thin ferromagnetic films, supported also with numerical results from PWM. Then we will describe in Section 8.4.4 the basic element of any magnonic device, it is the spin wave waveguide based on an antidot lattice. Finally the chapter is closed by the conclusion Section 8.5.

8.2 Spin Wave Dynamics in Homogeneous Ferromagnetic Film

The principal equation in magnonics is the Landau–Lifshitz (LL) equation, which describes time (t) and space ($\mathbf{r}$) dependence of the magnetization vector $\mathbf{M}(\mathbf{r}, t)$:

$$\frac{\partial \mathbf{M}(\mathbf{r}, t)}{\partial t} = \gamma \mu_0 \left(\mathbf{M}(\mathbf{r}, t) \times \mathbf{H}_{\text{eff}}(\mathbf{r}, t) \right), \tag{8.1}$$

where γ and μ_0 are gyromagnetic ratio and permittivity of vacuum, respectively. $\mathbf{H}_{\text{eff}}(\mathbf{r}, t)$ is an effective magnetic field, which is composed of external magnetic field, exchange field, and demagnetization field. In this chapter we limit our discussion only to these three terms of the effective magnetic field, while other possible components of $\mathbf{H}_{\text{eff}}$ will be omitted. This is an influence of an external radiofrequency field, electric currents, spin-torques, damping, and magnetocrystalline anisotropy field. By assuming harmonic in time plane wave solutions of LL equation, the dispersion relation of SW can be obtained.

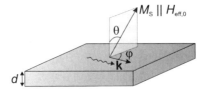

Figure 8.1 The thin-film geometry used in calculations.

At the beginning we make a reference to the dispersion relation of spin waves in homogeneous media. This is a textbook information; nevertheless, we will revise here some of this information, which will be useful for interpretation of the band structure in MCs presented in Sections 8.4.2 and 8.4.3. The analytical theory of spin waves in thin ferromagnetic films was developed by Kalinikos and Slavin (1986). We follow that approach and assume the homogeneous ferromagnetic thin film of thickness d, which is in magnetically saturated state with static part of the magnetization vector, **M** pointed along the direction of the static component of effective magnetic field, $\mathbf{H}_{\text{eff},0}$ (Fig. 8.1).[a] The spin wave of angular frequency ω and the wave vector **k** propagates in the plane of the film under the angle φ to the direction of the vector $\mathbf{H}_{\text{eff},0}$ projected onto the film plane. $\mathbf{H}_{\text{eff},0}$ and **M** make an angle θ with the normal to the film plane. In linear approximation the dispersion relation can be written in the following form (Kalinikos and Slavin, 1986):

$$\omega^2 = \omega_H(\omega_H + \omega_M F(\varphi, \theta)). \tag{8.2}$$

The function $F(\varphi, \theta)$ is defined as

$$F(\varphi, \theta) = P + \sin^2\theta \left[1 - P(1 + \cos^2\varphi) + \omega_M \frac{P(1-P)\sin^2\varphi}{\omega_H} \right], \tag{8.3}$$

where

$$P = 1 - \frac{1 - e^{-kd}}{kd}. \tag{8.4}$$

---

[a]The static effective field defines the equilibrium orientation of the magnetization and can be calculated by minimization of the self-energy usually composed of the demagnetizing term (see Section 8.3.1.2), exchange term (see Section 8.3.1.1), and, if present, anisotropy terms. Because we are concentrating on spin wave dynamics, the derivation of the equilibrium orientation of the magnetization is omitted in this chapter, and we refer the interested reader to Artman (1957a,b) and Aharoni (1991).

In Eq. (8.2) we have defined

$$\omega_H = \gamma \mu_0 \left(H_{\text{eff},0} + \frac{2A}{\mu_0 M_S} k^2 \right) \qquad (8.5)$$

and $\omega_M = \gamma \mu_0 M_S$, where M_S is a saturation magnetization, A is an exchange stiffness constant, and k is a spin wave wavenumber. The second term in the definition of ω_H, Eq. (8.5), is proportional to k^2 and describes the exchange contribution to the spin wave dispersion. The dynamic dipolar contribution is expressed by $F(\varphi, \theta)$ defined in Eq. (8.3).

The effective magnetic field, in this case, contains the external static magnetic field H_0 and the static demagnetizing field. For two configurations considered later in this chapter (i.e., for $\theta = 0$ and $\theta = 90°$), the static component of H_{eff} equals $H_{\text{eff},0} = H_0 - M_S$ and $H_{\text{eff},0} = H_0$, respectively. Equation (8.2) was obtained for the unpinned magnetization dynamics on the film surfaces and under assumption of a uniform excitation amplitude across the film thickness. These two assumptions are fulfilled to large extent in many recent experimental works realized on thin magnonic crystals (Bali et al., 2012; Gubbiotti et al., 2007; Mandal et al., 2012; Saha et al., 2013; Schwarze and Grundler, 2013; Semenova et al., 2013; Tacchi et al., 2011).

Compared with a parabolic and linear dispersion relations for electronic and electromagnetic waves in homogeneous media, respectively, the spin wave frequency $f(\mathbf{k})$ ($f = \omega/(2\pi)$) as described by Eq. (8.2) exhibits complex and nonmonotonous dependence on the in-plane wave vector. In particular, the spin wave frequency depends on (i) the wavenumber k, (ii) the strength, and (iii) orientation of the external magnetic field $\mathbf{H}_0$ with respect to $\mathbf{k}$ and to the film plane (angles φ and θ), (iv) the shape of the magnet affecting the demagnetization fields; (v) the magnetocrystalline anisotropy and (vi) magnetization configuration will add further important parameters for dependence $f(\mathbf{k})$ (Akhiezer et al., 1968; Gurevich and Melkov, 1996; Kabos and Stalmachov, 1994; Kalinikos and Slavin, 1986; Stancil and Prabhakar, 2009). Consequently, the band structure of magnonic crystals created from ferromagnetic film will be influenced by many additional factors apart from those ones that MCs have in common with other types of artificial crystals, i.e., the lattice symmetries and geometrical parameters given by the periodic patterning.

Usually three geometries of planar structures are distinguished: the Damon–Eshbach geometry (DE), backward volume magnetostatic wave geometry (BVMW), and forward volume magnetostatic wave geometry (FVMW) (Stancil and Prabhakar, 2009). In DE and BVMW geometries, the magnetic field $\mathbf{H}_0$ is in the plane of the film ($\theta = 90°$) and propagation of SWs is perpendicular ($\varphi = 90°$) and parallel ($\varphi = 0$) to the direction of this field, respectively. In Fig. 8.2d, these are directions y and z, respectively. The dispersion relation depends on the φ angle, thus is anisotropic. In the FVMW geometry, the static component of effective field $\mathbf{H}_{\mathrm{eff},0}$ is perpendicular to the film plane ($\theta = 0$), the φ dependence is dropped out from the dispersion relation Eq. (8.3) and the spin wave propagation in the plane of the film is isotropic.

8.2.1 Exchange and Dipole-Exchange Spin Waves

The propagation of spin waves is determined by two types of interactions. These are dipole and exchange interactions, which influence the magnetization dynamics at different ranges of length. The exchange interactions are usually limited only to the magnetic moments located in the nearest neighbor positions of the crystallographic lattice, while dipole interaction fades with lower rates— one over distance to the third power—thus the magnetic moments from the whole sample contribute to the effective magnetic field at every position.[a] This means that the relative strength of these two types of interactions will change with changes of the wavelength of spin oscillations ($\lambda = 2\pi/k$). Thus, the magnetization dynamics in the two regimes, with dominating dipole or exchange interactions, is different and therefore the spin wave spectrum does not follow a simple scaling rule known from electromagnetic waves (Joannopoulos et al., 2008).

The dispersion relation of spin waves in thin film with magnetization in the film plane is shown in Figs. 8.2a,b,c over the k_y–

---

[a]The detailed discussion of the exchange and dipole components of the effective magnetic field is presented in Section 8.3.1.

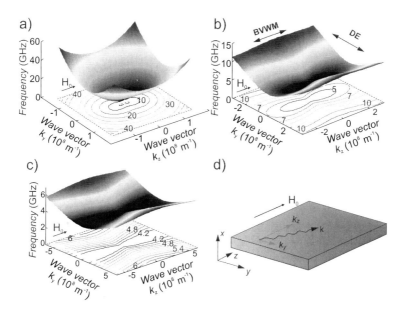

Figure 8.2 Dispersion relation of spin waves as a function of the in-plane wave vector (k_y and k_z components) in a thin film (d) of Py of thickness $d = 10$ nm magnetized by in-plane external magnetic field $\mu_0 H_0 = 0.02$ T for various range of wave vectors. In all figures k_y and k_z changes from $-\pi/a$ to π/a but in (a) $a = 20$ nm, (b) $a = 60$ nm, and (c) $a = 500$ nm. In calculations we have assumed $M_S = 0.8 \times 10^6$ A/m, $A = 1.3 \times 10^{-11}$ J/m and $\gamma = 175.9$ GHz/T. The insets in the bottom show contours of constant frequency in k-space (isofrequency plots). The almost circular contours in (a) are characteristic for the exchange-dominated spin waves at high frequencies. A strongly anisotropic dispersion of spin waves (b) and (c), clearly visibly at low frequencies, is characteristic for the magnetostatic part of the dispersion relation. The calculations were performed with the analytical formulas for spin wave dispersion relation—Eq. (8.2) taken from (Kalinikos and Slavin, 1986).

k_z plane. The calculations were performed for the thin Py film (of thickness 10 nm) with magnetic field $\mu_0 H_0 = 0.2$ T directed along the z-axis over various ranges of wave vector: (a) $k_y, k_z \in (-1.57; 1.57)$ [10^8 m^{-1}], (b) $k_y, k_z \in (-0.52; 0.52)$ [10^8 m^{-1}], and (c) $k_y, k_z \in (-0.06; 0.06)$ [10^8 m^{-1}]. The wave vector regime, where the exchange interaction dominates over the dipolar, can be easily extracted from these diagrams considering that exchange-

292 | Magnonic Crystals

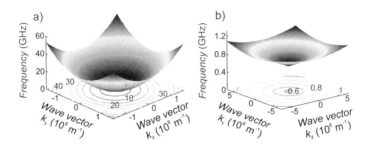

Figure 8.3 Dispersion relation of spin waves as a function of the in-plane wave vector (k_y and k_z components) in a thin film of Py of thickness $d = 10$ nm magnetized perpendicularly to the film plane; it is the FVMW geometry. In calculations we have assumed the same parameters as in Fig. 8.2 with static component of the effective magnetic field $\mu_0 H_{\text{eff},0} = 0.2$ T. (a) and (b) The dispersion relation for the wave vectors from $-\pi/a$ to π/a with a equal to 20 and 500 nm, respectively.

dominated spin waves follow a parabolic dispersion relation—this is for large wavenumbers. The exchange interactions dominate in the most area shown in Fig. 8.2a, while in Figs. 8.2b,c, the strongly anisotropic dispersion points at important contribution of dipole interactions. At small wavenumbers along k_y the dispersion relation is monotonous and linear with positive slope—this is a characteristic feature of the DE geometry (Fig. 8.2c). However, along k_z the $f(k_z)$ dependence is nonmonotonous, starting from the $k_z = 0$ the frequency decreases with increasing wavenumber, attains minimum (see Figs. 8.2b,c) and then increases with k_z. Usually, it is assumed that starting from k at the minimum, the exchange prevail the magnetostatic interactions. Such a nonmonotonic dispersion relation is a characteristic feature of BVMW geometry.

In the FVMW geometry the static component of the magnetic field $\mathbf{H}_{\text{eff},0}$ is perpendicular to the film plane and the dispersion relation is isotropic in this case. Nevertheless, the exchange and dipole-exchange regime can be still distinguished: Exchange dominating with parabolic dispersion is shown in Fig. 8.3a and the dipole-exchange with linear dependence between f and k is shown in Fig. 8.3b.

The results presented in Figs. 8.2 and 8.3 were calculated for some selected set of parameters. However, the dispersion relation will change with film thickness, amplitude, and direction (an angle θ in Fig. 8.1) of the magnetic field. For a given material and film thickness, the lattice constant a of the MC determines whether the first Brillouin zone (BZ) boundary (e.g., $k = \pi/a$ for a linear chain or square lattice of magnetic elements with lattice constant a) is in the magnetostatic or exchange-dominated regime. Thus for a relatively large lattice constant back-folding will happen in the magnetostatic regime of $f(k)$ (Fig. 8.2c and Fig. 8.3b). For small a BZ border will locate in the exchange regime (Fig. 8.2a and Fig. 8.3a). However, before presenting magnonic band structures in thin-film MCs, we introduce fundamentals of the spin wave theory and computational methods used in calculations of the magnonic band structure.

8.3 Theory of Spin Waves in Planar Magnonic Crystals

8.3.1 *Spin Waves in Inhomogeneous Medium*

The inhomogeneity of the medium is reflected in LL equation as a spatial dependence of the effective magnetic field. The effective magnetic field includes, with the exception of the external magnetic field, three important components, which depend on the magnetic properties and geometry of the magnetic material. These are (i) demagnetizing field, (ii) exchange field, and (iii) anisotropy field, if present in the material. All mentioned fields are introduced as spatially dependent parameters that do not remind their microscopic origin. However, it is instructive to discuss the relation between these macroscopic fields and microscopic parameters, because it will give us the limits for the applicability of the considered continuous model.

The magnetocrystalline anisotropy is a static magnetic field related to spin-orbit coupling in atomic lattice (Getzlaff, 2008; Stohr and Siegmann, 2006). At the distances much larger than interatomic distances the averaging of discrete atomic system to continuous medium allows us to relate the magnetocrystalline anisotropy field to the principal directions of the atomic lattice and assume its spatial independence in homogeneous sample. In heterostructures (or finite structures), where the interfaces (surfaces) between different materials appear, additionally the interface (surface) magnetic anisotropy can occur as a result of distortion of atomic orbitals in the vicinity of the interface. The field related to the interface (surface) anisotropy decays with the distance from the interface (surface), but this contribution is also often replaced with homogeneous bulk contribution (Vaz et al., 2008). Both bulk and surface magnetocrystalline anisotropy fields are local and depend only on the atomic structure of the system.

The demagnetizing field H_{dm}, being also the source of the magnetic anisotropy in the system, has both static and dynamic nature. This field is induced by the presence of nonuniform distribution of magnetization or by the presence of interfaces, including external surfaces (Aharoni, 1991; Kaczer and Murtinova, 1974). The spatial dependence of H_{dm} can be calculated from the Maxwell equations after stabilization of the magnetic configuration, which also depends on the H_{dm}. This procedure is nonlocal because it demands the integration of the all contributions to H_{dm} coming from the volume and all interfaces of the magnetic system.

The exchange field has a quantum origin and results in the presence of exchange interactions (White, 2007). The exchange interactions are related to the Coulomb interactions between electrons with preserving antisymmetry of the wave function of two indistinguishable fermions (electrons) with respect to their spatial exchange. This interaction is short range and can be considered as local one. In deep nanoscale the exchange interaction is the main factor responsible for magnetic ordering. In ferromagnetic (anitferromagnetic) materials the effect of exchange favorites energetically the parallel (antiparallel) alignments of the spins on the neighbor lattice points. However, the magnetic order observed at mesoscopic distances, such as the presence of magnetic domains and domain walls, is a result of exchange and classical magnetostatic interactions (and other fields like magnetic anisotropy if present). In the continuous model, like the one described by the LL equation (8.1), we have to relate the microscopic parameters describing the strength of exchange interaction to the effective parameters of the material and to find the relation between exchange energy and

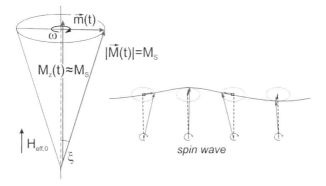

Figure 8.4 The linearized Landau–Lifshitz equation describes the precision of the magnetization vector around the direction of the static component of effective magnetic field, $\mathbf{H}_{\text{eff},0}$. For small procession angle ξ the magnetization component along direction of $\mathbf{H}_{\text{eff},0}$ can be considered as constant and equal to the saturation magnetization, $M_z(\mathbf{r}, t) \approx M_S(\mathbf{r})$. Then the dynamics can be assumed as harmonic oscillation with angular frequency ω of perpendicular component of the magnetization vector $\mathbf{m}(\mathbf{r}, t) = \mathbf{m}(\mathbf{r})e^{i\omega t}$, which can propagate as spin waves through the media.

exchange field. However, before doing that in Section 8.3.1.1 we introduce a linear approximation.

The LL equation (8.1) is nonlinear and gives the general description of magnetization dynamics, including nonlinear processes like magnetization switching. However, the spin waves are low energy excitations and to their description usually linear terms are sufficient. Thus, most of spin wave theories and variety of methods developed for magnonics are based on linear approximation. With this approximation, the description of wave effects becomes more transparent and its interpretation more intuitive. Therefore, it is useful to discuss the principles of the linearization of LL equation, which after that will be used in calculations.

For a small angle ξ of the precession of magnetization vector around the direction of the static component of effective field $H_{\text{eff},0}$ (we can assume that this precession is around the z-axis, as in Fig. 8.4), the M_z component of the magnetization vector is constant in time and we can assume that it is equal to the magnetization saturation: $M_z \approx M_S$. This assumption constitutes the bases for linear approximation. Linear regime of spin wave dynamics

296 | Magnonic Crystals

means that the dynamical components of magnetization $\mathbf{m}(\mathbf{r}, t)$ (the components perpendicular to the z-axis) change harmonically in time with angular frequency ω, $\mathbf{m}(\mathbf{r}, t) = \mathbf{m}(\mathbf{r})e^{i\omega t}$. In this case we can discuss the dynamics in terms of stationary modes and separate the time and space dependence of the spin waves. This linear approximation allows for neglecting nonlinear terms with respect to $\mathbf{m}$ in the LL equation. This is especially beneficial for the analysis of magnonic crystals, where the only in linear regime the spin wave excitations can have a form of Bloch waves (Chicone, 1999).

8.3.1.1 Exchange field in inhomogeneous media

The Heisenberg Hamiltonian describes exchange interaction between spins. The interaction of the one selected spin $\mathbf{S}_l$ on site l with the rest of spins is given by

$$H_l = \sum_{m \in \mathrm{NN}} J_{l,m} \mathbf{S}_l \cdot \mathbf{S}_m, \tag{8.6}$$

where the strength of this interaction is given by parameter $J_{l,m}$ called exchange integral. The summation here is limited only to the nearest neighbor (NN) spins of the spin from the site l. By introducing the unit vectors of the directions of individual spins: α_l, we can relate spins on discrete lattice to the magnetization being a continuous function of the position vector:

$$\alpha_l = \frac{\mathbf{S}_l}{|\mathbf{S}_l|} = \frac{\mathbf{M}(\mathbf{r}_l)}{|\mathbf{M}(\mathbf{r}_l)|}, \tag{8.7}$$

where $M(\mathbf{r}) = N\mu_B g S_l$ (N is the number of spins in the unit cell volume, μ_B is Bohr magneton, and g is g-factor). In the limit of small deviations of the neighboring spins, it gives us the expression for the exchange energy density (Chikazumi, 1997; Keffer, 1966):

$$\epsilon_{\mathrm{ex}} = \lambda_{\mathrm{w}} M^2 + A \sum_i \left(\frac{\partial \alpha}{\partial x_i} \right)^2, \tag{8.8}$$

where x_1, x_2, and x_3 are Cartesian components x, y, and z, respectively. The expression $\lambda_{\mathrm{w}} M$ have a meaning of so-called Weiss field. The material parameters λ_{w} and A are linked to the microscopic parameters by the following relations:

$$A = \frac{2J n S^2}{a}, \quad \lambda_{\mathrm{w}} = \frac{-2ZJ}{N\mu_B^2 g^2}, \tag{8.9}$$

where $n = 1$, 2, or 4 for sc, bcc, or fcc lattice, respectively, and Z is the number of the nearest neighbors. The integration of the energy density ϵ_{ex} over the volume of the material gives the exchange energy:

$$E_{ex} = \int_V \epsilon_{ex} dV. \tag{8.10}$$

To find the exchange field we have to calculate the functional derivative of E_{ex} with respect to the magnetization vector (Akhiezer et al., 1968):

$$\mathbf{H}_{ex} = -\frac{1}{\mu_0} \frac{\delta E_{ex}}{\delta \mathbf{M}}. \tag{8.11}$$

The other important issue is the contribution of the interfaces between the different materials, at which the magnetization is not continuous. This problem was discussed in detail by Krawczyk et al. (2012) where some ambiguity in the derivation of formula for the exchange field was discussed. According to the results presented in the paper (Krawczyk et al., 2012), the flowing formula gives reliable formulation of the exchange field:

$$\mathbf{H}_{ex} = \nabla l_{ex}^2 \nabla \mathbf{m}, \tag{8.12}$$

where l_{ex} is an exchange length:

$$l_{ex}^2(\mathbf{r}) = \frac{2A(\mathbf{r})}{\mu_0 M_S^2(\mathbf{r})}. \tag{8.13}$$

8.3.1.2 Demagnetizing field in inhomogeneous media

The magnetostatic self-energy is determined by the spatial distribution of magnetization in its own magnetic field called demagnetizing field:

$$\Delta E = -\frac{1}{2} \int_V \mathbf{H}_{dm}(\mathbf{r}) \cdot \mathbf{M}(\mathbf{r}) dV. \tag{8.14}$$

The demagnetizing field have to be calculated self-consistently. This means that the magnetic system has to distribute the magnetization to minimize magnetostatic self-energy, which in turn depends on the demagnetizing field produced by magnetization distribution. The demagnetizing field can be calculated directly from the Maxwell equations. For static fields and in the absence of electric currents,

298 | *Magnonic Crystals*

one of the Maxwell equations reduces to $\nabla \times \mathbf{H}_{dm} = 0$. It means that we can introduce the so-called magnetostatic potential Φ, the gradient of this potential is just a demagnetizing field:

$$\mathbf{H}_{dm} = -\nabla\Phi. \tag{8.15}$$

From the other Maxwell equation ($\nabla \cdot \mathbf{B}_{dm} = 0$, where $\mathbf{B}_{dm}$ is a magnetic induction field) and the definition of magnetostatic potential Eq. (8.15) we can obtain the Poisson (Laplace) equation for magnetostatic potential inside (outside) of the magnetic body:

$$\Delta\Phi_{in} = \nabla \cdot \mathbf{M},$$
$$\Delta\Phi_{out} = 0. \tag{8.16}$$

The requirement of the continuity of the normal component of $\mathbf{B}$ and tangential component of $\mathbf{H}$ on the interface leads to the following boundary conditions for magnetostatic potential:

$$\Phi_{in} = \Phi_{out},$$
$$\frac{\partial\Phi_{in}}{\partial\mathbf{n}} - \frac{\partial\Phi_{out}}{\partial\mathbf{n}} = \mathbf{M} \cdot \mathbf{n}, \tag{8.17}$$

where $\mathbf{n}$ is the unit vector normal to the interface. The magnetostatic potential mast be also regular at infinity (Aharoni, 1991). It means that both $|r\Phi|$ and $|r^2\Phi|$ must be limited for $r \to \infty$.

The general solution of Eqs. (8.16) has the following form:

$$\Phi(\mathbf{r}) = -\int_V \frac{\nabla' \cdot \mathbf{M}(\mathbf{r})}{|\mathbf{r} - \mathbf{r}'|} dV' + \int_S \frac{\mathbf{n} \cdot \mathbf{M}(\mathbf{r})}{|\mathbf{r} - \mathbf{r}'|} dS'. \tag{8.18}$$

The first and the second term are the contribution from volume and surface magnetic charges, respectively. The integrations are over the volume and surface of the magnetic body in the first and the second term, respectively. The superficial analysis of this terms concludes that for the homogeneous system in saturated state, the volume charges are absent and the only contribution from surface charges is present (Brown, 1963). When the sample consists of the homogeneous regions separated by sharp interfaces the boundary conditions can be applied also on those interfaces. This generates demagnetizing field related to interface magnonic charges, which appear in the solution of magnetostatic potential and can be written in the form: $\int_{S_i} \frac{\mathbf{n} \cdot (\mathbf{M}_1(\mathbf{r}) - \mathbf{M}_2(\mathbf{r}))}{|\mathbf{r} - \mathbf{r}'|} dS$, where the integration takes place over the i-th interface separating two homogeneous regions: M_1 and M_2.

The calculations of such integrals are time consuming, especially in the presence of volume chargers. Moreover, infinite systems like magnonic crystals are not suitable for such calculations. In the case of periodic systems, the quite efficient method of calculating demagnetizing field is based on the Fourier expansion (Krawczyk et al., 2012). This approach simplifies the solution of the Poisson and Laplace equation (8.16) for periodic structures. Solutions can be achieved by introducing fictitious potentials Ψ_x, Ψ_y, and Ψ_z related to the magnetostatic potential Φ by:

$$\Phi = - \left(\frac{\partial \Psi_x}{\partial x} + \frac{\partial \Psi_y}{\partial y} + \frac{\partial \Psi_z}{\partial z} \right),$$

(8.19)

which are related in turn to the corresponding components of magnetization in the same way as electrostatic potential is linked to the electric charge distribution:

$$\Delta \Psi_{x_i}(\mathbf{r}) = -M_{x_i}(\mathbf{r}),$$
$$\Psi_{x_i}(\mathbf{r}) = \int_V \frac{M_{x_i}}{|\mathbf{r} - \mathbf{r}'|} dV.$$

(8.20)

We can find independently the contribution to the magnetostatic potential coming from each component of the magnetization vector M_{x_i} separately. Having the fictitious potential Ψ calculated, the finding of the magnetic field is straightforward using the relations (8.19) and (8.15).

The magnetostatic approximation is also valid for the dynamical components of the demagnetizing field (Stancil and Prabhakar, 2009):

$$\nabla \times \mathbf{h} = 0.$$

(8.21)

Thus we can obtain both static $\mathbf{H}_{dm}(\mathbf{r})$ and dynamic $\mathbf{h}(\mathbf{r}, t)$ components of the demagnetizing field:

$$\mathbf{H}_{dm}(\mathbf{r}, t) = \mathbf{H}_{dm}(\mathbf{r}) + \mathbf{h}(\mathbf{r}, t), \qquad (8.22)$$

having a source in $M_S(\mathbf{r})$ and $\mathbf{m}(\mathbf{r}, t)$, respectively.

8.3.1.3 Boundary conditions for magnetization at interface with nonmagnetic material

The Landau–Lifshitz equation is valid only in magnetic medium and does not impose any boundary condition on the magnetization at the surfaces (i.e., at the interfaces between magnetic and nonmagnetic medium). Generally the magnetization is partially pinned with the strength depending on the surface magnetocrystalline anisotropy exchange and dipole fields (Guslienko and Slavin, 2002, 2005). To discuss the system with nonvanishing dipole field, let us consider a planar structure of infinite extension. The general form of the boundary conditions at external surfaces of the ferromagnetic film proposed by Guslienko in (Guslienko and Slavin, 2002, 2005) is

$$\mathbf{M} \times \left(l_{ex}^2 \frac{\partial \mathbf{M}}{\partial \mathbf{n}} + \frac{2K_s}{\mu_0 M_S^2} (\mathbf{M} \cdot \mathbf{n})\mathbf{n} + \mathbf{H}_{dm}d \right) = 0 \qquad (8.23)$$

and expresses the balance between torques acting on magnetization vector at film surface. This condition relates exchange interaction (first term), surface magnetocrystalline anisotropy (second term), and dipole field (third term). The strength of exchange term is mainly determined by exchange length l_{ex}. The term related to surface magnetic anisotropy (generated by breaking symmetry of atomic lattice on the surface, reconstruction or chemical processes in surface regions) is described by surface anisotropy constant K_s. The demagnetizing field $\mathbf{H}_{dm}$, appearing in the last term, depends on the geometry of the system (e.g., on its thickness d).

8.3.2 Bloch Theorem

Having defined the exchange (Eq. (8.12)) and demagnetizing (Eq. (8.22)) field in inhomogeneous ferromagnetic material, we can write explicitly LL equation (8.1) in a linear approximation, useful for spin wave calculations in MCs. If we assume that the magnetization saturation, although space dependent, is everywhere along the z-axis, the linearized LL equation takes the following form:

$$i\frac{\omega}{\gamma\mu_0}m_x(\mathbf{r}) = -M_S(\mathbf{r})[\nabla \cdot l_{ex}(\mathbf{r})\nabla]m_y(\mathbf{r})$$
$$+ m_y(\mathbf{r})(H_0 + H_{dm}(\mathbf{r})) - M_S(\mathbf{r})h_y(\mathbf{r}), \quad (8.24)$$

$$i\frac{\omega}{\gamma\mu_0}m_y(\mathbf{r}) = M_S(\mathbf{r})[\nabla \cdot l_{ex}(\mathbf{r})\nabla]m_x(\mathbf{r})$$
$$- m_x(\mathbf{r})(H_0 + H_{dm}(\mathbf{r})) + M_S(\mathbf{r})h_x(\mathbf{r}). \quad (8.25)$$

The solutions of this linearized wave equation in periodic system can have a form of Bloch waves. For spin waves propagating in magnonic crystals, the Bloch function describes two orthogonal dynamical components of the magnetization vectors $m_x(\mathbf{r})$ and $m_y(\mathbf{r})$:

$$m_{i,\mathbf{k}}(\mathbf{r}, t) = u_{i,\mathbf{k}}(\mathbf{r})e^{i(\mathbf{k}\cdot\mathbf{r})}, \quad (8.26)$$

where i stays for x or y, and $u_{i,\mathbf{k}}$ denotes the periodic part of the Bloch function for the wave vector $\mathbf{k}$: $u_{i,\mathbf{k}}(\mathbf{r}) = u_{i,\mathbf{k}}(\mathbf{r} + \mathbf{a})$, where $\mathbf{a}$ is a lattice vector.

The dispersion relation of spin waves in magnonic crystals manifests the all fundamental properties common for other periodic composites operating in linear regime. One of the most important features is the periodicity of the dispersion in a space of wave vector:

$$\omega(\mathbf{k}) = \omega(\mathbf{k} + \mathbf{G}), \quad (8.27)$$

where $\mathbf{G}$ is an arbitrarily reciprocal lattice vector. This leads to the back folding of the dispersion relation into the first Brillouin zone (BZ). In the first BZ the dispersion relation is then multivalued function consisting of frequency bands separated by frequency gaps. The width and the positions of the bands depend not only on the material but also on structural parameters. By appropriate adjustment of the structure in fabrication process, we can obtain the magnonic crystal of desirable spectrum. Therefore one of the tasks in magnonics is to obtain the relation between the structure and the dispersion relation of spin waves.

8.3.3 Plane Wave Method

In Vasseur et al. (1996) and Krawczyk and Puszkarski (2008), PWM was employed to model 2D and 3D magnonic crystals, respectively.

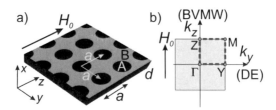

Figure 8.5 (a) Bi-component MC of square lattice formed by A circular dots of radius R included in B matrix. The thickness of the MC is d, the lattice constant is a. The magnetic field H_0 is oriented along the z-axis. (b) First Brillouin zone (BZ) for the structure shown in (a), with indicated high-symmetry points Γ, Y, M, and Z in the center and at the border of the first BZ. We distinguish the two directions of the wave vector **k** along the bias magnetic field and perpendicular to it, i.e., the BVMW and DE geometries, respectively.

Later, the method was modified and used to model 1D and 2D planar MCs (Sokolovsky and Krawczyk, 2011), including antidot lattices (Neusser et al., 2011; Tacchi et al., 2012b). With this new formulation the nonuniform static dipole field together with the finite thickness of the MC was also taken into account.

We will consider here a slab of 2D MC (Fig. 8.5) in the magnetically saturated state and use the linear approximation (described in Section 8.3.1). For this system the dynamics of the magnetization **m**(**r**, t) can be described with the equations (8.24) and (8.25), where the effective magnetic field $\mathbf{H}_{\text{eff}}$ consists of three terms:

$$\mathbf{H}_{\text{eff}}(\mathbf{r}, t) = \mathbf{H}_0 + \mathbf{H}_{\text{ex}}(\mathbf{r}, t) + \mathbf{H}_{\text{dm}}(\mathbf{r}, t). \quad (8.28)$$

The first term is the external static magnetic field, and we assume that this field is along the z-axis. The next component is the exchange field. In MCs we take into account the spatial inhomogeneity of both the exchange constant $A(\mathbf{r})$ and the saturation magnetization $M_S(\mathbf{r})$ and use the formulation defined in Eq. (8.12).

The last component in Eq. (8.28) is the demagnetizing field (Eq. (8.22)). According to the ideas presented in Kaczer and Murtinova (1974) and stressed already in Section 8.3.1.2, for a slab of a 2D MC with uniform magnetization along its thickness, the Maxwell equations can be solved in the magnetostatic approximation with electromagnetic boundary conditions properly taken into account on both surfaces of the 2D magnonic crystal slab, in our case located at $x = -d/2$ and $x = d/2$. For the considered structure, extended to the infinity in (y, z) plane, analytical solutions in the form of Fourier series for both static and dynamic magnetic fields are (Neusser et al., 2011; Tacchi et al., 2012a):

$$H_{dm}(\mathbf{r}_\parallel, x) = -\sum_{\mathbf{G}} M_S(\mathbf{G}) \frac{G_z^2}{G^2} A(\mathbf{G}, x) e^{i\mathbf{G}\cdot\mathbf{r}_\parallel},$$

$$h_x(\mathbf{r}_\parallel, x) = \sum_{\mathbf{G}} \left[i\, m_y(\mathbf{G}) \frac{k_y + G_y}{|\mathbf{k}+\mathbf{G}|} S(\mathbf{k}+\mathbf{G}, x) \right.$$
$$\left. - m_x(\mathbf{G}) C(\mathbf{k}+\mathbf{G}, x) \right] e^{i(\mathbf{k}+\mathbf{G})\cdot\mathbf{r}_\parallel},$$

$$h_y(\mathbf{r}_\parallel, x) = \sum_{\mathbf{G}} \left[i\, m_x(\mathbf{G}) \frac{k_y + G_y}{|\mathbf{k}+\mathbf{G}|} S(\mathbf{k}+\mathbf{G}, x) \right.$$
$$\left. - m_y(\mathbf{G}_\parallel) \frac{(k_y + G_y)^2}{|\mathbf{k}+\mathbf{G}|^2} A(\mathbf{k}+\mathbf{G}, x) \right] e^{i(\mathbf{k}+\mathbf{G})\cdot\mathbf{r}_\parallel}, \quad (8.29)$$

where $\mathbf{r}_\parallel = (y, z)$ is a position vector in the plane of periodicity, $\mathbf{G} = (G_y, G_z)$ denotes a reciprocal lattice vector of our structure, i.e., for square lattice of the lattice constant a: $\mathbf{G} = \frac{2\pi}{a}(n_y, n_z)$, n_y, and n_z are integers. $M_S(\mathbf{G}) = \int_V M_S(\mathbf{r}_\parallel) e^{-i\mathbf{G}\cdot\mathbf{r}} dV$ are Fourier components of the saturation magnetization, which can be calculated analytically for regular shape of dots. $\mathbf{k}_\parallel = (k_y, k_z)$ is a Bloch wave vector of spin waves that, according to the Bloch theorem, will be limited to the first Brillouin zone. The functions $S(\kappa, x)$, $C(\kappa, x)$, and $A(\kappa, x)$ are defined as $S(\kappa, x) = \sinh(|\kappa|x)/(\sinh(|\kappa|d/2) + \cosh(|\kappa|d/2))$, $C(|\kappa|, x) = \cosh(|\kappa|x)/(\sinh(|\kappa|d/2) + \cosh(|\kappa|d/2))$, and $A(\kappa, x) = 1 - S(\kappa, x)$. In derivation of Eqs. (8.29) we have used the Bloch theorem (Eq. (8.26)) to dynamical components of magnetization:

$$\mathbf{m}(\mathbf{r}) = \sum_{\mathbf{G}} \mathbf{m}(\mathbf{G}) e^{i(\mathbf{k}+\mathbf{G})\cdot\mathbf{r}_\parallel}, \quad (8.30)$$

and assumed the static demagnetizing field is parallel to the saturation magnetization.

The obtained formulas for the demagnetizing fields are represented in the reciprocal space for the in-plane components but depend also on the position across the thickness of the slab. However, when the slab is thin enough (which is the case for the latter discussed systems), the nonuniformity of the demagnetizing fields across its thickness can be neglected, and respective values of fields from Eqs. (8.29) with $x = d/2$ will be taken in calculations (Krawczyk et al., 2013). Because of their Fourier series form, the solution found for the demagnetizing fields can be used directly in the PWM technique. The PMW is based on the Fourier expansion of both material parameters and periodic parts of the Bloch functions describing the eigenmodes of magnonic crystal in linear regime. By applying these expansions to the (system of) differential equation(s), we transform it into the algebraic eigenproblem with the frequencies of modes being the eigenvalues and the sets of Fourier coefficients of their expansion as eigenvectors. After applying the PWM to the LL equation (8.1) and using the solution (8.29) for the demagnetizing field, one can obtain the algebraic eigenvalue problem in the following form:

$$\hat{M}(\mathbf{k})m_{\mathbf{k}} = i\Omega\mathbf{m}_{\mathbf{k}}, \tag{8.31}$$

where $\Omega = \omega/(\gamma\mu_0 H_0)$ is reduced frequency and matrix $\hat{M}$ is the block matrix:

$$\hat{M} = \begin{pmatrix} M^{xx} & M^{xy} \\ M^{yx} & M^{yy} \end{pmatrix}. \tag{8.32}$$

The elements of submatrices in (8.32) are defined as follows:

$$M_{ij}^{xx} = -i\frac{1}{H_0}\frac{k_y + G_{y,j}}{|\mathbf{k} + \mathbf{G_j}|} M_S(\mathbf{G_i} - \mathbf{G_j})S(\mathbf{k} + \mathbf{G_j}, x) = -M_{ij}^{yy},$$

$$M_{ij}^{xy} = \delta_{ij} + \frac{1}{H_0}\sum_{l}(\mathbf{k} + \mathbf{G_j}) \cdot (\mathbf{k} + \mathbf{G_l})\,l_{ex}^2(\mathbf{G_l} - \mathbf{G_j})M_S(\mathbf{G_i} - \mathbf{G_l})$$

$$+ \frac{1}{H_0}\frac{(k_y + G_{y,j})^2}{|\mathbf{k} + \mathbf{G_j}|^2} M_S(\mathbf{G_i} - \mathbf{G_j})A(\mathbf{k} + \mathbf{G_j}, x)$$

$$- \frac{1}{H_0}\frac{(G_{z,i} - G_{z,j})^2}{|\mathbf{G_i} - \mathbf{G_j}|^2} M_S(\mathbf{G_i} - \mathbf{G_j})A(\mathbf{G_i} - \mathbf{G_j}, x),$$

$$M_{ij}^{xy} = -\delta_{ij} - \frac{1}{H_0} \sum_{l} (\mathbf{k} + \mathbf{G_j}) \cdot (\mathbf{k} + \mathbf{G_l}) \, l_{ex}^2(\mathbf{G_l} - \mathbf{Gj}) M_S(\mathbf{G_i} - \mathbf{G_l})$$

$$- \frac{1}{H_0} M_S(\mathbf{G_i} - \mathbf{G_j}) C(\mathbf{k} + \mathbf{G_j}, x)$$

$$+ \frac{1}{H_0} \frac{(G_{z,i} - G_{z,j})^2}{|\mathbf{G_i} - \mathbf{G_j}|^2} M_S(\mathbf{G_i} - \mathbf{G_j}) A(\mathbf{G_i} - \mathbf{G_j}, x). \qquad (8.33)$$

The coefficients $l_{ex}(\mathbf{G})$ and $M_S(\mathbf{G})$ are the coefficients of Fourier expansion of periodic distributions of material parameters: exchange length $l_{ex}(\mathbf{r}_\parallel)$ and magnetization saturation $M_S(\mathbf{r}_\parallel)$, respectively. The eigenproblem (8.31) allows to find eigenvalues (SW frequencies ω) and eigenvectors (the sets of coefficients of Fourier expansion of periodic part of the Bloch functions) for fixed value of the wave vector. To find the dispersion relation $\omega(\mathbf{k})$, we have to solve eigenproblem (8.31) for successive values of $\mathbf{k}$ along the irreducible part in the first BZ, as shown for the square lattice in Fig. 8.5b.

8.4 Spin Wave Spectra in Magnonic Crystals

The dispersion relation of exchange spin waves is isotropic and quadratic with respect to the wave vector direction and amplitude, respectively. The spin wave spectrum in this case can be tuned by the change of structural and material parameters in similar way as in phononic and electronic systems. The behavior of SW dynamics in dipole regime is much more complex due to the anisotropy of the dispersion relation observed even in homogenous magnetic films (see Figs. 8.2 and 8.3). The change in the direction of spin wave propagation with respect to the magnetic field orientation alters significantly the spin wave dynamics. The isotropy of the system can be also broken by introducing the structuralization; in the case of periodic structure, this leads to the formation of the magnonic band structure. In the following sections, we will study formation of the magnonic band structure in various MCs at various length scales.

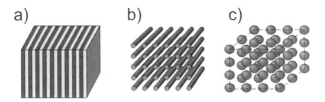

Figure 8.6 Magnonic crystals periodic in (a) 1D, (b) 2D, and (c) 3D. The elements forming the 1D, 2D, and 3D MCs have a form of infinitely extended slabs, infinitely long rods, and finite inclusions, respectively.

8.4.1 Periodicity

Periodic modulation of the structure or its material properties gives rise to the possibility of forming the material of desirable dynamic properties, which can be further tuned by the adjustment of the structure and the material parameters of constituent materials. Many interesting wave phenomena exhibit for the scattering of waves at wavelengths comparable to the period of the structure. For instance, due to coherent scattering the frequency band gaps can be opened or the group velocity of waves can be significantly altered and molded. The other feature is the anisotropy of dispersion relation along different crystallographic directions of the periodic structure. These general properties of periodic composites are also shared by magnonic crystals.

We can distinguish different forms of magnonic crystals with respect to their dimensionality and periodicity. The periodicity can be introduced along one direction, in two-dimensional space or in three dimensions, then we call these structures as 1D, 2D, and 3D MCs, respectively. The examples of 1D, 2D, and 3D MCs are shown in Fig. 8.6 in the form of multilayered structure, a square array of cylindrical rods and simple cubic lattice of spheres, respectively.

In ferromagnetic films the periodicity can be introduced only in 1D (Fig. 8.7) or 2D (Fig. 8.8); however, it is assumed to call planar MC as 3D structures whenever the inhomogeneity across the film thickness is important for SW dynamics. In ferromagnetic films periodicity (irrespective of their dimensionality) can be introduced in different ways. We can enumerate following methods of MC's formation, which have already been used:

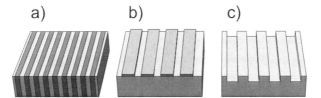

Figure 8.7 The examples of 1D planar magnonic crystal: (a) the sequence of uniform stripes of two kinds, (b) the magnetic slab with metal stripes on the top, modulating the boundary conditions for magnetostatic potential, and (c) the slab of magnetic material with the series of parallel grooves on the top face. The presented systems are confined in one dimension, periodic and extended in second one, and uniform and extended in third one.

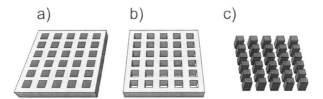

Figure 8.8 The different realizations of 2D planar magnonic crystals: (a) bi-component structures consisting of ferromagnetic inclusions embedded in ferromagnetic matrix, (b) periodically patterned magnetic slab—antidot lattice, and (c) array of magnetic dots.

- *Changes of the magnetic parameters.* Ferromagnetic material can be characterized by following magnetic parameters: magnetization saturation, exchange length or magnetocrystalline anisotropy. We can vary spatially the values of those parameters by changing the material composition of the system. In 1D MCs we can consider an array of stripes of two materials (Fig. 8.7a) where at least one of them is magnetic (Gubbiotti et al., 2005, 2007; Kostylev and Stashkevich, 2010; Wang et al., 2009, 2010; Zhang et al., 2012) whereas in 2D MCs the following structures can be investigated: bi-component MCs (Fig. 8.8a) (Duerr et al., 2011; Kłos et al., 2012b; Tacchi et al., 2012b), antidot lattices (Fig. 8.8b) (Gulyaev et al., 2003; Kumar et al., 2014; McPhail et al., 2005; Neusser et al., 2010; Saha et al.,

2013; Sklenar et al., 2013; Tacchi et al., 2010) or arrays of dots (Fig. 8.8c) (Barman and Barman, 2009; Barman et al., 2010; Tacchi et al., 2011; Zivieri et al., 2011). The other periodic structures, extended in all dimension, are: stack of layers (1D MC) (Barnaś, 1988; Dobrzynski et al., 1986; Grünberg and Mika, 1983; Hillebrands, 1990; Krawczyk et al., 2003; Kruglyak and Kuchko, 2003; Kruglyak et al., 2004), array of rods (2D MC) (Dmytriiev et al., 2013; Krawczyk and Puszkarski, 1998; Puszkarski and Krawczyk, 2003; Vasseur et al., 1996) or inclusions (3D MC) (Krawczyk and Puszkarski, 2008; Mamica et al., 2012b) embedded in the matrix of different magnetic properties (Figs. 8.6a,b,c).

- *Modulation of the external interfaces.* The periodic modulation of surfaces of the magnetic film (e.g., by creating the series of grooves (Chumak et al., 2008; Filimonov et al., 2012)) or by periodic change of the waveguide width (Chumak et al., 2009b; Kim, 2010) can also lead to the coherent scattering of spin waves and formation of the magnonic band structure.

- *Change of the boundary conditions on the surface of the ferromagnetic film.* The boundary conditions can be periodically altered by depositing the regular array of some material of different properties on the surface of the film. This can be done with metal, other ferromagnetic material, or antiferromagnetic material. This method of the MC formation is based on the impact of modified boundary conditions that affect the magnetization dynamics. In the literature so far only the evidence of magnonic band structure formation due to metallic stripes was demonstrated (Mruczkiewicz et al., 2013; Ustinov et al., 2010). On the interfaces with ideal metal (i.e., with perfect electric conductor) the magnetostatic potential is equal to zero, whereas in contact with nonmetallic medium (e.g., vacuum) the magnetostatic potential decay exponentially from the surface of the ferromagnetic film. The other approach is to modify some chemical or physical properties of the surface of the ferromagnetic film, which can lead to change of the surface magnetocrystalline anisotropy (Kisielewski et al., 2014) and alter the exchange boundary condition for magnetization dynamics (Eq. 8.23). This can be done by change of chemical composition of the surface (e.g., by oxidation or ion implantation (Obry et al., 2013)) or by proximity with an antiferromagnetic or other ferromagnetic material (Vysotskii et al., 2013).

- *Application of the external magnetic field.* The homogeneous ferromagnetic material can be placed in the external magnetic field variable in the space (Demokritov et al., 2004). Hence, the magnonic crystal can be formed by periodic modulation of such field. The grid of conducting wires producing periodic Oersted field in the thin ferromagnetic field was demonstrated to be sufficient for formation magnonic band gaps (Bai et al., 2011). Moreover, it was shown that such periodic magnetic field can be changed in time of the spin wave propagation, providing the platform for formation of the dynamical magnonic crystals (Chumak et al., 2009a, 2010).

- *Anisotropy induced by periodic bending of the film or waveguide.* The magnetic wire (or plane) with the constant width (or thickness) can be periodically bended (Fig. 8.13c). The demagnetizing field in such structures will prefer the alignment of the magnetization tangential to the external surfaces. The exchange spin waves propagating in the system with such curved geometry shall experience the effective anisotropy field proportional to the square of curvature (Tkachenko et al., 2012). Therefore the considered system can be regarded as MC in which the spin waves feel the periodic anisotropy field induced by curvature.

The fabrication techniques of periodic nanostructures have been improved in recent years. However, not every geometry presented in the discussion above is equally feasible for implementation. The techniques for multilayered structures are well developed and allow the fabrication of the 1D MC with atomic resolution. Also, the techniques for periodic modulation of the surface of ferromagnetic layer by creating grooves or by placing on it metallic nonmagnetic stripes were the subject of intensive studies in the past decades. However, the fabrication of 2D MC is more challenging. The current technology allows the fabrication of the antidot MCs or antidot periodic waveguide of planar structure with resolution in the range to tens of nanometers. The formation of bi-component planar magnonic crystals is more difficult but feasible for slightly large modulations, i.e., with resolution in the range of hundreds of nm. The 3D MCs with periodicity in nm scale are extremely difficult to fabricate using top-down approaches, but there are promising methods based on the bottom-up approach, which will be discussed in another chapter of this book.

8.4.2 Dipole-Exchange Spin Waves

We will consider 2D bi-component MCs of thickness $d = 10$ nm with a square lattice of circular Co dots of radius $R = 155$ nm (material A) embedded in Py (material B) as shown schematically in Fig. 8.5a. The saturation magnetizations and the exchange constants in Py and Co have been fixed to the values used in (Tacchi et al., 2012b): $M_{S,Py} = 0.78 \times 10^6$ A/m, $A_{Py} = 1.3 \times 10^{-11}$ J/m and $M_{S,Co} = 1.0 \times 10^6$ A/m, $A_{Co} = 2.0 \times 10^{-11}$ J/m, respectively. The magnetic field is in the film plane along the z-axis. The lattice constant is a, thus the first Brillouin zone has a square shape covering wave vectors with in-plane components k_y and k_z between $-\pi/a$ to π/a as shown in Fig. 8.5b. We have already shown in Section 8.2 that in dependence on the wavenumber the dispersion relation is determined by the exchange or dipole-exchange interactions. In this chapter, we will focus on the formation of the magnonic band structure for the dipole-exchange spin waves. To emphasize the effects of dipole interactions, we will consider lattice constant $a = 600$ nm, when at the first Brillouin zone border $k \approx 5 \times 10^6$ m^{-1} (Fig. 8.2c).

We start with the discussion of the model of homogeneous ferromagnetic film with an artificial periodicity. In solid-state physics this model is known as the *empty lattice model* (ELM). The results of the PWM calculations done in the ELM for a 10 nm-thick film with a square lattice periodicity of $a = 600$ nm

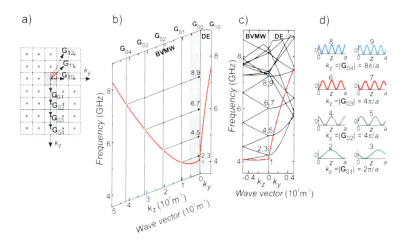

Figure 8.9 (a) Periodic representation of the reciprocal space with a few reciprocal lattices vectors (G_{ij}) for the 2D MC with a square lattice. The gray-shaded area marks the first BZ shown in Fig. 8.5b. (b) Dispersion relation (dashed lines) of SWs in a homogeneous ferromagnetic thin film with in-plane magnetization and external magnetic field ($\mu_0 H_0 = 0.02$ T). The dispersion is shown along the k_z–k_y direction, i.e., in BVMW and DE geometry, respectively. Vertical dotted lines show positions of the reciprocal lattice vectors **G** along the z-axis as introduced by a 2D periodic lattice in (a). The gray-shaded area marks the first Brillouin zone (BZ) that would follow from a periodicity of 600 nm. (c) Magnonic band structure in the first BZ for the homogeneous film with the artificial periodicity introduced (the lattice constant is $a = 600$ nm), i.e., the empty lattice model (ELM). The part of the dispersion from (a) in the first BZ is marked by a bold dashed line (red online). The low-frequency branches along k_z are formed directly by shifts (back-folding) of the SW dispersion relation from higher-order BZs. The frequencies in the BZ center ($k = 0$) are obtained by shifts according to the arrows shown in (a). (d) The spin-precessional amplitudes of SWs at the center of the BZ as derived from the ELM (modes 2 to 9). The respective value of the wave vector is written below each mode.

and averaged values of magnetic parameters ($\bar{l}_{ex} = l_{ex}(\mathbf{G} = 0)$, $\overline{M}_S = M_S(\mathbf{G} = 0)$) are shown in Fig. 8.9c. There is a dense spectrum of magnonic bands, which is a direct consequence of the periodicity. The analytical dispersion relation of spin wave in a homogeneous thin film is marked by bold dashed line (red online). We see a perfect coincidence of the analytical and numerical results.

Magnonic Crystals

According to the Bloch theorem (Section 8.3.2), the discrete translational symmetry in real space introduces the periodicity of the dispersion relation $f(\mathbf{k}) = f(\mathbf{k} + \mathbf{G})$ with a period equal to the reciprocal lattice vector $\mathbf{G}$ (for a square lattice it is $2\pi/a$ along y and z) (Ashcroft and Mermin, 1976). In the center of the BZ ($k = 0$), additional solutions (frequencies 2–9) occur, as compared with the single solution (frequency 1) in the homogeneous film due to back-folding effect. These solutions are exactly the same as for spin waves with wave vectors equal to the reciprocal lattice vectors. In the case of the ferromagnetic film considered here, SWs with wave vectors $\mathbf{k} = \mathbf{G}_{01}$, $\mathbf{G}_{02}$, $\mathbf{G}_{03}$, and $\mathbf{G}_{04}$ have a frequencies below 7.57 GHz (the indexes for reciprocal lattice vectors refer to centers of successive BZs: $\mathbf{G}_{n,m} = 2\pi/a\,(n, m)$, where lattice constant a is equal to 600 nm—see Fig. 8.9a). It means that these frequencies are below the frequency of the DE mode at the first BZ boundary (right edge of Fig. 8.9b). The new mode frequencies appear at the BZ center (frequencies 2–9) once the periodicity of 600 nm is introduced, as it is schematically shown in Fig. 8.9b by arrows. The same will happen for other wave vectors with $\mathbf{k} \neq 0$ in the first BZ.

Because in ELM the periodicity is artificial, the solutions should be consistent with the plane waves, being solutions for the homogeneous film. The profiles of SW modes calculated within the ELM in the BZ center are shown in Fig. 8.9d and are just the same as plane-wave solutions of the homogeneous thin film with wave vectors being equal to the respective reciprocal lattice vectors. Not shown in Fig. 8.9c is the profile of mode 1 because it is a uniform excitation. The pairwise degeneracy of the modes 2 and 3, 4 and 5, 6 and 7, and 8 and 9, seen in Fig. 8.9c, is a direct consequence of the artificial periodicity. The profiles of these couples of degenerate modes have the same shape but are shifted by $a/4$ along the z-axis; thus, in the homogeneous film, their energies are degenerate.

The magnonic band structure calculated along the path Z-Γ-Y-M-Z of the first BZ (Fig. 8.5b) for a 2D bi-component MC composed of Co dots in Py is shown in Fig. 8.10. We consider a lattice constant $a = 600$ nm and thickness $d = 20$ nm. The MC is saturated in the periodicity plane along the z-axis by an external magnetic field $\mu_0 H_0 = 0.02$ T. The magnonic band structure in Fig. 8.10 according with expectations from ELM is composed of many bands. In the

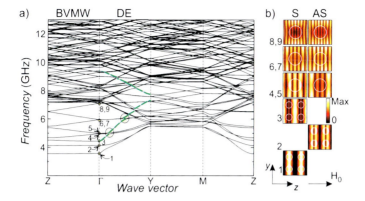

Figure 8.10 (a) Magnonic band structure of a 2D bi-component MC consisting of Co dots in Py with $a = 600$ nm calculated along different directions of the first BZ of Fig. 8.5b. The sample is saturated along the z-axis by the static external magnetic field $\mu_0 H_0 = 0.02$ T. The bold dashed line (green online) marks the DE-like mode in the first BZ. (b) Modulus of the spin-precessional amplitude for low frequency modes with wave vectors from the center of the BZ (Γ point). The numbers 1 to 9 in (a) and (b) indicate the mode number. In (b) the SW profiles are grouped into symmetric (S) and antisymmetric (AS) modes with respect to a line parallel to the y-axis and crossing the Co dots in the middle.

experiments by Duerr et al. (2011) and Tacchi et al. (2012b), the two branches with largest group velocities near the BZ center and with the highest scattering cross section are due to SW excitations where maximum spin-precessional amplitudes exist mainly in channels either crossing the Co dots (mode 1, c.f. Fig. 8.10b) or laying in Py in-between neighboring columns of Co dots (mode 3 in Fig. 8.10b) (Duerr et al., 2011; Tacchi et al., 2012b). Modes with higher frequencies have nodal lines along the direction of the external magnetic field, i.e., they have small scattering cross section.

We have pointed at the obvious similarity between the magnonic band structure of bi-component MC shown in Fig. 8.10a and the SW dispersion calculated in the ELM (Fig. 8.9c), but there are also discrepancies that have different origins. We distinguish four main differences:

    i. *The low frequency of the first mode in the bi-component MC.* This is an effect of the static demagnetizing field, which lowers the effective magnetic field in Co at the interfaces with Py and creates potential wells for SWs. This property was already found in micromagnetic simulations, and the resulting mode channeling was confirmed by micro-focused BLS measurements (Duerr et al., 2011).

ii. *The larger splitting of the bands 2 and 3 and small splittings between the bands 4 and 5, 6 and 7, and 8 and 9.* Near the Γ point, the spin wave modes 1 and 3 concentrate their maxima in two complementary channels parallel to the y-axis comprising the Co dots and in between them, respectively (Fig. 8.10b). In comparison to the spectrum of the ELM (Fig. 8.9c), the dispersive branch of mode 3 in the bi-component MC is shifted to higher frequencies near the Γ point with respect to mode 2. This effect is also due to the inhomogeneous demagnetizing field not considered in the ELM.

The dispersions $f(k)$ of the modes 4 to 9 in the spectrum of both the bi-component MC and ELM are very similar. These modes in the center of the BZ are directly related to the BVMWs of the homogeneous film folded back into the first BZ due to periodicity. The profiles of the plane waves from the ELM (Fig. 8.9d) and amplitudes of modes in the bi-component MC along the z-axis (Fig. 8.10b) are closely related to each other. In the unit-cell center (i.e., in the Co dots) there are maxima of the spin-precessional amplitudes for symmetric modes (modes 4, 6, and 8) and nodal lines for antisymmetric modes (5, 7, and 9). These differences in the position of amplitude maxima are responsible for the band splittings at the BZ boundary and center. The splittings between bands in the bi-component MC decrease with increasing mode number.

iii. *The crossing and anti-crossing between DE mode and other SW excitations in the bi-component MC.* We have already shown that with increasing wave vector k_y the DE mode (marked by a bold dashed line (red online) in Fig. 8.10b) crosses other bands in the ELM. Because folding to the first BZ is due to the artificial periodicity introduced in the ELM, interaction between modes is absent. However, in the bi-component MC the DE-like mode interacts with other modes.

iv. *Magnonic band splitting at the BZ boundary or center.* This splitting result from the destructive interference of Bragg reflected DE-like SWs at the BZ boundary (Tacchi et al., 2012b).

For SWs in thin films with out-of-plane magnetization the isofrequency contours are isotropic, mimicking the case of electromagnetic waves in an isotropic dielectric medium (Bali et al., 2012; Schwarze et al., 2012). Moreover, in the dipole-exchange part of the spectra, ω is a linear function of k (see bottom part of Fig. 8.3b). The magnonic band structure of homogeneous Py film magnetized perpendicular to its plane (by the external magnetic field 1.2 T) with the artificial periodicity $a = 600$ nm (the ELM) is shown in Fig. 8.11a. We can see the linear dispersion relation of spin waves and the clear degeneracy of the first and the second magnonic band around of the first BZ marked by dashed line. Compared with the geometry with the in-plane magnetic field, the BZ border is at similar frequencies (it is between 6.6 and 6.8 GHz) in every direction of $\mathbf{k}$. This means that the magnonic band gap shall be easier to open in FVMW geometry than in BVMW-DE geometry.

In Fig. 8.11b, we show the magnonic band structure for MC composed of Co dots in square lattice included in Py thin film (10 nm thick). Both materials are magnetized perpendicular to the film plane. Due to the narrow bands and large magnonic band gaps, we split the figure into two separate pictures showing the first band (bottom picture) and the second and third bands together (top picture). The narrow bands point at localized character of spin wave excitations and in fact this is confirmed by spin wave amplitude shown in Fig. 8.11c. In all three bands the spin waves are concentrated in Co dots and are just a normal mode excitation of the Co dot, only slightly broadened by interactions. This strong localization is an effect of static demagnetizing field, which is much larger in Co (due to large magnetization saturation value) than in Py matrix.

The further discussion of the magnonic band structure for dipole-exchange spin waves in bi-component MCs can be found in Krawczyk et al. (2013).

316 | *Magnonic Crystals*

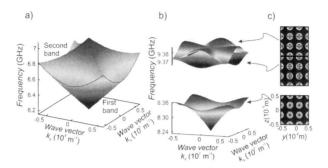

Figure 8.11 (a) Magnonic band structure of a homogeneous Py film magnetized perpendicular to the film plane with the artificial periodicity of square lattice. External magnetic field $\mu_0 H_0 = 1.2$ T and the lattice constant $a = 600$ nm. (b) Magnonic band structure of bi-component magnonic crystal (Co dots of 155 nm radius in Py matrix, thickness 10 nm) in FVMW geometry with perpendicular magnetic field 1.5 T. The different frequency scale is assumed for the first and second with third band. (c) Modulus of the amplitude of the dynamical component of magnetization vector for spin wave modes from the center of the BZ.

8.4.3 Exchange Waves

For investigation the magnonic band structure in exchange dominated regime, we chose Ni as the material of inclusions and Fe as matrix in a planar MC with a small lattice constant $a = 50$ nm of square lattice and the film thickness 10 nm. For this lattice constant at BZ boundary, we expect to obtain the band structure similar to the one shown in Fig. 8.2a, with almost isotropic dispersion relation. In such a composite the spin waves with the lowest frequency (corresponding to the lowest magnonic band) should concentrate in Ni, due to the lower FMR frequency of Ni as compared with Fe. At those values the static demagnetizing effects should not change this relation, although it will decrease effective field in matrix and increase in Ni. In inclusions (having a confined geometry) the localization of the spin waves is strong (stronger than it is in the matrix) and this results in narrower bands and wider gaps in the spin wave spectrum of the MC with Ni inclusions in an Fe matrix (than in the reverse structure Fe inclusions in Ni matrix).

The PWM results are shown in Fig. 8.12. We assume external magnetic field 50 mT in the MC plane. The flat band (with a

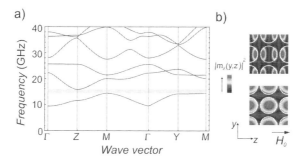

Figure 8.12 (Color online) (a) Spin wave spectrum for a square lattice of circular Ni inclusions in an Fe matrix in the small lattice constant ($a = 50$ nm). The assumed values of structural parameters are as follows: filling fraction $f = 0.55$ and slab thickness $d = 10$ nm. An external magnetic field $\mu_0 H_0 = 50$ mT is applied in the film plane along the z-axis. (b) Spatial distribution of the z-component of the amplitude of dynamic magnetization for the first (bottom row) and second (top row) magnonic band in the vicinity of point Γ.

little negative slope near the Γ point) along the direction of the magnetic field (Γ-Z) is a signature of the BVMW geometry, while the linear slope along orthogonal direction is due to DE geometry. The anisotropy in the dispersion is still present at the Brillouin zone boundary (at Y it is 14 GHz while at Z it is 12 GHz) but the full magnonic band gap is opened. This is an indirect gap as the maximum of the first band is in M and minimum of the second band is in X. As expected the modes from the first two bands have an amplitude concentrated in Ni. This makes the magnonic bands (and magnonic gap) robust on the distortions of the shape of inclusions, until the filling fraction is fixed.

To investigate the impact of the lattice symmetry and the inclusion shape on the spin-wave spectrum in Kłos et al. (2012b), circular, square, and hexagonal inclusions arranged in a square lattice and a triangular one were considered with fixed filling fraction. It was shown that the shape of inclusions does not play a significant role in the range of small lattice constant values. In the same paper, it was shown that the symmetry of the lattice is of much importance for the spectrum. In the spectrum of the MC with the triangular lattice the band gap is significantly wider. The same conclusions were obtained for photonic crystals, pointed at that the smaller differences in the wavenumbers at the Brillouin zone border between different directions favor the band gaps. The symmetry reduction by rotating inclusions of noncircular shape or additional scatterer in the unit cell were also investigated in 2D MC (Mamica et al., 2012a; Wang et al., 2014). It was shown that this approach offer additional possibility for tuning magnonic band gaps.

8.4.4 *Magnonic Waveguides*

The waveguides are structures where the modes can propagate only in one direction due to confinement in the two remaining dimensions. One of the most popular realizations of a waveguide for magnonics has a form of planar structure with the ratio of thickness to width of the waveguide being a small number. For the thin waveguide the assumption of a spin wave unpinning on the top and bottom surfaces of the waveguide is fulfilled to a large extent. The strength of the pinning of the spin wave modes along the width of the waveguide (i.e., in the transferral in-plane direction) depends on the aspect ratio of the waveguide cross section, the absolute width and thickness of the waveguide, and magnetocrystalline anisotropy at the waveguide edges (Guslienko and Slavin, 2005) (see the discussion in Section 8.3.1.3). In low frequency range we can then assume that spectrum of propagating modes will consist of the modes quantized across the width of the waveguides and uniform across its thickness.

The waveguides with the molded dispersion relation can be interesting for magnonic applications due to existence of the magnonic band gap and thus its filtering properties. This can be achieved by periodic patterning, waveguide corrugation or waveguide banding, as schematically shown in Fig. 8.13. Any periodic modulation will result in the back-folding of the spin wave dispersion relation to the first BZ (see Eq. (8.27)). At the BZ edges the band splitting is then expected. If the periodic modulation is big enough and the successive standing modes quantized along the waveguide width are well separated in the frequency, the magnonic band gap opening in the whole range of wavenumbers is possible.

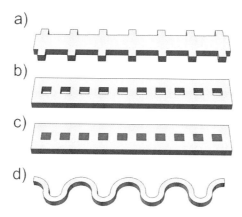

Figure 8.13 Magnonic crystals waveguides. The periodicity can be introduced by (a) modulation of the width, (b) periodic patterning, (c) periodic arrangement of inclusions, and (d) curvature induced anisotropy.

Here, we consider the system presented in Fig. 8.13b (Kłos et al., 2012a), where the row of antidots splits the waveguide into two coupled subwaveguides. For strong pinning at the edges of the antidots, the coupling between the spin waves propagating in two subwaveguides (separated by the row of antidots) will be weak. In Fig. 8.14 the dispersion relation of spin waves for such a waveguide (made of Permalloy) is shown. The magnonic gaps are marked by gray bars. The dispersion branches of periodic systems (black lines) that are close to the dispersion branches of homogeneous subwaveguides of the width equal to the halve of width of the whole patterned waveguide (dashed lines) represent the propagating modes guides by two channels delimited by the row of antidots. These modes form doublets that are split in dependence on the strength of coupling. For a small system (Fig. 8.14a), dipole interaction cannot compete with short-range exchange interactions, whereas for large systems (Fig. 8.14b), the long-range dipole interaction dominate and the coupling is stronger. The strong coupling is manifested in Fig. 8.14b by significant splitting between modes. Another signature of dipole interactions is negative value of the group velocity observed at small values of wavenumbers in Fig. 8.14b. We can notice there the negative slope of dispersion characteristic for BVMW (Kłos et al., 2014).

320 | *Magnonic Crystals*

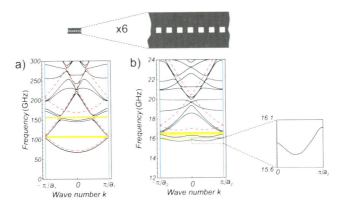

Figure 8.14 (Color online) Spin wave dispersion relation for periodically patterned magnonic waveguides (slid black lines). The waveguide is formed by the thin Py stripe with the square antidots placed equidistantly in the center. Two ranges of sizes are considered: (a) exchange regime where the system is small (width of the waveguide is 45 nm, size of antidots is 6 × 6 nm, the lattice constant $a = 15$ nm, and thickness is 1 nm) and (b) a crossover of dipole-exchange regime where the system was magnified by the factor of six with respect to (a). The same value of the external magnetic field $\mu_0 H_0 = 1$ T was applied along the axis of waveguides in both cases. The inset in (b) shows the signature of dipole effects where for small values of the wave vector a negative group velocity, a characteristic for BVMW geometry, is observed. The red dashed lines show the dispersion for uniform waveguide of the width equal to the halves of patterned ones, i.e., 19.5 nm in (a) and 27 nm in (b).

8.5 Concluding Remarks

Spin wave propagation in ferromagnetic material is a complex phenomenon. The spin wave dispersion is highly anisotropic even in homogeneous thin film, where the direction of external magnetic field with respect to the orientation of external faces of the film nodify spin wave propagation.

The other unique feature of magnonic systems is the presence of two competitive interactions: exchange and dipole interactions. The relative strength of these interactions varies with the extension (or contraction) of the sizes of the system or its structuralization. It means that spin wave spectrum will not scale, as it is for photonics, with the sizes of the system. For small systems (with the dimension in the range up to tens of nm) the short-range exchange interaction dominate. In this regime, intrinsic anisotropy of the dispersion relation is weak and the dispersion can be considered as isotropic and parabolic in many cases. The magnonic systems of the characteristic length in the range of beyond hundreds of nanometers are dominated by dipole interactions. The long-range dipole interactions are highly anisotropic and give the origin to the variety of the unusual effects, such as caustic effect or nonreciprocity.

Magnonic crystals are periodic magnetic structures supporting propagation of spin waves. Periodicity in magnonic system can be introduced in many different ways, by (i) modulation of material parameters, (ii) periodic corrugation of surfaces (modulation of width or thickness), (iii) periodic modulation of boundary condition on the interfaces with nonmagnetic medium, and (iv) application of periodic external magnetic field.

The spin waves propagating in magnonic crystal, excited in linear regime, have a form of Bloch waves. The dispersion relation can be back-folded into limited range of space of wave vector called first Brillouin zone. Due to the folding and intersection of the branches of dispersion relation at the Brillouin zone border, the frequency band gaps forbidden for spin wave propagation can be opened. By adjusting structural and material parameters of the magnonic crystal, we can tune the spectrum of spin waves. The propagation for spin waves in magnonic crystals can have additional anisotropy related to the distinguishability of different crystallographic directions.

For the description of spin wave dynamics in the nanometer scale and larger scale, we can use description based on classical equation of motion for magnetic moment in an effective magnetic field called Landau–Lifshitz equation. In this approach, the magnetic material is described as a medium with continuous distribution of the magnetization. Also, the exchange interaction and magnetocrystalline anisotropy are expressed as continuous in space magnetic fields, giving the contribution to the total effective magnetic field.

One of the most efficient semi-analytical methods used to find the dispersion relation $\omega(k)$ and spatial profiles of the dynamical magnetization $m_\alpha(r)$ for periodic structures is PWM. The starting point for this method is the set of two equations for dynamical components of the magnetization obtained from the linearized Landau–Lifshitz equation. In this method the differential equation is converted into the algebraic eigenproblem for the Fourier components of Bloch functions and the frequencies of spin waves being eigenvalues, that can be solved numerically. To illustrate the potential of this method and to show the influence of the exchange and dipole interactions on the magnonic band structure, we focused on the planar 2D bi-component magnonic crystal and planar periodic antidot waveguides.

Acknowledgments

We acknowledge the financial support from the National Science Centre of Poland project DEC-2-12/07/E/ST3/00538. Part of calculations presented in this chapter were performed at the Poznan Supercomputing and Networking Center.

References

Aharoni, A. (1991). Magnetostatic energy calculations, *IEEE Trans. Magn.* **27**, p. 3539.

Akhiezer, A. I., Bar'yakhtar, V. G., and Peletminskii, S. V. (1968). *Spin Waves* (North Holland, Amsterdam).

Artman, J. O. (1957a). Ferromagnetic resonance in metal single crystals, *Phys. Rev.* **105**, p. 74.

Artman, J. O. (1957b). Microwave resonance relations in anisotropic single-crystal ferrites, *Phys. Rev.* **105**, p. 62.

Ashcroft, N. W., and Mermin, N. D. (1976). *Solid State Physics* (Holt, Rinehart and Winston).

Bai, L., Kohda, M., and Nitta, J. (2011). Observation of spin wave modes depending on a tunable periodic magnetic field, *Appl. Phys. Lett.* **98**, 17, p. 172508.

Bali, R., Kostylev, M., Tripathy, D., Adeyeye, A. O., and Samarin, S. (2012). High-symmetry magnonic modes in antidot lattices magnetized perpendicular to the lattice plane, *Phys. Rev. B* **85**, p. 104414.

Barman, A., and Barman, S. (2009). Dynamic dephasing of magnetization precession in arrays of thin magnetic elements, *Phys. Rev. B* **79**, p. 144415.

Barman, A., Barman, S., Kimura, T., Fukuma, Y., and Otani, Y. (2010). Gyration mode splitting in magnetostatically coupled magnetic vortices in an array, *J. Phys. D: Appl. Phys.* **43**, p. 422001.

Barnaś, J. (1988). Spin waves in superlattices. I general dispersion equations for exchange, magnetostatic and retarded modes, *J. Phys. C* **21**, p. 1021.

Bloch, F. (1929). Über die quantenmechanik der elektronen in kristallgittern, *Zeitschrift fr Physik* **52**, p. 555.

Bloch, F. (1930). Zür theorie des ferromagnetismus, *Z. Phys.* **61**, p. 206.

Bloch, F. (1932). Zür theorie des austauschproblems und der remanenzerscheinung der ferromagnetika, *Z. Phys.* **74**, p. 295.

Brataas, A., Kent, A. D., and Ohno, H. (2012). Current-induced torques in magnetic materials, *Nature Mat.* **11**, p. 372.

Brillouin, L. (1953). *Wave propagation in periodic structures. Electric filters and crystal lattices*, 2nd edn. (Dover Publications).

Brown, W. F. (1963). *Micromagnetics* (Interscience Publishers, New York).

Cherepov, S., Khalili Amiri, P., Alzate, J. G., Wong, K., Lewis, M., Upadhyaya, P., Nath, J., Bao, M., Bur, A., Wu, T., Carman, G. P., Khitun, A., and Wang, K. L. (2014). Electric-field-induced spin wave generation using multiferroic magnetoelectric cells, *Appl. Phys. Lett.* **104**, p. 082403.

Chicone, C. (1999). *Ordinary Differential Equations with Applications* (Springer).

Chikazumi, S. (1997). *Physics of Ferromagnetism* (Oxford University Press).

Chumak, A. V., Neumann, T., Serga, A. A., Hillebrands, B., and Kostylev, M. P. (2009a). A current-controlled, dynamic magnonic crystal, *J. Phys. D: Appl. Phys.* **42**, 20, p. 205005.

Chumak, A. V., Pirro, P., Serga, A. A., Kostylev, M. P., Stamps, R. L., Schultheiss, H., Vogt, K., Hermsdoerfer, S. J., Laegel, B., Beck, P. A., and Hillebrands, B. (2009b). Spin-wave propagation in a microstructured magnonic crystal, *App. Phys. Lett.* **95**, p. 262508.

Chumak, A. V., Serga, A. A., Hillebrands, B., and Kostylev, M. P. (2008). Scattering of backward spin waves in a one-dimensional magnonic crystal, *Appl. Phys. Lett.* **93**, 2, p. 022508.

Chumak, A. V., Tiberkevich, V. S., Karenowska, A. D., Serga, A. A., Gregg, J. F., Slavin, A. N., and Hillebrands, B. (2010). All-linear time reversal by a dynamic artificial crystal, *Nature Commun.* **1**, p. 141.

Demokritov, S. O., Serga, A. A., André, A., Demidov, V. E., Kostylev, M. P., Hillebrands, B., and Slavin, A. N. (2004). Tunneling of dipolar spin waves through a region of inhomogeneous magnetic field, *Phys. Rev. Lett.* **93**, p. 047201.

Demokritov, S. O., and Slavin, A. N. (eds.) (2013). *Magnonics from Fundamentals to Applications* (Springer).

Ding, J., Kostylev, M., and Adeyeye, A. O. (2012). Realization of a mesoscopic reprogrammable magnetic logic based on a nanoscale reconfigurable magnonic crystal, *Appl. Phys. Lett.* **100**, 7, p. 073114.

Dmytriiev, O., Al-Jarah, U. A. S., Gangmei, P., Kruglyak, V. V., Hicken, R. J., Mahato, B. K., Rana, B., Agrawal, M., Barman, A., Mátéfi-Tempfli, M., Piraux, L., and Mátéfi-Tempfli, S. (2013). Static and dynamic magnetic properties of densely packed magnetic nanowire arrays, *Phys. Rev. B* **87**, p. 174429.

Dobrzynski, L., Djafari-Rouhani, B., and Puszkarski, H. (1986). Theory of bulk and surface magnons in Heisenberg ferromagnetic superlattices, *Phys. Rev. B* **33**, 5, p. 3251.

Dreher, L., Weiler, M., Pernpeintner, M., Huebl, H., Gross, R., Brandt, M. S. and Goennenwein, S. T. B. (2012). Surface acoustic wave driven ferromagnetic resonance in nickel thin films: Theory and experiment, *Phys. Rev. B* **86**, p. 134415.

Duerr, G., Madami, M., Neusser, S., Tacchi, S., Gubbiotti, G., Carlotti, G., and Grundler, D. (2011). Spatial control of spin-wave modes in $Ni_{80}Fe_{20}$ antidot lattices by embedded Co nanodisks, *Appl. Phys. Lett.* **99**, 20, p. 202502.

Edwards, E. R. J., Ulrichs, H., Demidov, V. E., Demokritov, S. O., and Urazhdin, S. (2012). Parametric excitation of magnetization oscillations controlled by pure spin current, *Phys. Rev. B* **86**, p. 134420.

Elachi, C. (1976). Waves in active and passive periodic structures: A review, *Proceedings of the IEEE* **64**, p. 1666.

Filimonov, Y., Pavlov, E., Vystostkii, S., and Nikitov, S. (2012). Magnetostatic surface wave propagation in a one-dimensional magnonic crystal with broken translational symmetry, *Appl. Phys. Lett.* **101**, p. 242408.

Floquet, G. (1883). Sur les équations différentielles linéaires à coefficients périodiques, *Annales de l'École Normale Supérieure* **12**, p. 47.

Getzlaff, M. (2008). *Fundamentals of Magnetism* (Springer).

Grünberg, P., and Mika, K. (1983). Magnetostatic spin-wave modes of a ferromagnetic multilayer, *Phys. Rev. B* **27**, 5, p. 2955.

Gubbiotti, G., Tacchi, S., Carlotti, G., Vavassori, P., Singh, N., Goolaup, S., Adeyeye, A. O., Stashkevich, A., and Kostylev, M. (2005). Magnetostatic interaction in arrays of nanometric permalloy wires: A magneto-optic Kerr effect and a Brillouin light scattering study, *Phys. Rev. B* **72**, p. 224413.

Gubbiotti, G., Tacchi, S., Carlotti, G., Singh, N., Goolaup, S., Adeyeye, A. O., and Kostylev, M. (2007). Collective spin modes in monodimensional magnonic crystals consisting of dipolarly coupled nanowires, *Appl. Phys. Lett.* **90**, p. 092503.

Gubbiotti, G., Tacchi, S., Madami, M., Carlotti, G., Adeyeye, A. O., and Kostylev, M. (2010). Brillouin light scattering studies of planar metallic magnonic crystals, *J. Phys. D: Appl. Phys.* **43**, p. 264003.

Gulyaev, Y., Nikitov, S., Zhivotovskii, L., Klimov, A., Tailhades, P., Presmanes, L., Bonningue, C., Tsai, C., Vysotskii, S., and Filimonov, Y. (2003). Ferromagnetic films with magnon bandgap periodic structures: Magnon crystals, *J. Exp. Theor. Phys. Lett.* **77**, p. 567.

Gulyaev, Y. V., and Nikitov, S. A. (2001). Magnonic crystals and spin waves in periodic structures, *Dokl. Physics* **46**, p. 687.

Gurevich, A., and Melkov, G. (1996). *Magnetization Oscillations and Waves* (CRC Press, Boca Raton).

Guslienko, K. Yu., Demokritov, S. O., Hillebrands, B., and SlavinSlavin, A. N. (2002). Effective dipolar boundary conditions for dynamic magnetization in thin magnetic stripes, *Phys. Rev. B* **66**, p. 132402.

Guslienko, K. Y., and Slavin, A. N. (2005). Boundary conditions for magnetization in magnetic nanoelements, *Phys. Rev. B* **72**, p. 014463.

Hartemann, P. (1987). Reflection of magnetostatic forward volume waves by ion-implanted periodic arrays, *J. Appl. Phys.* **62**, p. 2111.

Hill, G. W. (1886). On the part of the motion of lunar perigee which is a function of the mean motions of the sun and moon, *Acta Math.* **8**, p. 1.

Hillebrands, B. (1990). Spin-wave calculations for multilayered structures, *Phys. Rev. B* **41**, 1, p. 530.

Huber, R., Schwarze, T., and Grundler, D. (2013). Nanostripe of subwavelength width as a switchable semitransparent mirror for spin waves in a magnonic crystal, *Phys. Rev. B* **88**, p. 100405(R).

Joannopoulos, J. D., Johnson, S. G., Winn, J. N., and Meade, R. D. (2008). *Photonic Crystals: Molding the Flow of Light*, 2nd edn. (Princeton University Press).

John, S. (1987). Strong localization of photons in certain disordered dielectric superlattices, *Phys. Rev. Lett.* **58**, p. 2486.

Kabos, P., and Stalmachov, V. S. (1994). *Magnetostatic Waves and Their Applications* (Chapman Hall and Ister Science Press).

Kaczer, J., and Murtinova, L. (1974). On the demagnetizing energy of periodic magnetic distributions, *phys. stat. sol. (a)* **23**, p. 79.

Kadic, M., B uckmann, T., Schittny, R., and Wegner, M. (2013). Metamaterials beyond electromagnetism, *Rep. Prog. Phys.* **76**, p. 126501.

Kalinikos, B., and Slavin, A. (1986). Theory of dipole-exchange spin wave spectrum for ferromagnetic films with mixed exchange boundary conditions, *J. Phys. C* **19**, p. 7013.

Keffer, F. (1966). *Spin Waves* (Springer-Verlag, Berlin).

Khitun, A., Bao, M., and Wang, K. L. (2008). Spin wave magnetic nanofabric: A new approach to spin-based logic circuitry, *IEEE Trans. Magn.* **44**, 9, p. 2141.

Kim, S.-K. (2010). Micromagnetic computer simulations of spin waves in nanometre-scale patterned magnetic elements, *J. Phys. D: Appl. Phys* **43**, p. 264004.

Kisielewski, J., Dobrogowski, W., Kurant, Z., Stupakiewicz, A., Tekielak, M., Kirilyuk, A., Kimel, A., Rasing, T., Baczewski, L. T., Wawro, A., Balin, K., Szade, J., and Maziewski, A. (2014). Irreversible modification of magnetic properties of Pt/Co/Pt ultrathin films by femtosecond laser pulses, *J. App. Phys.* **115**, p. 053906.

Kłos, J. W., Kumar, D., Krawczyk, M., and Barman, A. (2014). Influence of structural changes in a periodic antidot waveguide on the spin-wave spectra, *Phys. Rev. B* **89**, p. 014406.

Kłos, J. W., Kumar, D., Romero-Vivas, J., Fangohr, H., Franchin, M., Krawczyk, M., and Barman, A. (2012a). Effect of magnetization pinning on the spectrum of spin waves in magnonic antidot waveguides, *Phys. Rev. B* **86**, p. 184433.

Kłos, J. W., Sokolovskyy, M., Mamica, S., and Krawczyk, M. (2012b). The impact of the lattice symmetry and the inclusion shape on the spectrum of 2D magnonic crystals, *J. Appl. Phys.* **111**, p. 123910.

Kostylev, M. P., and Stashkevich, A. A. (2010). Stochastic properties and Brillouin light scattering response of thermally driven collective magnonic modes on the arrays of dipole coupled nanostripes, *Phys. Rev. B* **81**, p. 054418.

Krawczyk, M., and Grundler, G. (2014). Review and prospects of magnonic crystals and devices with reprogrammable band structure, *J. Phys.: Condens. Matter* **26**, p. 123202.

Krawczyk, M., Mamica, S., Mruczkiewicz, M., Kłos, J. W., Tacchi, S., Madami, M., Gubbiotti, G., Duerr, G., and Grundler, D. (2013). Magnonic band structures in two-dimensional bi-component magnonic crystals with in-plane magnetization, *J. Phys. D: Appl. Phys.* **46**, p. 495003.

Krawczyk, M., and Puszkarski, H. (1998). Magnonic spectra of ferromagnetic composites versus magnetization contrast, *Acta Physica Polonica A* **93**, p. 805.

Krawczyk, M., and Puszkarski, H. (2008). Plane-wave theory of three-dimensional magnonic crystals, *Phys. Rev. B* **77**, p. 054437.

Krawczyk, M., Puszkarski, H., Levy, J.-C. S., and Mercier, D. (2003). Spin-wave mode profiles versus surface/interface conditions in ferromagnetic Fe/Ni layered composites, *J. Phys.: Condens. Matter* **15**, p. 2449.

Krawczyk, M., Sokolovskyy, M., Kłos, J. W., and Mamica, S. (2012). On the Formulation of the Exchange Field in the Landau-Lifshitz Equation for Spin-Wave Calculation in Magnonic Crystals, *Advances in Condensed Matter Physics* **2012**.

Kruglyak, V., and Kuchko, A. (2003). Spectrum of spin waves propagating in a periodic magnetic structure, *Physica B* **339**, p. 130.

Kruglyak, V., Kuchko, A., and Finokhin, V. (2004). Spin-wave spectrum of an ideal multilayer magnet upon modulation of all parameters of the Landau-Lifshitz equation, *Physics of the Solid State* **46**, p. 867.

Kruglyak, V. V., Demokritov, S. O., and Grundler, D. (2010). Magnonics, *J. Phys. D: Appl. Phys.* **43**, 26, p. 264001.

Kumar, D., Kłos, J. W., Krawczyk, M., and Barman, A. (2014). Magnonic band structure, complete bandgap, and collective spin wave excitation in nanoscale two-dimensional magnonic crystals, *J. Appl. Phys.* **115**, p. 043917.

Lenk, B., Ulrichs, H., Garbs, F., and Münzenberg, M. (2011). The building blocks of magnonics, *Phys. Rep.* **507**, p. 107.

Mamica, S., Krawczyk, M., and Kłos, J. W. (2012a). Spin-wave band structure in 2D magnonic crystals with elliptically shaped scattering centres, *Adv. Cond. Matt. Phys.* **2012**, p. 1.

Mamica, S., Krawczyk, M., Sokolovskyy, M. L., and Romero-Vivas, J. (2012b). Large magnonic band gaps and spectra evolution in three-dimensional magnonic crystals based on magnetoferritin nanoparticles, *Phys. Rev. B* **86**, p. 144402.

Mandal, R., Saha, S., Kumar, D., Barman, S., Pal, S., Kaustuv, D., Raychaudhuri, A. K., Fukuma, Y., Otani, Y., and Barman, A. (2012). Optically induced tunable magnetization dynamics in nanoscale Co antidot lattices, *ACS Nano* **6**, p. 3397, doi:10.1021/nn300421c.

McPhail, S., Gürtler, C. M., Shilton, J. M., Curson, N. J., and Bland, J. A. C. (2005). Coupling of spin-wave modes in extended ferromagnetic thin film antidot arrays, *Phys. Rev. B* **72**, p. 094414.

Mruczkiewicz, M., Krawczyk, M., Sakharov, V. K., Khivintsev, Y. V., Filimonov, Y. A., and Nikitov, S. A. (2013). Standing spin waves in magnonic crystals, *J. Appl. Phys.* **113**, 9, p. 093908.

Neusser, S., Bauer, H. G., Duerr, G., Huber, R., Mamica, S., Woltersdorf, G., Krawczyk, M., Back, C. H., and Grundler, D. (2011). Tunable metamaterial response of a $Ni_{80}Fe_{20}$ antidot lattice for spin waves, *Phys. Rev. B* **84**, 18, p. 184411.

Neusser, S., Duerr, G., Bauer, S., H. G. amd Tacchi, Madami, M., Woltersdorf, G., Gubbiotti, G., Back, C. H., and Grundler, D. (2010). Anisotropic propagation and damping of spin waves in a nanopatterned antidot lattice, *Phys. Rev. Lett.* **105**, p. 067208.

Neusser, S., and Grundler, D. (2009). Magnonics: Spin waves on the nanoscale, *Adv. Mater.* **21**, p. 2927.

Obry, B., Pirro, P., Brcher, T., Chumak, A. V., Osten, J., Ciubotaru, F., Serga, A. A., Fassbender, J., and Hillebrands, B. (2013). A micro-structured ion-implanted magnonic crystal, *App. Phys. Lett.* **102**, p. 202403.

Owens, J. M., Collins, J. H., Smith, C. V., and Chiang, I. I. (1977). Oblique incidence of magnetostatic waves on a metallic grating, *Appl. Phys. Lett.* **31**, p. 781.

Owens, J., J.M., Smith, C., Lee, S., and Collins, J. (1978). Magnetostatic wave propagation through periodic metallic gratings, *IEEE Trans. Magn.* **14**, p. 820.

Parekh, J. P., and Tuan, H. S. (1977). Magnetostatic surface wave reflectivity of a shallow groove on a yig film, *Appl. Phys. Lett.* **30**, 12, p. 667.

Pirro, P., Brcher, T., Chumak, A. V., Lgel, B., Dubs, C., Surzhenko, O., Grnert, P., Leven, B., and Hillebrands, B. (2014). Spin-wave excitation and propagation in microstructured waveguides of yttrium iron garnet/Pt bilayers, *Appl. Phys. Lett.* **104**, p. 012402.

Puszkarski, H., and Krawczyk, M. (2003). Magnonic crystals – the magnetic counterpart of photonic crystals, *Solid State Phenomena* **94**, p. 125.

Saha, S., Mandal, R., Barman, S., Kumar, D., Rana, B., Fukuma, Y., Sugimoto, S., Otani, Y., and Barman, A. (2013). Tunable magnonic spectra in two-dimensional magnonic crystals with variable lattice symmetry, *Adv. Funct. Mater.* **23**, 19, p. 2378.

Schwarze, T., and Grundler, D. (2013). Magnonic crystal wave guide with large spin-wave propagation velocity in cofeb, *Appl. Phys. Lett.* **102**, p. 222412.

Schwarze, T., Huber, R., Duerr, G., and Grundler, D. (2012). Complete band gaps for magnetostatic forward volume waves in a two-dimensional magnonic crystal, *Phys. Rev. B* **85**, p. 134448.

Semenova, E. K., Montoncello, F., Tacchi, S., Dürr, G., Sirotkin, E., Ahmad, E., Madami, M., Gubbiotti, G., Neusser, S., Grundler, D., Ogrin, F. Y., Hicken, R. J., Kruglyak, V. V., Berkov, D. V., Gorn, N. L., and Giovannini, L. (2013). Magnetodynamical response of large-area close-packed arrays of circular dots fabricated by nanosphere lithography, *Phys. Rev. B* **87**, p. 174432.

Sklenar, J., Bhat, V. S., DeLong, L. E., Heinonen, O., and Ketterson, J. B. (2013). Strongly localized magnetization modes in permalloy antidot lattices, *Appl. Phys. Lett.* **102**, p. 152412.

Sokolovsky, M., and Krawczyk, M. (2011). The magnetostatic modes in planar one-dimensional magnonic crystals with nanoscale sizes, *J. Nanopart. Res.* **13**, p. 6085.

Soukoulis, C. M., and Wegener, M. (2011). Past achievements and future challenges in the development of three-dimensional photonic metamaterials, *Nature Phot.* **5**, p. 523.

Stancil, D. D., and Prabhakar, A. (2009). *Spin Waves: Theory and Applications* (Springer).

Stohr, J., and Siegmann, H. (2006). *Magnetism: From Fundamentals to Nanoscale Dynamics* (Springer).

Strutt, J. W. (1887). On the maintenance of vibrations by forces of double frequency, and on the propagation of waves through a medium endowed with a periodic structure, *Phil. Mag. S5* **24**, 147, p. 145.

Sykes, C. G., Adam, J. D., and Collins, J. H. (1976). Magnetostatic wave propagation in a periodic structure, *Appl. Phys. Lett.* **29**, p. 388.

Tacchi, S., Botters, B., Madami, M., Kłos, J. W., Sokolovskyy, M. L., Krawczyk, M., Gubbiotti, G., Carlotti, G., Adeyeye, A. O., Neusser, S., and Grundler, D. (2012a). Mode conversion from quantized to propagating spin waves in a rhombic antidot lattice supporting spin wave nanochannels, *Phys. Rev. B* **86**, p. 014417.

Tacchi, S., Duerr, G., Kłos, J. W., Madami, M., Neusser, S., Gubbiotti, G., Carlotti, G., Krawczyk, M., and Grundler, D. (2012b). Forbidden band gaps in the spin-wave spectrum of a two-dimensional bicomponent magnonic crystal, *Phys. Rev. Lett.* **109**, p. 137202.

Tacchi, S., Madami, M., Gubbiotti, G., Carlotti, G., Adeyeye, A. O., Neusser, S., Botters, B., and Grundler, D. (2010). Angular dependence of magnetic normal modes in NiFe antidot lattices with different lattice symmetry, *IEEE Trans. Mag.* **46**, p. 1440.

Tacchi, S., Montoncello, F., Madami, M., Gubbiotti, G., Carlotti, G., Giovannini, L., Zivieri, R., Nizzoli, F., Jain, S., Adeyeye, A. O., and Singh, N. (2011). Band diagram of spin waves in a two-dimensional magnonic crystal, *Phys. Rev. Lett.* **107**, p. 127204.

Tkachenko, V. S., Kuchko, A. N., Dvornik, M., and Kruglyak, V. V. (2012). Propagation and scattering of spin waves in curved magnonic waveguides, *Appl. Phys. Lett.* **101**, p. 152402.

Topp, J., Duerr, G., Thurner, K., and Grundler, D. (2011). Reprogrammable magnonic crystals formed by interacting ferromagnetic nanowires, *Pure Appl. Chem.* **83**, 11, p. 1989.

Ustinov, A. B., Kalinikos, B. A., Demidov, V. E., and Demokritov, S. O. (2010). Formation of gap solitons in ferromagnetic films with a periodic metal grating, *Phys. Rev. B* **81**, p. 180406.

Vasseur, J. O., Dobrzynski, L., Djafari-Rouhani, B., and Puszkarski, H. (1996). Magnon band structure of periodic composites, *Phys. Rev. B* **54**, p. 1043.

Vaz, C. A. F., Bland, J. A. C., and Lauhoff, G. (2008). Magnetism in ultrathin film structures, *Rep. Prog. Phys.* **71**, 5, p. 056501, URL http://stacks.iop.org/0034-4885/71/i=5/a=056501.

Vysotskii, S., Nikitov, S., Pavlov, E., and Filimonov, Y. (2013). Bragg resonances of magnetostatic surface waves in a ferrite-magnonic-crystal-dielectric-metal structure, *J. Commun. Technol. El.* **58**, p. 347.

Wang, Q., Zhang, H., Tang, X., Su, H., Bai, F., Jing, Y., and Zhong, Z. (2014). Effects of symmetry reduction on magnon band gaps in two-dimensional magnonic crystals, *J. Phys. D: Appl. Phys.* **47**, p. 065004.

Wang, Z. K., Zhang, V. L., Lim, H. S., Ng, S. C., Kuok, M. H., Jain, S., and Adeyeye, A. O. (2009). Observation of frequency band gaps in a one-dimensional nanostructured magnonic crystal, *Appl. Phys. Lett.* **94**, p. 083112.

Wang, Z. K., Zhang, V. L., Lim, H. S., Ng, S. C., Kuok, M. H., Jain, S., and Adeyeye, A. O. (2010). Nanostructured magnonic crystals with size-tunable bandgaps, *ACS Nano* **4**, p. 643.

White, R. M. (2007). *Quantum Theory of Magnetism* (Springer).

Yablonovitch, E. (1987). Inhibited spontaneous emission in solid-state physics and electronics, *Phys. Rev. Lett.* **58**, 20, p. 2059.

Yeh, P. (1977). Electromagnetic propagation in periodic stratified media. I. general theory, *J. Opt. Soc. Am.* **67**, 4, p. 423.

Yu, H., Duerr, D., Huber, R., Bahr, M., Schwarze, T., Brandl, F., and Grundler, D. (2013). Omnidirectional spin-wave nanograting coupler, *Nature Commun.* **4**, p. 2702.

Zhang, V. L., Ma, F. S., Pan, H. H., Lin, C. S., Lim, H. S., Ng, S. C., Kuok, M. H., Jain, S., and Adeyeye, A. O. (2012). Observation of dual magnonic and phononic bandgaps in bi-component nanostructured crystals, *Appl. Phys. Lett.* **100**, p. 163118.

Zighem, F., Faurie, D., Mercone, S., Belmeguenai, M., and Haddadi, H. (2013). Voltage-induced strain control of the magnetic anisotropy in a ni thin film on flexible substrate, *J. Appl. Phys.* **114**, 7, p. 073902.

Zivieri, R., Montoncello, F., Giovannini, L., Nizzoli, F., Tacchi, S., Madami, M., Gubbiotti, G., Carlotti, G., and Adeyeye, A. O. (2011). Collective spin modes in chains of dipolarly interacting rectangular magnetic dots, *Phys. Rev. B* **83**, p. 054431.

Chapter 9

Physical Properties of 2D Spin-Crossover Solids from an Electro-Elastic Description: Effect of Shape, Size, and Spin-Distortion Interactions

Kamel Boukheddaden,[a] Ahmed Slimani,[b] Mouhamadou Sy,[a] François Varret,[a] Hassane Oubouchou,[a,c] and Rachid Traiche[a,d]

[a]*Groupe d'Etudes de la Matière Condensée, Université de Versailles-CNRS UMR 8635, 45 Avenue des Etats Unis 78035 Versailles cedex, France*
[b]*Department of Chemical System Engineering, The University of Tokyo, 7-3-1, Hongo, Bunkyo-Ku, Tokyo 113-8656, Japan*
[c]*Centre de Recherche Scientifique et Technique en Soudage et Contrôle, Division des Procédés Electriques et Magnétiques, Route de Dély-Ibrahim - BP 64 Chéraga, Alger, Algeria*
[d]*Université Hassiba Benbouali-Chlef, 13 rue de la révolution, Ténes, Chlef, Algeria*
kbo@physique.uvsq.fr

9.1 Introduction

Today, materials science is an interdisciplinary area at the interface between physics, chemistry, and biology focusing on new materials with novel and/or multi-functionalities that are quite promising

---

Magnetic Structures of 2D and 3D Nanoparticles: Properties and Applications
Edited by Jean-Claude Levy
Copyright © 2016 Pan Stanford Publishing Pte. Ltd.
ISBN 978-981-4613-67-5 (Hardcover), 978-981-4613-68-2 (eBook)
www.panstanford.com

for the future development of the high-tech applications. Indeed, based on the Moore law, predicting that the capacity of the electronic devices doubles every two years on an average, it is not difficult to conceive that the usual miniaturization techniques of the devices will reach their limitation in few decades. That is why the molecular electronic emerges as a powerful research area to propose alternative solutions well adapted to the constant need for highly efficient devices at very small scales. Unlike the usual top down approaches constituted by lithography techniques, the molecular electronic is based on a bottom up approach going from the molecule to the functionalized solid. Such research area aims to replace the current technologies based on silicon by new ones based on molecules or an ensemble of molecules. However, the integration of molecular materials in miniaturized electronic devices presents the most exciting and challenging goal for scientists as they should be able to reproduce the functionalities at low dimensionality. In this context, the spin-crossover (SC) materials (although many of their fundamentals aspects remain to be clarified) present serious potentialities in the development of new generation of electronic devices, but also as very accurate sensors, displays as well as optical memories. These materials combine thermo-, piezo, magneto-, and photo-switching features, leading to original physical properties. At the solid state, however, all of these properties remain very slow and complicated to control and therefore incite scientists to the development of a fertilized area devoted to the investigations of the physical properties of SC nanoparticles. Prior to the development of the associated chemistry, the idea was that very small particles will present very short switching times when addressing the information as well as when reading it. However, this naïve picture was faced with the serious problem of the non-linear changes produced on the physical (magnetic, elastic, and optical) properties of SC solids when reducing the size. The transition from the solid state to a nanoparticle is indeed not continuous, and the nanoparticle itself is an intriguing object that merits to be studied for its own rights.

The molecular and thermodynamic properties of SC compounds are well documented in literature, and there is no need to recall them from the basis of the ligand theory. The reader can consult reference books and articles on the subject [1–10]. In contrast, we will recall the global aspects of the phenomenon to be accessible for a non-specialized reader. In particular, the magnetic properties of SC solids will be briefly reviewed and the connection with ordered magnetic systems will be discussed, in order to avoid the usual confusion between both topics. However, we do not pay much attention to the several types of spin-crossover transition reported in the wide literature and somehow jungle of experimental data, in which a tremendous number of ambiguous or pathological behaviors can be found. Instead, we will focus mainly on SC solids showing first-order or gradual phase transitions as a function of temperature. In this chapter, we will also tackle the problem of nonequilibrium properties of spin-crossover solids. In particular, since these systems are photo-sensitive; adequate light populates excited molecular levels which relax back to some long-lived metastable state; the relaxation properties from these states are full of information about the nature of the interactions in these systems and on the free energy or dynamical potential landscape which drive the thermal relaxation.

In octahedral symmetry, transition metal complexes with $3d^4$-$3d^7$, as well as $3d^8$ when molecular symmetry is lower than O_h, may present an electronic configurations adopting two electronic ground states according to the splitting of the d orbitals in the e_g and t_{2g} subsets. When the energy gap between them, called the ligand field, Δ, is greater than the electronic repulsion energy, Π, the electrons tend to occupy the lower energy orbitals, t_{2g} and the metal complex adopts the low-spin (LS) state. In contrast, when $\Delta \ll \Pi$, the d electrons follow Hund's first rule and the metal complex adopts the high-spin (HS) state. When Δ and Π have the similar values, the energy difference between the HS and the LS states, ΔE, become of the same order of magnitude of the thermal energy, $k_B T$. Consequently, when $\Delta E \simeq k_B T$, the application of an external stimulus (temperature, light, pressure) may induce a change in the spin state. However, this view is limited, since it only explains the intra-molecular process of the spin-crossover transition, which also appears at the solid state.

Historically, in the 1930s Cambi et al. [11] observed for the first time anomalous magnetic properties of some Fe(III) complexes.

Concomitantly, Pauling et al. [12] were discussing similar anomalous magnetic behavior in ferrihemoprotein hydroxides. Two decades later, the foundations of the ligand theory were solidly established, Orgel [13] suggested the possible occurrence of the spin equilibrium as a plausible explanation for the observed anomalous magnetic properties. Later, in 1964, Baker and Bobonich [14] reported the first unusual cooperative spin transition behavior in the complex [Fe(Phen)$_2$(NCX)$_2$] (X = S, Se), which represented the first spin-crossover based on Fe(II). Some years later, König and Majeda clarified definitively the physical nature of the spin-crossover transition in Fe(II) derivatives from detailed magnetic and Mössbauer investigations. Soon, after the 1970s, the number of spin crossover compounds tremendously increased and several new complexes based on cobalt [15–23], nickel [24–26], chromium [17, 18, 27, 28], and manganese [15, 29–38] appeared.

Now, it is well admitted that the thermally induced spin crossover (SC) transition between the low spin (LS) and the high spin (HS) states of Fe(II) complexes with suitable ligands is a typical example of switchable molecular solids (SMSs). Such materials have been studied [3, 4, 39] for many years due to their promising applications as materials for information storage. The bistability of SMSs originates from an intra-molecular vibronic coupling [40, 41] between the electronic structure and the vibrational one. The latter can be enhanced at the solid state by inter-molecular interactions. Indeed, elastic interactions [83–85] are recognized as one of the basic ingredients of the SC transition and lead to various behaviors: from gradual, with a transition corresponding to the simple Boltzmann distribution between two states, which is generally obtained in highly diluted crystals (i.e., in noncooperative systems), to rather abrupt thermal spin transitions, and even up to hysteretic behavior denoting a first-order phase transition [42–44] above a threshold value of the interaction strength. The occurrence of two-step transitions has been assigned to the coexistence of interactions with opposite signs [44, 45].

Most of models historically developed (regular solutions [46], Ising-like [47, 48]) are based on two-level approaches which totally discarded the volume change at the transition and consequently the effect of mechanical stresses on the transition mechanism. In such models, the first-order thermal transition is obtained through a phenomenological interaction parameter, electronic in nature. At variance from these electronic models, the continuous medium model developed by Spiering [84] showed that an elastic interaction could as well give rise to the observed first-order transitions. However, one of the most severe limitations of the previous models arises with the inherent mean-field approximation, which of course reproduces the experimentally observed first-order transitions but ignores the local fluctuations and correlations. More recently, discrete atomistic models based on deformable lattices [49–56] were introduced so as to mimic the spatio-temporal features of the nucleation and growth process of spin phases, which were revealed by optical microscopy investigations [57–65] at the thermal transition of SC single crystals. These models combine electronic and elastic degrees of freedom, which opens the possibility to discriminate the contribution of these two important aspects of the spin transition that were treated separately in the past.

The present work aims to conciliate the continuous medium and discrete lattice approaches, with purely elastic interactions, so as to reproduce both the lattice and spin transformations of the system, thus giving access to spatio-temporal properties. The relevance of model will be established through its ability to reproduce, at least qualitatively, the spatio-temporal effects reported as follows in the well-documented case of the HS $\rightarrow$ LS transition of fresh single crystals: (i) the transition usually starts from a corner or/and an edge of the crystal (depending on its shape); (ii) the LS phase spreads over the whole crystal with a well-defined frontline, the shape of which depends on the interplay with the edges of the crystal, (iii) the frontline propagates at a very slow velocity ($\approx$2–10 μm/s); (iv) the existence of large mechanical stresses is evidenced by an irreversible damage; (v) the stresses (after thermal cycling) strongly impact the subsequent transition temperatures. By proceeding to a local analysis of the kinetics of the transformation, we deduced that the nucleation and growth (NG) process at the thermal transition is a multi-scale process driven by the propagation of mechanical stresses ahead of the transformation frontline [58]. These results pointed out the importance of the coupling between the spin state change and the local deformation of the lattice, opening the way to very interesting theoretical problems in which electronic (spin) and structural (lattice parameter) degrees of freedom are coupled. The present problem is similar to that of miscibility and superstructure formation in binary solid alloys, which in the past was studied in the frame of lattice-gas statistical models. However, these historical lattice-gas models involve rigid lattices and actually belong to the class of two-level models previously quoted [46–48]. More recent models were proposed on the basis of deformable lattices with various interaction potentials [49–55]. Indeed, the choices of a relevant interaction potential and degrees of freedom of sites are crucial issues. Long-range potentials may be very time consuming, since they necessitate the recalculation of the energy of the whole lattice for each visited site during the simulation. In addition, spin-crossover materials are molecular crystals, and so most probably short-ranged potentials can already describe most of their important physical properties in a relevant way, as we will show in this work. Therefore, we used here a short-range anharmonic quartic potential, recently introduced and exactly solved for the 1D case by the transfer integral method [66] some years ago. Here, we restrict the investigations to the case of a 2D square lattice, with fixed topology. It is solved by a Monte Carlo algorithm which alternatively changes the spin states and moves the atomic positions in the framework of a canonical ensemble. The choice of the input parameters aimed to match the experimental values of the bulk modulus and Debye temperatures, as well as the structure variations at the transition. This chapter is organized as follows: In Section 9.2, we summarize the recent results of optical microscopy visualizing the spin transition in a spin-crossover single crystal; the model and the simulation method are described in detail in Section 9.3. Section 9.4 is devoted to the choice of interaction parameter values, which are derived from the available literature on elastic properties of SC solids. Section 9.5 presents the results of the simulations (phase diagram, spatio-temporal properties of the thermal transition) with a focus on the atomic displacements and the propagation of mechanical stresses; in Section 9.6 we summarize the main conclusions and draw some possible developments of this work.

9.2 Some Recent Experimental Results of Optical Microscopy: Visualization of the Spin Transition

Imaging spin-transition in single crystals was initiated by Jeftic and Hauser in the 1990s [67]. These authors showed optical microscope images of $[Fe(ptz)_6](BF4)_2$ (Fe-ptz) large single crystals at low temperature during low-temperature photo-excitation. They already outlined a major problem inherent to volume change upon transformation, which creates inhomogeneous stresses and finally generates cracks and irreversible damage to the crystals. Further images were reported by Ogawa et al. during the photo-excitation of $[Fe(2\text{-pic})_3]Cl_2 \cdot EtOH$ by a femtosecond laser pulse [68]. This second work showed inhomogeneous features occurring transiently during the photo-transformation or subsequent relaxation of the metastable state, leading to spin patterns the organization of which remained far from clear. The real start of optical microscope investigations occurred later, when the quantitative character of the method was recognized in terms of colorimetric analysis. A decisive step was reached when an accurate temperature control was obtained, which is an obvious prerequisite of a true physical approach of the first-order transition, as recently exemplified by the Hauser group, through an optical absorption investigation of the parented spin transition crystals $[Fe(bbtr)_3](ClO_4)_2$ (Fe-bbtr). Of course the latter technique only provides average information on the crystal state, while optical microscopy provides access to the full spatio-temporal properties of the system. Till recently such properties remained almost unknown in the case of the spin transitions and, more generally, of switchable molecular solids. It is worth mentioning here that a recent microscope investigation of the same Fe(bbtr) system which illustrated the major role of stresses at the thermal spin transition in large crystals (the larger the crystal, the larger the stresses). In parallel, micro-Raman experiments on bulk samples were developed in order to characterize the presence of like spin domains (LSD), that is, the domains made of mainly HS or LS molecules expected at the thermal transition range. Such LSDs were not detected at the resolution limit of the technique, for the compound $[Fe(pz)[Ni(CN)_4] \cdot 2H_2O$, (pz = pyrazine). This established

The molecular structure

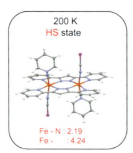
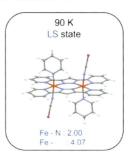

Figure 9.1 Molecular structure of the Fe(NCSe) molecule in the LS and HS states. The red and blue bonds are the most affected during the spin transition. Red dots represent Fe atoms, while the blue (gray) ones are N (C). The white (Purple) dots are hydrogen (Se) atoms. The Fe-N distance increases by about 10% at the transition.

that the domain size should be sub-micronic (for this compound, at least). In another complex, [Fe(bapbpy)(NCS)$_2$], macroscopic structures, in shape of stripes, were observed on the microscope images of the crystal under transformation and assigned to the presence of inhomogeneous elastic strains, while the point-by-point Raman spectral mapping revealed macroscopic structures. However, videos based on optical microscope observations of the thermal spin transition of the same compound previously revealed the regular progression of the transformation front. The Versailles group also reported the more or less regular propagation of the transformation front in various systems: Fe-ptz [69], Fe(bbtr) [59], Fe(btr)2(NCS)2]·H$_2$O [57, 58] here abbreviated Fe(btr).

[Fe(NCSe)(py)$_{22}$(m-bpypz)], where py = pyridine and bpypz = 3,5-bis(2-pyridyl)-pyrazolate, hereafter abbreviated Fe(NCSe), is an archetype of cooperative spin crossover compounds, that is, it exhibits a first-order transition between low-temperature LS and HS states. The molecular structure of the Fe(NCSe) is presented in Fig. 9.1. The molecule is constituted of two metal centers connected by bridging ligands. Upon conversion from LS to HS, the Fe-ligands distances increase by ∼10% and the distance between the two iron centers within the binuclear unit by 5.5%, which shows that even at the intramolecular level, the structure absorbs a part of the

Some Recent Experimental Results of Optical Microscopy | 341

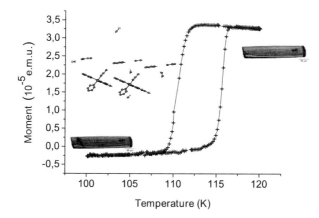

Figure 9.2 Magnetic response of 20 selected micro-crystals of Fe(NCSe) performed in a PPMS device under external field 5000 Oe. A clear thermal hysteresis loop is obtained as a result of a first-order LS–HS transition. The transition is accompanied by a change in the crystal color, which transform from purple (LS) to orange (HS). Temperature sweep rate was 1 K/min.

Fe-ligand expansion. The occurrence of first-order transition with hysteresis is assigned to the presence of elastic interactions between the spin-crossover molecular units. The switching between spin states can be triggered by several means such as temperature and pressure variations, irradiation in the visible range, pulsed magnetic field, ligand photo-isomerization [1–10]. It can be monitored by various techniques due the large changes in the physical properties (structure, volume, color, magnetization, dielectric moment, specific heat, etc.) and spectroscopic properties (optical, Raman, Mössbauer) under the effect of the spin transformation at the molecular and lattice levels. Detailed and comprehensive information can be found in some recent reviews [1]. The transformation from the diamagnetic LS state to the paramagnetic HS state of Fe(NCSe) is illustrated in Fig. 9.2, with a color change consistent with the reported optical properties. The structural changes have been analyzed in detail by X-ray diffraction on single crystals showing a volume increase of ~10% upon the LS to HS transformation. However, a fine analysis showed that the change in the cell parameters upon transition was highly anisotropic. Indeed, while b and c increase at the transition from LS to HS, a parameter shrinks.

9.2.1 *A Front Propagation Slower Than the Snail*

We performed optical microscopy studies on single crystals of Fe(NCSe). These crystals revealed to be rather robust to the repeated thermal cycling unlike the previous spin crossover crystals which display severe cracks upon the transition. Indeed, the Fe(NCSe) spin crossover compound display undergoes an incomplete spin transition ($\approx 2/3$ of molecules switch to the LS state). Such fact reduces the volume change accompanying the transformation and consequently the mechanical stresses. The preparation of the crystals was described in Nakano et al. [70, 71]. The results shown here are obtained with elongated platelets, typically a few 100 μm long. To start with, we selected 20 of them for the magnetic measurements performed in a PPMS device, which are reported in Fig. 9.2. The data evidence a square-shaped hysteresis loop associated with the first-order spin transition, ~ 5 K wide.

We show in Fig. 9.3 typical images obtained during the on-cooling and subsequent on-heating transitions of the single crystal. The nucleation of the LT phase (on cooling) started in the right part of the crystal, while that of the HT phase (on heating) at the left tip, and the orientation of the macroscopic interface remained constant for both processes. It was possible to repeat such a behavior several times in high-quality fresh single crystals before major alterations (cracks) of the crystal. On a large assembly of crystals (~ 100), the observation of interfaces propagating in various directions but always with the same orientation rules out the existence of sizable temperature gradients in the sample cell, as the driving force of the present transformations. We observed on all good-quality crystals a preferred orientation of the front line, at an angle ~ 120–$125°$, as shown in Fig. 9.3. This stable orientation obviously is the one that minimizes the elastic energy associated with the structural mismatch between the HS and LS phases.

We report in Fig. 9.4 the propagation data (interface position vs. time), recorded during the on-cooling and on-heating transitions, located in the [107.7–107.5] K and [115.9–116.0] K temperature ranges, respectively. These propagation data determine fairly constant velocities, ~ 6.1 and 26.2 μms^{-1}, for the on-cooling and on-heating transformations, respectively. This velocity is slower than

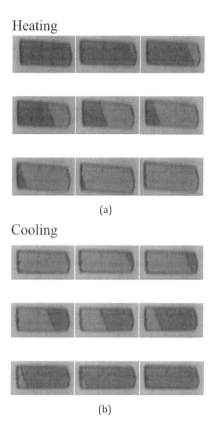

Figure 9.3 Snapshots of the interface position during the thermally induced transition of Fe(NCSe) single crystal. The interface propagates in the heating (a) and cooling (b) transitions with the respective velocities, 6.1 and 26.2 μms^{-1}.

that of a garden snake, whose velocity is ∼1 mm·s^{-1}. It is postulated that the origin of the difference in velocities is related to the 7–8 K difference between the upper and the lower temperatures associated with the thermal hysteresis loop, as well as to the difference in elastic properties of the HS and LS phases. It seems easier to grow nuclei of a soft phase inside a stiffer one than the reverse.

We have followed the experimental interface motion inside the hysteresis loop at temperature $T = 115.04$ K where the HS (resp. LS)

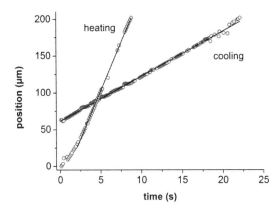

Figure 9.4 (Color online) Propagation plots for HT ⇌ LT interface during the first on-cooling and on-heating transitions of the Fe(NCSe) selected crystal, leading to the propagation velocity values (along the long axis of the crystals): 6.1 and 26.2 μm s^{-1} for the on-cooling and on-heating transitions, respectively.

state is stable (resp. metastable), above the experimental transition temperature, $T_{eq} = 112.6$ K of the sample.

In Fig. 9.5, we present simultaneously the experimental (crosses) and the theoretical (dashed curve) time dependences of the HS fraction during the nucleation and propagation process of the front transformation from LS to HS inside the hysteresis loop of Fig. 9.2. The HS fraction is here averaged over the whole just by following the relative surface of the HS spin phase. The results depict a clear linear character with time, indicating that the velocity of the front propagation is almost constant.

For the present sample, it is clearly seen that the hyperbolic tangent law, of type

$$n_{HS}(x) \simeq \frac{n^*_{HS} + n^*_{LS}}{2} + \frac{n^*_{HS} - n^*_{LS}}{2} \tanh\left(\frac{x - x_0}{\delta}\right) \quad (9.1)$$

where $x_0 = vt$ is the position of the interface center, δ the interface width and n^*_{HS} and n^*_{LS}, well reproduce the interface shape and its propagation over the crystal. This is, however, the solution of the well-known Fisher-Kolmogorov equation [72] which describes well the diffusion and propagation of infectious diseases.

Next, we have also analyzed the local kinetics of the HS fraction on a point inside the crystal. Here, the idea is to follow the time-

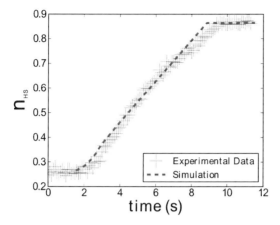

Figure 9.5 Experimental (crosses) kinetic behavior of the HS fraction during LS to HS transformation of the crystal Fe(NCSe) inside the hysteresis loop at temperature $T = 115.04$ K. The dashed curve is derived from the resolution of a reaction diffusion equation. The parameter values are given in the text. An excellent agreement is found between theory and experiment.

dependence of the HS fraction in a fixed small region of the crystal (one pixel at the scale of optical microscopy measurements, i.e., around 1 μm^2) during the front propagation. As long as the front is far from this point, the HS fraction is that of the stationary LS state, i.e., $n_{HS} \simeq 0.3$. When the front comes around the considered point, the HS fraction increases in time until to reach the value of the stationary HS state (i.e., $n_{HS} \simeq 0.9$). The experimental and theoretical results are depicted in Fig. 9.7 and are in very good agreement. In particular, interesting information arising from this local kinetics analysis concerns the time of local nucleation of the HS fraction, obtained from the width of the curve of Fig. 9.7, as $\Delta t \simeq 1$s. This value is an excellent agreement with other experimental lifetime values obtained on different spin-crossover compounds using pump probe reflectivity measurements allowing to monitor the relaxation curves of the HS fraction at different temperatures [73]. On the other hand, it is useful to mention that in a recent experimental work, based on optical microscopy investigations on the spin-crossover sample [Fe(btr)$_2$(NCS)$_2$].H$_2$O (btr=bis-triazole) [58], a similar plot was obtained and phenomenologically described in terms of a

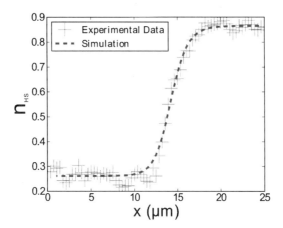

Figure 9.6 Experimental (crosses) spatial dependence of the HS fraction profile along the propagation direction during the LS to HS transformation of the compound Fe(NCSe). The dashed curve is the theoretical curve, obtained with the same parameter values as those of Fig. 9.5. The two curves show the spatial cross section of the interface at time $t = 6$s.

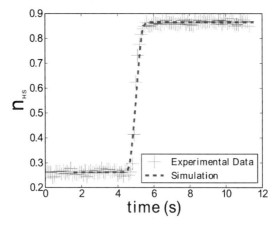

Figure 9.7 Local kinetics of the LS to HS transformation on a local point of the crystal Fe(NCSe) during the front propagation. Crosses (dashed line) stand for experimental data (theoretical curve). The parameter values of the calculated curve are the same as those of Figs. 9.5 and 9.6.

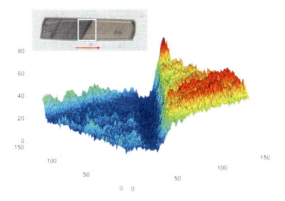

Figure 9.8 Map of the emerging intensity, during the course of the HS to LS transition of a selected crystal. The white-blue area corresponds to the LS phase, while the red-yellow area corresponds to the HS phase.

nucleation and growth process, through the Kolmogorov–Johnson–Mehl–Avrami-type law [72], $n_{HS}(t) = C\{1 - \exp[-k(t-t_0)^\gamma]\}$, with an exponent value $\gamma \simeq 2$.

9.2.2 The HS–LS Interface Deduced from Optical Microscopy Measurements

To make the HS–LS interface much more visible, we have transformed the selected region of Fig. 9.8 in an optical density matrix, $DO(x, y, t)$, illustrated in Fig. 9.8. The front line presents an oscillatory behavior with a ridge line in the LS state and a hollow in the HS state, which are also seen in the interface cross sections along the direction of propagation. The oscillations should not be associated to those of the HS fraction, which can be assumed as homogeneous in the LS and HS phases far from the interface region. They can be interpreted as due to the variation of the refractive index of the crystal near the interface region as a result of the local volume change induced by the spin transition. Such local variations of the refractive index may deflect the light beam, thus producing gray areas as well as mirage effects across the interface.

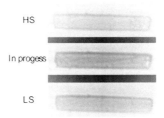

Figure 9.9 Homogeneous transformation of the Fe(NCSe) single crystal from LS to HS after photo-excitation at 4 K using the light beam of the microscope.

9.2.3 Photo-Induced Effects and Relaxation of the Metastable HS State at Low-Temperature

As mentioned in the introduction, SC solids can be photoexcited at low temperature using an adequate wavelength. Usually, a white lamp can be sufficient to produce a sizeable HS metastable fraction at very low temperature. We aimed to visualize for the first time the photoexcitation of a spin-crossover single crystal under the microscope. We then photoexcited the present system Fe(NCSe) at 4 K using the light beam of the microscope. We have found a homogeneous transformation of the crystal from LS to HS without the presence of any macroscopic domain, as seen in Fig. 9.9. For moderate intensities, the transformation from LS to HS follows a single exponential, which denotes that the process is noncooperative. Once we get the photo-induced metastable HS state, we increase the temperature so as to reach the thermally activated regime, and then we let the system relaxing back to LS state in an isothermal way. We did several cycles of photo-excitations and relaxations, whose results are summarized in Fig. 9.10. The curves show a sigmoidal shape, contrasting with those of the photoexcitation regime. This non-linear behavior is known as due to the elastic interactions in the HS phase, which delay the relaxation lifetimes and affect the shape of the relaxation curves in a sigmoidal way. An interesting issue of these curves relates with the lifetime of the metastable HS state which is less than 5 min at 62 K, and which increases almost exponentially when we decrease the temperature (see Fig. 9.11). It is remarkable to see that these curves follow nicely

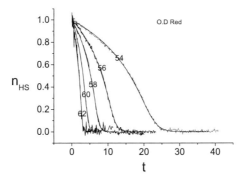

Figure 9.10 Several relaxations of the photo-induced metastable state.

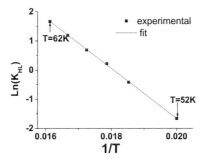

Figure 9.11 Relaxations of the photo-induced metastable state performed at different temperatures (from 54 to 62 K, every 2 K; after each relaxation, the crystal is cooled down until 10 K, then photoexcited till saturation).

a simple mean-field kinetic model [74–77] which predicts that the HS fraction follows the simple law

$$\frac{dn}{dt} = -k_{HL} n \exp\left(-\frac{\alpha n}{k_B T}\right) \qquad (9.2)$$

where n is the HS fraction, $k_{HL} = k_0 \exp(-E_0/k_B T)$ is the Boltzmann frequency factor, α is a parameter proportional to the strength of the interactions, k_B the Boltzmann constant and E_0 is an intrinsic energy barrier. We have checked the validity of this simple description by drawing the rate of the transition $\ln(-dn/ndt)$ as function of n and found that the experimental results follow the expected linear law. All relaxation curves (see Fig. 9.12) lead to a linear behavior, the slope of which gives the factor $-\alpha/k_B T$. The energy barrier E_0

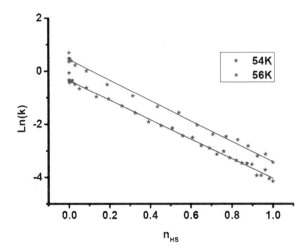

Figure 9.12 The rate of the transition as a function of the HS fraction n_{HS}.

is determined from the intersection with the y axis. The values, $\alpha \sim 210$ **K** and $E_0 = -859.45$ **K** are derived from this analysis.

Let us now discuss the physical reasons leading to the absence of macroscopic domain during the photo-transformation. In the case of weak intensities, the molecular switching proceeds as a single site process. The photon is absorbed by each with some probability which depends of the state of the molecule in the lattice. This stochastic process has also some intrinsic probability, related to the quantum yield of the process, which is temperature dependent. At the end, the photo-conversion takes place as a random stochastic process. Each molecule can then switch under light with some probability, and since the switching does not concern several molecules at the same time (that is the case of the domino effect) the mechanism is that of independent sites, and then it follows a natural exponential growth. The proportion of the switched molecules depends on the fraction of remaining LS molecules inside the lattice. In the case where a strong density of photons is used, the physics becomes different, since a macroscopic fraction of molecules can be switched in a short time, thus causing a shock wave, produced by the subsequent volume expansion, which can drive the collective switching of the other molecules in the crystal, that is the domino effect [78, 79, 81, 82].

9.3 Theoretical Description of the Spin-Crossover Problem

The spin-crossover phenomenon has been described in the past by several types of models [49–51, 51–55, 83–86, 88, 89], aiming to mimic the macroscopic experimental data (magnetic, Mössbauer, heat capacity, ...) derived for most of them from measurements on powder samples. These models were based on Ising-like descriptions, homogeneous elastic medium or regular solutions approaches, which are mean-field theories by construction. Monte Carlo simulations were also conducted on Ising-like models, but with the goal to reproduce thermal hysteresis dependences on model parameters, like interaction and ligand-field. Optical microscopy measurements revealed that the true nature of the spin-transition is spatio-temporal, and then the space and time must be explicitly accounted for to get realistic descriptions. In addition, in the past, the Ising-like models did not include the volume change at the transition, which is admitted today as one of the most important aspects of the spin-crossover phenomenon. Indeed, the expansion of the molecular volume upon transformation from LS to HS delocalizes this local excitation among several unit cells; it is then at the origin of the long-ranged interaction between SC molecules. However, this does not exclude the existence of short-range interactions too, which can be elastic or electronic.

In this section, we describe the present phenomenon through a minimal anelastic model. It can be defined as a hybrid approach mixing an Ising-like model coupled to an elastic lattice. Such models are usually called compressible Ising models [90, 91]. The model consists on a discrete deformable elastic lattice made of two-level units representing the HS and LS states. The symmetry and shape of the lattice are so flexible and could fit different situation. Each lattice's node represents a SC molecule, whose electronic state is described by a fictitious spin $S = +1$ (HS) or -1 (LS). The two spin states have different degeneracies as the diamagnetic LS state has a total spin quantum number $S = 0$, and then it is a non-degenerate state unlike the paramagnetic HS state whose the total spin quantum number is $S = 2$ leading to an electronic degeneracy of 5 that becomes 15 when the orbital moment is considered. Additionally, the volume of the both spin states are as well different due to the antibonding character of the e_g orbitals which makes the metal ligand larger in the HS state (up $\approx$ 10% for Fe(II) molecular systems). Such fact results a rise in the density of levels during the transformation from LS to HS state via supplying an additional effective degeneracy and consequently enhances the multiplicity of the HS state. Such facts entail a rigid LS state with larger vibrational frequencies when compared to those of the HS state. Indeed, Raman spectroscopy investigations [92] combined with DFT studies [93], have shown a net decrease of the intramolecular vibrational frequencies when the molecule experiences the transition from the LS to the HS. In view of the above observations, in the present model we will consider $g_{HS} \gg g_{LS}$, where g_{HS} (g_{LS}) is the degeneracy of the HS (LS) state. In fact, we can demonstrate that the ratio $g^{HS}/g^{LS} = g$ is enough to describe the effect of these degeneracy differences between the HS and the LS states, and for simplicity, such degeneracy will be considered independent of the temperature. Besides, the energy gap between the two spin states at 0 K will be denoted Δ and defined as $\Delta = E^{HS} - E^{LS}$. Such term will be assumed as temperature independent. Indeed, in a previous work [75, 95], we have demonstrated that the present Ising-like system with $g^{LS} \neq g^{HS}$, is isomorphic to a pure two-state Ising system under a temperature-dependent field $-\frac{1}{2}k_B T \ln(g)$. We then obtained an effective temperature-dependent energy gap, $\frac{1}{2}(\Delta - k_B T \ln(g))$, where the second contribution has an entropic origin.

The interactions between the molecules in the lattice can be conceived as springs, where each molecule has four nearest neighbors and four next nearest neighbors as illustrated in (see Fig. 9.13).

We assume here that the lattice deformations remain inside the plane. For a more general study, including the buckling and the crumpling of the lattice, the reader can refer to [94]. It is worth mentioning that the inter-site distances used here do not directly refer to the bond lengths between the metal ion and the ligands, which is the Fe-N distance in iron SC materials, for example. The inter-sites distances actually correspond to those connecting SC metal centers, and the first-neighbors distance will

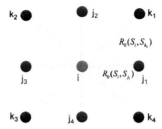

Figure 9.13 Topology of the bonds in the square lattice model. Each site i interacts with its nearest neighbors (nn) j and next nearest neighbors (nnn) k. The equilibrium bond length between nn (nnn) is denoted as $R_0(S_i, S_j)$ ($R'_0(S_i, S_k)$) and depends on the spin states of the linked atoms.

be called hereafter lattice parameter or intermolecular distance. In the present case of a square planar lattice, the presence of second-neighbor interactions is needed for ensuring mechanical stability with respect to shear distortion. Our simulations are performed on a system with open boundary conditions, so as to reveal surface effects which are expected to be large in the case of nano-size systems. Moreover, the simulations performed in the canonical ensemble, do not explicitly involve pressure effects.

9.3.1 The Hamiltonian

The equilibrium distances between the nodes depend on their spin states and are depicted by $R_0(S_i, S_j)$ and $R'_0(S_i, S_k)$ for the nearest neighbors and next nearest neighbors, respectively. The corresponding bond stiffness constants are denoted $A(r_{ij})$, $B(r_{ik})$, where r_{ij} (r_{ik}) is the distance between the sites i and j (i and k). In this problem, we assume that the topology of the bond network remain invariant during the simulation, irrespective of the actual values of the lattice parameters. The total Hamiltonian writes,

$$H = H_{\text{elec}} + H_{\text{elas}} \qquad (9.3)$$

where H_{elec} is the electronic contribution:

$$H_{\text{elec}} = \sum_i \frac{1}{2} \left(\Delta - k_\text{B} T \ln(g) \right) S_i \qquad (9.4)$$

and H_{elas} accounts for the elastic part:

$$H_{elas} = \sum_{i-j} \frac{A_{ij}(r_{ij})\,(r_{ij} - R_0(S_i, S_j))^2}{2}$$

$$+ \sum_{i-k} \frac{B_{ik}(r_{ik})\,(r_{ik} - R_0'(S_i, S_k))^2}{2}. \qquad (9.5)$$

Here, the elastic constants account for anharmonic contributions as follows

$$A_{ij}(r_{ij}) = A_0 + A_1(r_{ij} - R_0^{HH})^2 \text{ and } B_{ik}(r_{ik}) = B_0 + B_1(r_{ik} - \sqrt{2}R_0^{HH})^2,$$
$$(9.6)$$

where $i - j$ and $i - k$, respectively, denote the first- and second-neighbor bonds. The bond vectors and the intermolecular distances are written $\vec{r}_{ij} = \vec{r}_j - \vec{r}_i$ and $\vec{r}_{ik} = \vec{r}_k - \vec{r}_i$, respectively, and $R_0^{HH} \equiv R_0(1, 1)$. The variations of the equilibrium intermolecular distance and bond stiffness upon the LS $\rightarrow$ HS transition result in volume and bulk modulus changes, which are major experimental features of the spin transition. Anharmonic contributions, such as those written in Eq. 9.6 (in terms of even degree polynomials for ensuring stability of the system with respect to large distortions), can be introduced so as to generate the thermal dependences of the bulk modulus and of the Debye temperature, which can be determined experimentally. We neglected here the third order contributions to the total elastic constant, which are responsible for the thermal expansion of the lattice. In fact, the anharmonic form of the elastic constant, given in Eq. 9.6, should be seen as an empirical form, due to the lack of appropriate studies on the elastic constants of SC single crystals, the anisotropy of which makes the Brillouin scattering investigations very complicated.

Let's assume for simplicity that $A_1/A_0 = B_1/B_0$, then the stiffness constant of the lattice only depends on the ratio A_1/A_0. In the present work, we took $A_0 = 4 \times 10^3$ K/nm^2, $A_1 \times a^2 = 4 \times 10^4$ K/nm^2 where $a = 1$ nm is the lattice parameter of the LS state. Namely, $A_{ij}(R_0^{HH}) = 4 \times 10^3$ K/nm^2 and $A_{ij}(R_0^{LL}) = 5.6 \times 10^3$ K/nm^2 and $A_0 = B_0$ and $A_1 = B_1$.

The ratio of the elastic constants A_0 and A_1 can be derived from the changes in the Debye temperatures of the HS and LS phases as expressed in references [66, 96]:

$$\frac{\Theta_D^{LL}}{\Theta_D^{HH}} = \sqrt{\frac{A(R_0^{LL})}{A(R_0^{HH})}} = \sqrt{1 + \frac{A_1}{A_0}\left(R_0^{LL} - R_0^{HH}\right)^2}, \qquad (9.7)$$

where

$$R_0^{LL} \equiv R_0(-1, -1). \qquad (9.8)$$

As a matter of fact, the anharmonic terms are not needed for obtaining the phase transition, and their effects were finally found negligible.

9.3.2 Structure of the Hamiltonian and the Analogy with the Ising-Like Model

To derive the analytical dependence of the total potential energy on the stiffness constant and the lattice misfit, we investigated the structure of the Hamiltonian. For simplicity reason, we only considered in the present section the harmonic part and the nearest neighbor interactions. Let us first use the formal transformation of the equilibrium distances, as follows:

$$R_0(S_i, S_j) = \rho_0 + \rho_1(S_i + S_j) + \rho_2 S_i S_j. \qquad (9.9)$$

Simple identification to the equilibrium distances of HH, LH, LL bonds leads to

$$\rho_0 = \frac{1}{4}\left(R_0^{HH} + R_0^{LL} + 2R_0^{HL}\right), \rho_1 = \frac{1}{4}\left(R_0^{HH} - R_0^{LL}\right),$$

$$\text{and } \rho_2 = \frac{1}{4}\left(R_0^{HH} + R_0^{LL} - 2R_0^{HL}\right). \qquad (9.10)$$

The parameter ρ_0 is obviously associated with the average lattice parameter of the system, termed a in the following, while the parameters ρ_1 and ρ_2 (see Fig. 9.14) are associated with the mismatch of the HS and LS lattices.

Then, the total Hamiltonian (9.3) is accordingly re-expressed in terms of an Ising-like Hamiltonian with space-dependent effective interactions and effective field:

$$H = \sum_{i-j} J_{ij} S_i S_j + \sum_i h_i S_i + \frac{A_0}{2} \sum_{i-j} \left(r_{ij} - \rho_0\right)^2 + C, \qquad (9.11)$$

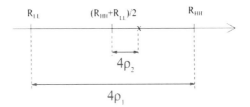

Figure 9.14 Schematic representation of the different equilibrium bond-lengths and their relation with ρ_0, ρ_1, and ρ_2.

where the parameters J_{ij} and h_i are the local exchange-like interactions and the local field-like contributions, respectively. The expressions of J_{ij} and h_i are given by

$$J_{ij} = A_0 \left[\rho_1^2 - \rho_2 \left(r_{ij} - \rho_0 \right) \right] \tag{9.12}$$

and

$$h_i = \frac{1}{2}(\Delta - k_B T \ln g) + \frac{z}{2} A_0 \rho_1 \rho_2 - A_0 \rho_1 \sum_{j=1}^{z} \left(r_{ij} - \rho_0 \right), \tag{9.13}$$

where the index j runs over the neighbors of a given site i, and $z (= 4)$ is the coordination number. The constant C is given by

$$C = \frac{zN}{4} A_0 (2\rho_1^2 + \rho_2^2). \tag{9.14}$$

The third term of Eq. 9.11 is related to the cohesion energy of the elastic lattice whose the equilibrium distance is given by ρ_0 when the spin variables are absent. The constant term C will be omitted hereafter, since it does not play any role in the thermodynamic properties of the system.

Equation 9.11 shows here that the previous "elastic" Hamiltonian can be mapped under the form of an Ising-like Hamiltonian in which the interaction and the field-like parameters are local and depend on the stress field. Interestingly, a meticulous inspection of the expression of J_{ij} shows that this contribution includes both short-range and long-range contributions. The short-range part, $A_0 \rho_1^2 S_i S_j$ which emerges from the structure of the Hamiltonian is naturally antiferro-elastic, while the long-range part, $= -A_0 \rho_2 (r_{ij} - \rho_0) S_i S_j$, is much more complex. If ρ_0 is not the center of mass of R_0^{HH} and R_0^{LL} $\left(\rho_0 \neq \frac{(R_0^{HH} + R_0^{LL})}{2} \right)$, then $\rho_2 \neq 0$. Moreover, when $\rho_2 <$

$\frac{\left(R_0^{\text{HH}}-R_0^{\text{LL}}\right)}{2}\left(\rho_2 > \frac{\left(R_0^{\text{HH}}-R_0^{\text{LL}}\right)}{2}\right)$, the long-range elastic contribution in J_{ij} will stabilize the HS (LS) sate. So, J_{ij} appears clearly as a complex term in which all the misfit lattice parameters (ρ_1 and ρ_2) accounting for the volume change at the transition are included. In the current simulations, however, we have used $R_0^{\text{HL}} = \frac{\left(R_0^{\text{HH}}+R_0^{\text{LL}}\right)}{2}$, and therefore ρ_2 vanishes, which leads to the following simplified expressions of J_{ij} and h_i:

$$J_{ij} = A_0 \rho_1^2 \tag{9.15}$$

and

$$h_i = \frac{1}{2}\left(\Delta - k_{\text{B}}T \ln g\right) - A_0\rho_1 \sum_{j=1}^{z}\left(r_{ij} - \rho_0\right). \tag{9.16}$$

Equation 9.15 clearly shows a spatially invariant effective interaction J_{ij} that consists on the product of the elastic constant and the lattice misfit squared. It also expresses a bilinear coupling between displacements and spins, proportional to ρ_1^2, in agreement with the continuous medium model of H. Spiering [84, 85]. In addition, Eq. 9.16 evidences the direct synergy between the effective "field", $\frac{1}{2}\left(\Delta - k_{\text{B}}T \ln g\right)$, which stabilizes the LS (HS) state at low (high) temperature and the elastic contribution in the local field-like h_i, which acts in the same way. Indeed, the sign of the quantity $-A_0\rho_1 \sum_{j=1}^{z}\left[\left(r_{ij} - \rho_0\right)\right]$ reverses from positive to negative when the lattice goes from LS ($r_{ij} = R_0^{\text{LL}}$) to HS ($r_{ij} = R_0^{\text{HH}}$) states. Thus, the local field-like h_i constitutes one of the driving forces of the SC transition. So, one of the important issues of this model is the direct identification of the intimate nature of the effective interaction term J_{ij} to the elastic energy associated with the volume misfit between the HS and the LS lattices. As said previously, the short-range effective interaction $J_{ij} = A_0\rho_1^2$ is strictly positive and then it leads to stabilize energetically the HS–LS configurations at short range. However, the elastic interaction arising from the field-like contribution, h_i, is long ranged and favors the appearance of HS or LS domains. In a recent work we reported on the coexistence of antiferroelastic correlation for the nearest neighbors and ferroelastic correlation with long range impact [87]. A competition between these two contributions may lead to unexpected results, like self-organized structures, reminiscent of two-step transitions. Such aspects are too heavy to be included here and will be addressed elsewhere.

9.3.3 The Monte Carlo Procedure

The way the total Hamiltonian is solved may be the most crucial issue for the physical relevance of the model, and it involves the different time-scales of the spin state switching and of the lattice dynamics. Representative values of the spin state lifetimes upon the thermal transition typically are in the ms range and may increase to several hours at lower temperatures ($\sim$10 K). On the other hand, the lattice distortions are expected to propagate at the velocity of sound in the crystal, typically at some 10^3 m/s, that means within less than 1 ps through a unit cell (10^{-9} m). Therefore, we assumed that after each spin switching the lattice has enough time for (almost) full relaxation from the excess of elastic energy generated by the spin switching. Consequently, we adopted a two-step iterative strategy. The Monte Carlo (MC) procedure is performed on spins variables and atomic positions as follows: we chose randomly a node and flip its spin state using the usual MC Metropolis algorithm. In any case (spin flip accepted or rejected), we visit randomly the lattice and attempt to move each site by $\delta u = 0.05$ nm in randomly chosen direction. If the new position is permitted by the usual Monte Carlo Metropolis algorithm, then we updated the position of the node, otherwise the node is left in its original position. Then we randomly a chose a new site for which execute the same procedure. This procedure is repeated N times for the spins, where N represents the number of nodes. Since for each spin flip, we visit the lattice 100 times, and if we define the "single MC time unit," on the average, as one attempt of position change made for each atom, then we arrive to 3 10^5 MCS on position updates for each atom when we have visited all the spins one time. During the elastic relaxation process of the lattice, at constant spin configuration, we recorded the lattice configuration every $\tau_0 = N/\delta r^2 = 1200$ MC times units for the total time $t_0 \simeq 100\tau_0$. We checked that these times are sufficient to guarantee both statistical independence of the reached configurations and good averages. It is important to remark that the motion of the atoms is here purely diffusive, so that their equations of motions can be roughly written as

$$\frac{k_B T}{D}\frac{\partial u}{\partial t} = K\nabla^2 u, \tag{9.17}$$

where u and K are the displacement and the bulk modulus, and D is the diffusion constant of a unit surface area. The diffusion of a single atom in the present case is $4\delta r^2$. The diffusion constant per unit area D is then given by $4\delta r^2/a^2$, where a is the lattice parameter (the mean distance between the atoms). Solids governed by the equation of diffusion 9.17 are characterized by spatially normal modes decaying exponentially in time with a time constant depending on the length scale of the oscillation. For a solid of linear size Na, the lowest mode has a wavelength $2Na$, and it relaxation will be

$$\tau \sim \frac{Nk_B T a^4}{\pi^2 \delta r^2 K} \simeq \frac{k_B T a^4}{\pi^2 K} \tau_0.$$
(9.18)

In our calculations, K is lying in the range $[100 - 10000]$ K·nm^{-2} and $a = 1.2$ nm, which gives $\tau \sim 0.01\tau_0$, ensuring that time intervals between the measurements of the positions of the atoms were long enough to ensure statistically independent configurations.

9.3.4 *The Parameter Values*

The relationships between electronic parameters and thermodynamic data of the system are well known for Ising-like models, see for example [45, 47, 75]. The LS–HS energy gap Δ is related to the molar enthalpy variation upon complete transition, according to $\Delta H = N_A \Delta$, where N_A is Avogadro number. The values of ΔH are in the range $5 - 20$ kJ/mol [98]. The degeneracy ratio is related to the molar entropy change upon complete transition ΔS, according to $\Delta S = R \ln g$, where R is the perfect gas constant. The values of ΔS are in the range $35 - 80$ J/K/mol. In the present work we used $\Delta = 900$ K, $\ln g \simeq 10$, respectively, leading to $\Delta H \approx 7, 5$ kJ/mol, and $\Delta S \approx 83$ J/K/mol. The corresponding value of the transition temperature is

$$T_{eq} = \frac{\Delta H}{\Delta S} = \frac{\Delta}{k_B \ln g},$$
(9.19)

which is equal to ≈ 90 K for the used parameter values. Some of the elastic parameter values of the model can be derived from the known properties of the pure phases (LS or HS). For example, the equilibrium Fe(II)-ligand distances increase by about 10% upon the LS–HS transition [3], but the variations of the lattice parameters usually are closer to a few percent. We actually used for the first- neighbor pairs (lattice parameter distance): $R_0^{LL} = 1$ nm, $R_0^{HH} = 1.2$ nm, and for the second-neighbor pairs $R_0'^{LL} = \sqrt{2}R_0^{LL}$, $R_0'^{HH} = \sqrt{2}R_0^{HH}$, based on the idea that the angular distortions [99] may be neglected. Within, these values, we have $\rho_0 = 1.1$ nm, $\rho_1 = 0.2$ nm and $\rho_2 = 0$.

The choice of such a large lattice parameter variation is rather arbitrary since the relationship between the lattice parameter and the metal–ligand bond distance variations is certainly complex, and it is so far unknown. In addition, the use of a large value of the lattice mismatch is expected to balance the reduction of the stress effects due to the small size of the lattice. Moreover, the possible overestimate of the lattice mismatch parameter can be merely compensated by a decrease in the bulk modulus value, as shown by Eqs. 9.12 and 9.15. In other words, we believe that the qualitative nature of the phenomenon is universal.

The values of the harmonic interaction parameters A_{ij} in the HS and LS phases can be derived from the bulk modulus E, which in spin-crossover solids amounts to ≈ 5–20 GPa [100, 101]. An order of magnitude of the stiffness constant A_0 is obtained by considering the elongation of a cubic cell using a 3D lattice, with lattice parameter a, submitted to an uniaxial stress, and neglecting the transversal effects. This simplified model results in the (approximate) relationship $A_0 + 2B_0 \approx Ea$. Following these observations, we took $A_0 \approx 4000$ K·nm^2 ≈ 4 meV·Å^2, leading to a bulk modulus value, $E = 6$ GPa, which is in good agreement with the experimental data reported in the literature [100]. The ratio A_1/A_0 can be estimated by the knowledge of the Debye temperature ratio (Eq. 9.7), and we took $A_1 = 4 \times 10^4$ K/nm^2, leading to $A_1\rho_1^2 = 100$ K. The parameters associated with the HL mixed bonds cannot be determined by similar a priori considerations. We assumed simple arithmetic mean values such as $R_0^{HL} = \frac{R_0^{HH}+R_0^{LL}}{2}$, $A_0^{HL} = \frac{A_0^{HH}+A_0^{LL}}{2}$, and identical relations for the second-neighbor interactions. This simple working assumption concerned terms which apparently did not play a leading role at the transition, and it was not further questioned in the course of the present work.

9.4 Results and Analysis

We investigated here the thermal properties of two kind of systems: (i) a planar square-shaped system and (ii) a planar rectangular system, with free boundary conditions for both cases. The objectives of this section are

- the determination of the thermal properties of the system, including the temperature dependences of the HS fraction;
- the study of the nucleation, growth, and propagation of the HS fraction inside the lattice;
- the analysis of the spatio-temporal behavior of the system and its relation with the crystal shape.

We worked on a square lattice of size ($L^2 = 50 \times 50$) with the aim to understand its equilibrium and nonequilibrium properties. First, we started by simulating the thermally induced hysteresis loop for the square system for various interaction parameter values. The temperature was increased from $T = 1$ to 200 K and then decreased to 1 K, by 1 K increments. As described before, the stochastic procedure was alternatively performed on spin and lattice variables, with 1000 lattice MCS after each eventual spin state switching. Random application of the two-step MC procedure over all sites of the lattice—termed here a "spin-lattice MCS"—was repeated 2.0×10^3 times at each temperature. The first 10^3 times were used for reaching the equilibrium of the system, and the second 10^3 ones for the statistical analysis of the physical quantities of interest. First, we determined the phase diagram of the thermal transition of the system in order to validate the model (with respect to the well-known properties of switchable molecular solids), and in a second step we analyzed the spatio-temporal features of the thermal transition, from both electronic and structural viewpoints.

9.4.1 *The Thermally Induced First-Order Transition*

The average thermodynamical properties of the system are investigated through the thermal dependence of the HS fraction, n_{HS}, and

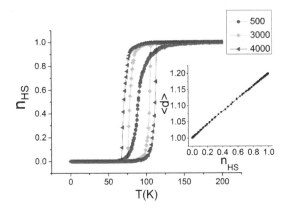

Figure 9.15 (Color online) Thermal dependence of the HS fraction for different values of the elastic constant A_0. Inset: correlation between the lattice parameter $<d>$ and the HS fraction (computed for $A_0 = 4000$ K/nm^2).

the average distance, d, between the nearest neighbors given by

$$n_{HS} = \frac{(1 + \langle S_i \rangle)}{2} \text{ and } \langle d \rangle = \frac{\sum_{ij} \sqrt{(x_j - x_i)^2 + (y_j - y_i)^2}}{N(N-1)/2}, \quad (9.20)$$

where i, j run over $[1, N]$ and are restricted to nn pairs.

9.4.2 Square-Shaped Lattice

The simulation results are shown in Fig. 9.15 for different values of the harmonic elastic constant A_0 in the range of the significant values determined in the previous section. By raising A_0, the transition transforms from gradual to first-order.

As shown in Fig. 9.15, the model exhibits a first-order phase transition due to the elastic interaction, which is now discussed by analogy to the usual phenomenological description [45, 75]. Indeed, A_0 can be scaled to the phenomenological interaction parameter of the usual Ising-like models through Eq. 9.15. It is then legitimate to find the physical criteria to fulfill in order to observe this first order phase transition. For the needs of the coming discussion, it is important to understand that the present first-order transition takes place between two ordered states, namely HS and LS states.

The presence of the HS state is mainly stabilized by the existence of the thermally dependent effective field, Δ_{eff}, and the first-order transition occurs at a temperature, T_{eq} at which the field-like h_i becomes equal to zero, which gives

$$T_{\text{eq}} = \frac{\Delta}{k_{\text{B}} \ln g} \simeq 90\text{K}. \tag{9.21}$$

T_{eq} does not depend on the misfit elastic energy A_0, mainly due to the fact that the average contribution of the elastic "field", $A_0 \rho_1 \sum_1^z (r_{ij} - \rho_0)$, to the local field-like contribution, h_i (given in Eq. 9.16), is negligible around the transition temperature, for which $\langle r_{ij} \rangle \simeq \rho_0$.

One of the important prerequisites of the existence of the first-order transition is that T_{eq} should be smaller than the order-disorder temperature, T_{C}, resulting from the Hamiltonian (9.11) in the case where $\Delta = 0, g = 1$ (non-degenerate states). In other words, the transition between the two states must occur in the ordered phase of this model with equal HS and LS degeneracies and without ligand field. So for comparative purpose, we have considered an Ising model with an exchange interaction parameter $A_0 \rho_1^2$ (antiferro-type) and a long-range ferroelastic field-like interaction $h_i = -A_0 \rho_1 \sum_{j=1}^z (r_{ij} - \rho_0)$, for which we performed MC simulations to investigate the thermal dependence of the HS fraction $_{\text{HS}} = (1 + \langle S \rangle)/2$ for several values of the elastic energy $A_0 \rho_1^2$. The results are illustrated in Fig. 9.16a and reveal a clear indication of a second order-like transition. Indeed, we found that the order-disorder transition temperature, T_{C}, behaves linearly with the misfit elastic energy, as $T_{\text{C}} = \alpha A_0 \rho_1^2$, with α 25 [56]. The results are reported in Fig. 9.16b, where the T_{C} data (red symbols) are compared to the equilibrium temperature values T_{eq} (for which $n_{\text{HS}} = 0.5$, red symbols).

As a general property, the criterion for the existence of a first-order transition is given by $T_{\text{eq}} < T_{\text{C}}$. For the values of the elastic misfit energy such as $T_{\text{eq}} < T_{\text{C}}$, a gradual spin transition is obtained, as seen in Fig. 9.15. It is worth mentioning the previous conditions on the occurrence of the first order transition related to Hamiltonian 9.5 are very general and can be applied to any kind of lattice, irrespective of its shape and symmetry. Indeed, the lattice symmetry does not play any role on the value of the first-order transition

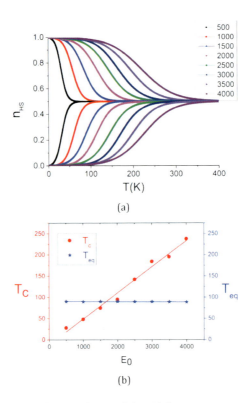

Figure 9.16 Thermal dependence of the HS fraction corresponding to the Ising-like model in an elastic lattice without degeneracy and ligand-field ($g = 1$, $\Delta = 0$), with competing short- (antiferro) and long-range (ferro) interactions. A second order transition is obtained. Parameter values are given in the text.

temperature, T_{eq}, but it clearly affects the second order transition T_C due to the change of connectivity. In addition, the thermal hysteresis width, which results from the existence of finite lifetime metastable states, and which strongly depends on the Monte Carlo kinetics, the temperature sweep rate, etc., may also depend on the shape of the lattice.

In the next section, we report on the elastic properties of the lattice at some positions on the hysteresis loop shown in Fig. 9.15 for the two kind of lattices.

9.5 Lattice Configurations upon the SC Transition of the Square Lattice

This section is devoted to the study of the spatio-temporal properties of the spin transition along the thermal hysteresis transition (see Fig. 9.17) of square-shaped lattice. Previous models and simulations [50, 56, 86] have shown that the transition in such lattice starts from the four corners, which constitute fragile points at which the first nucleus of the new phase can be initiated. In Fig. 9.17b, we show the corresponding snapshots of the system. Here we represent the true site positions and each spin state $+1$ or -1 is identified by its own color, red or blue, respectively. In Fig. 9.17c, we plotted the distribution of the local pressure $P(i)$ defined as

$$P(i) = -\sum_j A_{ij}(r_{ij})\,(r_{ij} - R_0(S_i, S_j)) - \sum_k B_{ik}(r_{ik})\,(r_{ik} - R_0(S_i, S_k)),$$

(9.22)

where j (k) runs over nearest (next nearest) neighbors of site i.

Figure 9.17b gives evidence for a domain growth process starting from all corners, for both the HS $\rightarrow$ LS and the LS $\rightarrow$ HS transformations, as already reported by previous Molecular Dynamics and Monte Carlo studies [80, 88]. The growing domains extend toward the center of the crystal and then collapse. This behavior is reminiscent of experimental observations obtained by optical microscopy [58], for which the nucleation and growth process starts from a single point (edge, corner or defect) and can propagate through the whole crystal in isothermal conditions. However, the simultaneous nucleation at all corners followed by propagation toward the center was not experimentally observed so far. A possible reason for this difference may be a shape effect, because all crystals investigated till now always departed from the perfect square shape of the present model. An alternative explanation might be the presence of defects in real systems.

The maps of the local pressure, shown in Fig. 9.17c, mainly show short-range inhomogeneous distributions of pressures resulting from local expansions and shrinkages. Along the cooling branch (from B to C), the peaks of positive pressure progressively take over, due the cooperative effect of the local shrinkages resulting from the growth of LS domains, and the reverse is observed

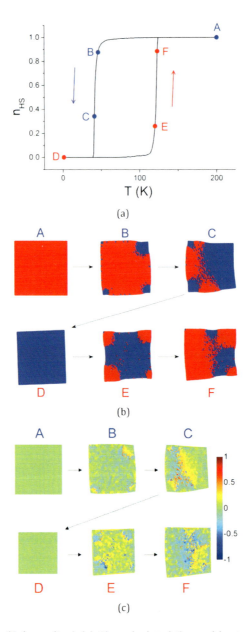

Figure 9.17 (Color online) (a) The calculated thermal hysteresis loop for $A_0 \rho_1^2 = 220$ K. The corresponding snapshots (b) and pressure map (c) of the system at positions A to F along the hysteresis loop.

(mutatis mutandis) along the heating branch. The onset of such inhomogeneous structures was reported during the domain wall propagation observed by optical microscopy [58].

We can explain why nucleation in the model starts from the corners, by simple energetic considerations. Let us start from a lattice in a saturated HS state and consider a nucleus made of a single LS site. The energy cost associated with the creation of a LS nucleus at the corner, edge, center of the HS lattice is written:

$$\Delta E_{\text{corner}} = E(LS) - E(HS) = -(\Delta - k_{\text{B}} T \ln g) + 4(A_0 + B_0)\rho_1^2$$
$$(9.23)$$

$$\Delta E_{\text{edge}} = -(\Delta - k_{\text{B}} T \ln g) + 2(3A_0 + 4B_0)\rho_1^2 \qquad (9.24)$$

$$\Delta E_{\text{center}} = -(\Delta - k_{\text{B}} T \ln g) + 4(2A_0 + 4B_0)\rho_1^2 \qquad (9.25)$$

The first contribution to Eqs. 9.23–9.25, that is, the temperature-dependant field, is negative due to the situation of the system in the metastable state. The second term is positive, and it is obviously minimized at the corner position.

On heating, similar considerations can be developed, with the only difference that larger thermal fluctuations may increase a little bit the (low) probabilities of nucleation at the center or edges, with respect to the center position.

In addition, we calculated the total elastic energy of the HS lattice with a LS domain embedded at the corner, at the edge, and in the center. We show in Fig. 9.18 how the energy relaxes in these cases. Clearly, the elastic energy is best relaxed in the corner case [80]. This should make further growth easier. The relaxation process of the elastic energy is an important issue which will be detailed separately.

9.5.1 Derivation of the Elastic Stress from the Displacement Field

Stresses generated by the nucleation and growth process were found of prior importance in the experiments. The elastic stresses at the various positions (A–F) of Fig. 9.17 can be determined from the knowledge of the displacement field $\vec{u}(i_x, i_y)$ associated with the lattice site (i_x, i_y), defined as

$$\vec{u}(i_x, i_y) = \vec{r}(i_x, i_y) - \vec{r}_0(i_x, i_y), \qquad (9.26)$$

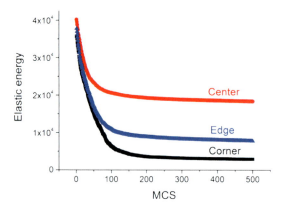

Figure 9.18 (Color online): Spontaneous evolution of the elastic energy of a LS lattice through mechanical relaxation when one node of the lattice has been set in the HS state, in a corner (black curve), edge (blue curve), and center (red curve). The time scale is expressed in Monte Carlo steps, and calculation was performed at low temperature (1 K).

where $\vec{r}_0(i_x, i_y)$ and $\vec{r}(i_x, i_y)$ are the initial and final atomic positions of the site (i_x, i_y). Here, we used the positions of the HS state as a reference, i.e.,

$$\vec{r}_0(i_x, i_y) = \left(i_x R_0^{HH}, i_y R_0^{HH}\right). \tag{9.27}$$

The deformations are deduced in the continuum limit, as follows:

$$\frac{\partial u_x}{\partial x} = \frac{u_x(i_x + 1, i_y) - u_x(i_x, i_y)}{R_0^{HH}}, \tag{9.28}$$

where $\frac{\partial u_y}{\partial y}, \frac{\partial u_x}{\partial y}$ and $\frac{\partial u_y}{\partial x}$ are also given in the same way.

We have calculated the displacement field and its spatial distribution (see Fig. 9.19a), at the positions A–F along the hysteresis loop of Fig. 9.17, from which we derived the spatial distribution of the strain tensors, ϵ_{ij}. Here, ϵ_{xx} and ϵ_{yy} bring information about the local relative volume change and the last two terms ϵ_{xy} and ϵ_{yx} connect to the pure shear strains. The "dilatation" and "distortion" fields, which respectively describe the pure relative volume expansion and shear stress, are given by

$$div\,[\vec{u}\,(\vec{r})] = \epsilon_{xx}\,(\vec{r}) + \epsilon_{yy}\,(\vec{r}), \text{ and } \vec{rot}\,[\vec{u}\,(\vec{r})] = \epsilon_{xy}\,(\vec{r}) - \epsilon_{yx}\,(\vec{r}). \tag{9.29}$$

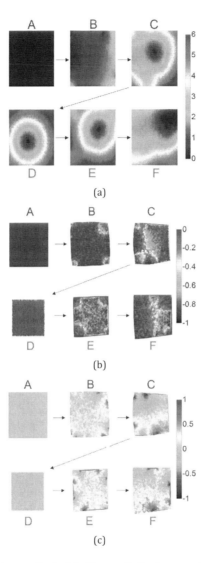

Figure 9.19 (Color online) (a) Maps of displacement field at several positions along the hysteresis loop of Fig. 9.17. (b) The corresponding maps of divergence of displacement field. The reference state is the HS phase, thus leading to a negative value of the divergence, which reaches its maximum value in the saturated LS state. (c) The corresponding maps of rotational of displacement field. The maximum values are observed at the edges of the system and/or at the interfaces between the LS and the HS domains, as depicted in images B, C, E, and F. They outline the presence of shear strains.

Their spatial distributions are given in Fig. 9.19. An evident correlation between the dilatation maps of Fig. 9.19 and the spin configurations of Fig. 9.17 is found, although the maps of the cooling and the heating branches clearly differ, due to the choice of the HS state as the reference state. $\text{div}[\vec{u}\,(\vec{r})]$ starts from zero in the HS state (Fig. 9.19b) and turns negative in the cooling branch, due to the shrinking of the system. The detailed inspection of the maps shows that the latter is mainly concentrated around the corners (nucleation regions) of the lattice, while it is weak at the center and on the edges. The rotational fields (see Fig. 9.19c) present a different feature. An enhancement of the shear stresses is observed at the intersects of the HS–LS interface and the edges of the lattice. These regions constitute brittle points at which dislocations and/or fractures may be initiated.

9.6 Spatio-Temporal Aspects of the HS to LS Relaxation Process: Case of the Square Lattice

We aim here to study the propagation of the HS–LS interfaces in isothermal conditions. For that, we prepare initially the system in the HS state (all spins are set as $S = +1$ and all atomic distances R_{ij} are set equal to $R_{ij} = R_{\text{HH}}$. The temperature is 10 K and the Monte Carlo procedure for relaxing the system is exactly the same as that described in the previous sections.

The present metastable state can be obtained by a rapid thermal quenching [112, 113] of the HS state at low temperature, although we do include in the preparation of the initial state, the fluctuations of positions. The parameter values are those of the thermal hysteresis loop illustrated in Fig. 9.15. It is well known that elastic effects generate long-range interactions. For example, a classical behavior has been observed in the gas-liquid transition of hydrogen in metals [104], in structural phase transitions [105], and in ferroelectrics [106] long time ago. The theoretical side of the problem was considered in [107], and other pioneering works [108]. It is also quite well established that an elastic interaction between dipolar elastic defects [107] (i.e., localized stresses) decays as r_{ij}^{-3}. However, the defect-defect interaction is strongly modified

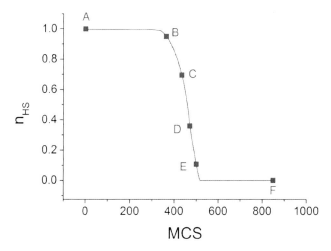

Figure 9.20 (Color online) Evolution of the HS fraction as a function of MCS upon the relaxation of a trapped HS state at 1 K.

by the shape of the lattice and by finite size effects [110]. Much more recently, Miyashita et al. [87, 102] have demonstrated that models including spin-distortion interactions belong to the mean-field universality class. Moreover, recent investigations on circular lattices have evidenced the macroscopic character of the nucleation and growth phenomena involved in this class of models [53], as well as the importance of shape effects.

The study of the stochastic relaxation from HS to LS states for lattices having different shape is then an interesting issue. The relaxation curve, showing the time dependence of the HS fraction is presented in Fig. 9.20. Its variation shows a clear existence of an incubation time (∼400 MCS) during which the HS fraction remains constant, but the elastic lattice not. We have followed the spatio-temporal properties of the relaxation process, through by storing the electro-elastic configurations (electronic states and the atomic positions) over the course of the relaxation process. The configurations at points A, B, C, D, E, and F are shown in Fig. 9.21. An evidence of single macroscopic domain nucleation is obtained in this square lattice. This result contrasts with the available data in literature which show nucleations from the four corners of the square. This is also obtained with this model when changing the size

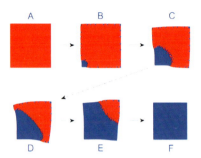

Figure 9.21 (Color online) Snapshots of the lattice along the relaxation curve of Fig. 9.20.

or the elastic constants, as already presented in a previous work [56]. The single domain nucleation in a perfect symmetric system must obey to the subtle balance between the nucleation time (time necessary to the appearance of a single droplet of LS phase) and the time scale, $\tau = L/v$ of the propagation of the macroscopic domain over the crystal of size L at the velocity v.

The first process (from A to B) is clearly stochastic and the escape from the metastable state is dominated by rare events, localized at the surface of the lattice and preferentially at the corners where the energy cost for the formation of the HS nucleus is the lowest. The size of the critical droplet depends on the geometry of the lattice (and so on its size), which is different from the nucleation process in short-range interaction systems, for which the size of critical nuclei does not scale with the system size. This is the concept of macroscopic nucleation introduced in [53]. The second (from B to E) is purely deterministic and thus independent of the system size. In the case where the timescale of the second process is much more smaller than that of the stochastic process, it is remarked that the single domain nucleation dominates the system behavior, as shown in Fig. 9.21. However, as one can see in snapshot B of Fig. 9.21, the nucleation began at several sites at the lattice border (surface), but once the left bottom corner grew the process was stopped for the other nucleus, due to the propagation of the strain field, induced by the macroscopic deformation of the lattice, whose shape accompanies the propagation of the HS–LS interface. Indeed, the lattice volume (here surface) decreases during the relaxation

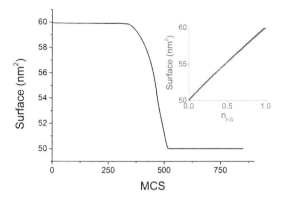

Figure 9.22 (Color online) Distribution of the elastic energy corresponding to the snapshots shown in Fig. 9.20.

process as clearly indicated by Fig. 9.22, in which we have also inserted the correlation volume-HS fraction.

It is of particular relevance to focus as well on the spatio-temporal aspects involved in the latter transformation. Indeed, as stated previously the relaxation of the trapped HS sate was done at low temperature 1 K where the elastic properties clearly stand out since the thermal fluctuation are significantly reduced unlike the thermal spin transition reported earlier. Figure 9.23 shows the distribution of the elastic energy over the lattice at different instants along the relaxation process. By taking the snapshots of Fig. 9.21 as reference, it is easily seen that the distribution of the elastic energy peaks around the HS/LS interface and vanishes far away. On the other hand, one can as well observe that such distribution has more extend in the center of the lattice unlike around the edge where the elastic energy becomes more localized. Such fact is assigned to the surface effect which result a quick relaxation of the excess in the elastic energy unlike the central part of the lattice where the relaxation extends over some lattice parameters. Additionally, we judged it will be interesting to report here as well on the "dilatation" and "distortion" fields shown in Fig. 9.24. These results agree well with the earlier observations and stress more clearly the relative volume change during the transition.

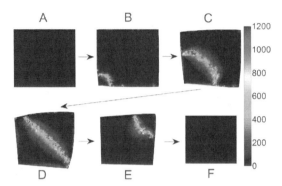

Figure 9.23 (Color online) Distribution of the elastic energy corresponding to the snapshots shown in Fig. 9.20.

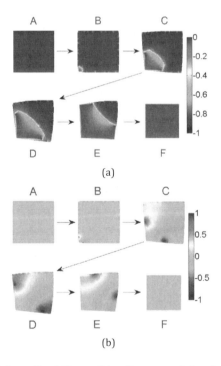

Figure 9.24 (Color online) Maps of the divergence (a) and rotational (b) of the displacement field derived from the snapshots of Fig. 9.20.

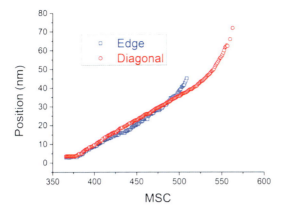

Figure 9.25 (Color online) Position of the HS/LS interface derived from the snapshots of Fig. 9.20 as a function of the MCS along the edge (blue) and the diagonal (red) of the lattice.

9.6.1 The Front Shape and the Interface Velocity

We could follow the interface position on time along several directions, as shown in Fig. 9.25. Here we selected two directions corresponding to the edge of the lattice and the diagonal. Obviously, the square shape of the lattice imposes very close velocity values along the two edges. Indeed, the front has a curved shape but it is not circular due to the global deformation of the lattice which shrinks upon relaxation. During the growth of the LS phase, the curvature the front changes when the transformation along the edge reaches the extremities of the lattice. This behavior is well evidenced on the snapshots C and E of the divergence of the displacement field, which also expresses the spatial distribution of the relative volume expansion. To understand the driving force of this curvature change, we have calculated, at each time during the relaxation process, the interface length, and the associated elastic energy, as well as the HS fraction. Figure 9.26 shows that the elastic energy, E_{elas} scales linearly with the interface length, Γ, following the law $\Gamma \simeq \frac{3}{4} E_{elas}$. This linear relation proves that (i) most of the elastic energy of the system is stored in thin region around the HS–LS interface and (ii) the interface width does not vary significantly during the movement,

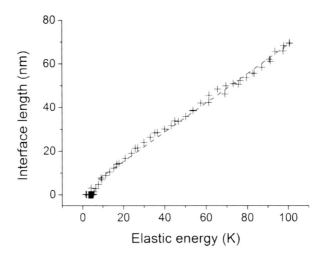

Figure 9.26 Black crosses: Interface length as a function of the total elastic energy associated with the relaxation of Fig. 9.20, showing a linear dependence with the density of the elastic energy, $E_{elas}/\Gamma \sim 3/4$. As a result, $E_{elas} \propto L$, where L is the lattice size, this confirming the concept of macroscopic nucleation.

which is also in good agreement with the constant propagation velocity of the front.

To investigate the effect of the lattice geometry on the interface shape, we consider an analytical interface (radial) for a comparative purpose with that derived from above MC simulations. The radial interface (not shown here) is represented by a portion of a disk of a given radius ρ, centered in the nucleation point (left bottom corner) intersecting the lattice's edge with a contact angle of $\pi/2$, except for a small region around $\rho \sim L$, where the contact angle was maintained equal to $\pi/4$, because the interface is straight. The former assumption means that the projection of the gradient of the HS fraction on the lattice boundary vanishes, which corresponds to the well known Neumann boundary conditions.[a] Obviously such assumption does not take into account the edge deformation during

---

[a] The Neumann boundary condition is one of the three types of boundary conditions commonly encountered in the solution of partial differential equations. It specifies the normal derivative of the function, U, for example, on a surface. It writes, $\vec{n}.\vec{\nabla}U = f(\vec{r}, t)$, where $\vec{n}$ is the normal vector to the surface at $\vec{r}$.

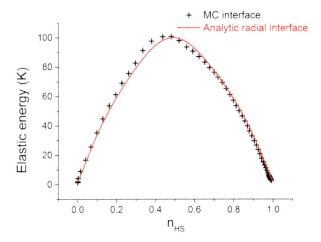

Figure 9.27 Total elastic energy (black crosses) versus HS fraction corresponding to the relaxation of the thermally quenched HS fraction of Fig. 9.20, resulting from MC simulations. The maximum of the elastic energy corresponds to $n_{HS} = 1/2$, and reveals that the elastic energy connects directly to the length of the interface, as proved in Fig. 9.26. The red line is the predicted elastic energy calculated for a curved interface constructed as the shortest trajectory between two points on the border of the lattice satisfying the condition of a wetting angle equal to $\pi/2$. A good agreement is obtained with the simulation.

the front propagation. The MC simulations have been performed on these analytic interfaces as follows: (i) a LS domain with radial shape is formed at the border of the HS lattice and (ii) the system is relaxed mechanically with locking the spin variables of the nodes. To examine the energy distribution and analyze the shape variation of the interface, we monitor the evolution of the relaxed elastic energy, E_{elas}, of the fully relaxed lattice as function of n_{HS} for different LS domain sizes. An excellent agreement is found between MC simulations performed with analytic interfaces at fixed spin configuration and the full MC simulations including spin and lattice positions, as illustrated in Fig. 9.27.

To make the discussion more substantial, we should mention at this stage that the elastic energy of the MC data of Fig. 9.27 has been corrected from elastic energy contribution of the LS atoms located at the surface of the lattice. Indeed, during the relaxation process of the

HS fraction (see Fig. 9.21), the HS surface atoms switch very early to the LS state, due to their smaller elastic energy barrier, compared to that of bulky atoms. Thus, the HS fraction and the total elastic energy must be corrected from this residual contribution, which is expected to be negligible for large size lattices. One can also see that the two set of data of Fig. 9.27 are almost superimposed, which proves the consistency of our geometric analytical description of the interface properties. The linear relation between the elastic energy and the interface length means that most of the elastic energy induced by the lattice parameter misfit is stored in a thick stripe around the interface region. Consequently, *at fixed HS fraction, the most stable interface is the shortest one which respects the boundary conditions, which are reflected here by the fixed contact angle.* So, finally the present problem is reduced to an interface optimization under the constraints imposed by the presence of a free surface (boundary).

Let us try to have a comprehensive view of this problem. Let us neglect the lattice distortion at the surface, and then we keep the lattice shape invariant. This assumption is somewhat true at the macroscopic scale (or for large size systems). If we assume that the system reaches its stable mechanical state in a short time scale, then the interface dynamics follows instantaneously the high-spin fraction motion. As a result, for each HS fraction value (that is the surface of the blue area in Fig. 9.21, for example) the system optimizes the shortest interface in order to minimize the interfacial energy. This problem is somehow similar (up lattice distortions) of interfacial energy in soap bubbles. Suppose A is the surface of the growing new phase (the LS fraction in the HS to LS relaxation) and Γ, the length of the interface. The problem is then to find the best trajectory $y(x)$ (called also the stationary solution) which minimizes Γ, that is

$$\frac{\partial \Gamma}{\partial [y(x)]} = \frac{\partial}{\partial [y(x)]} \int_A^B \sqrt{dx^2 + dy^2} = 0 \qquad (9.30)$$

with the constraint,

$$\iint_\Omega dx\,dy = A. \qquad (9.31)$$

The physical problem turns into a variational problem of mathematics which can be solved either analytically (using Lagrange multiplier tests) or numerically by Monte Carlo method. It is interesting to mention that except the constraints imposed on the contour Ω of the lattice and the surface A, the present resolution of this problem will produce the best wetting angle of the interface at the lattice border. We are currently working on this problem using different lattice geometries to test the validity of the arguments developed here, as well as their limits.

9.7 Lattice Configurations upon the SC Transition of the Rectangular Lattice

Since the nucleation, growth, and propagations processes involved in the first-order transformation and/or in the relaxation between the LS and the HS states have macroscopic features and depend on the shape of the material, it is legitimate to investigate different kind of lattice shapes. Here, we study the case of a 2D lattice with a square symmetry but having a rectangular shape. The lattice has the dimensions $L_X = 70$ and $L_Y = 20$, and the parameter values are exactly the same as those of the square lattice, studied before. The thermal dependence of the HS fraction (see Fig. 9.28(a)) shows that the transition temperature is still around 90 K, but the hysteresis width is of $\sim$80 K. The lattice configuration corresponding to different temperatures in the heating and cooling modes, as well incorporated along the thermal hysteresis, are shown in Fig. 9.28(b), where the blue and red parts represent the LS and HS phases, respectively. The snapshots show that on cooling the LS phase started from three corners (snapshot 39 K), with the existence of a clear HS/LS interface propagating from left to right. On heating, however, the HS phase appeared around the four corners of the lattice leading to the formation of two fronts propagating in opposite ways. These different behaviors are mainly due to a significant contribution of thermal fluctuations at upper temperature of the hysteresis loop, and also because the nucleation of a phase with smaller volume into a phase with bigger volume and inversely are not equivalent from the elastic point of view. The latter aspect is not yet developed in the elastic models devoted to the SC phenomenon though it is of particular relevance to well

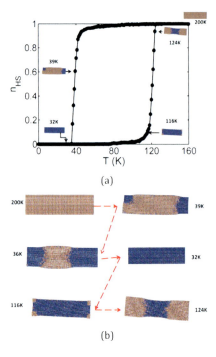

Figure 9.28 (a) Computed thermal dependence of the HS fraction on a 2D rectangular lattice with square symmetry, showing a first-order transition accompanied with a hysteresis loop of 83 K width. (b) Snapshots of electro-elastic state of the lattice showing the domain structures along the cooling and heating branches of the thermal hysteresis loop. The lattice size is $L_X \times L_Y = 70 \times 20$ and the temperature sweep rate was 0.1 K/MCS. The parameter values are $g = 150$ for the degeneracy ratio between the HS and the LS states, the harmonic elastic constant $A_0 = 22000$ K/nm^2 (resp. $B_0 = 0.3 \times 22000$ K/nm^2) for the nn (resp. nnn), $\Delta = 450$ K, and the angle between the unit cell parameters $\vec{a}$ and $\vec{b}$ is $\theta = \pi/2$.

understand the mechanical properties involved in the transition such as the deformations, cracks, elasticity, etc. Indeed, the current elastonic model investigated here has proven its ability to exhibit some interesting mechanical features that arise during the spin transition such as remarkable deformations of the lattice in both branches of the hysteresis loop which are enhanced by the finite size effects.

9.7.1 Relaxation of the Metastable HS Fraction at Low-Temperature

To avoid the thermal fluctuations on the interface dynamic, we study the relaxation of the HS state at low temperature (10 K) for the rectangular lattice. Initially, the system was set in the HS state from the electronic (all spins are +1) and the elastic (all nn distances are equal to R_{HH}) point view. Such state is metastable since at low temperature, the ground state is the LS state (see Fig. 9.28(a)) and thus the HS state which is thermodynamically unstable at low temperature will be forced to relax. The corresponding relaxation curve is shown in Fig. 9.29 and presents typical features of sigmoidal relaxation with the existence of an incubation time ([0, 300] MCS) and a deterministic regime ([300, 360] MCS) of rapid change of the HS fraction. Let us now examine in details the pattern formation during this isothermal relaxation. Snapshots of electro-elastic configurations associated to points A–F of Fig. 9.29 are presented in Fig. 9.30, where the blue (red) part denotes the LS (HS) phase. The nucleation of LS domains takes place at the corner of the lattice and spread over the whole system which agree well with the experimental optical microscopy observations. Such behavior evidences the effect of the elastic interactions which generate long-range strains and hinder the nucleation of the LS state from other

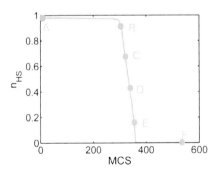

Figure 9.29 Computed thermal dependence of the HS fraction on a 2D rectangular lattice with square symmetry, showing the relaxation from HS to LS states at 10 K. The parameter values are those of Fig. 9.28(a). The stochastic (incubation) and the deterministic (unstable) regimes are easily identified.

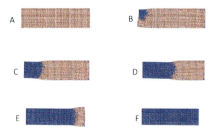

Figure 9.30 (Color online) Snapshots of the rectangular lattice at different instants ($A \rightarrow F$) along the relaxation process from (HS $\rightarrow$ LS) shown in Fig. 9.29.

parts of the lattice. Furthermore, it should be noted that the interface shape drastically changes during the relaxation process; one can see that briefly after the onset of the LS domain, the HS–LS interface becomes circular, and this shape is maintained as long as its radius, ρ, is smaller than the width, L_y, of the lattice. This circular shape of the interface results from the minimization of its length (the interface energy cost) under the constraint of a contact angle of $\sim \pi/2$, imposed by the open boundary conditions of the problem. Afterward, once $\rho = L_y$, the interface changes gradually through the propagation. After a while, $t \simeq 350$ MCS, the interface tends to a stable shape, almost straight and perpendicular to the lattice side borders. It is worth to notice that while these results are new in the spin-crossover topic, similar behaviors of nucleation and propagation have been already reported in literature of magnetic materials [114, 115], where the propagation of the magnetic domain wall is driven by an applied magnetic field. However, the present situation is much more similar to the nucleation and propagation of domains in ferroelectric systems in the vicinity of the thermal instabilities caused by the presence of first-order phase transition.

9.7.2 On the Interface Propagation

We now focus on the interface travelling along the x axis, which can be followed by selecting a typical atomic line, for example at $y = L_Y/2$, and by plotting the position of its center with MC time. Two types of interfaces exist, namely an electronic and an elastic interface. The electronic interface is defined by the spatial profile of the HS fraction along the chosen atomic line by plotting the $n_{HS}(x)$-dependence for different MC time, t, values. The center of the interface is then located at $n_{HS} = 1/2$, for which the fictitious magnetization, $\langle S \rangle = 0$. Due to the discrete nature of the model, the spin state jumps at the HS–LS interface; however, a continuous electronic interface can be obtained by averaging over several lines or over a set of MC simulations using different sequences of random numbers, as it was done in [116].

It is important to mention here that the HS–LS interface can be characterized from both electronic $n_{HS}(x, y)$ and elastic $d_{nn}(x, y)$ point of view. Here, the electronic interface is sharp due to the discrete nature of the spin states $S = +1, -1$ associated with the LS and the HS states. However, in optical microscopy experiments, for example, the resolution is about 0.3 µm, which means that the HS fraction is the spatially averaged over a cell containing millions of atoms. Indeed, the size of a unit cell is about 1 nm and thus the number of atoms present in one pixel is about 0.3 µm^2/1 nm$^2 \sim 10^5$. Consequently, in a rigorous approach, to avoid this sharp electronic interface resulting from the discrete nature of the lattice, the interface should be calculated by averaging over a number of Monte Carlo simulations using different seeds. This statistical average will smooth out the curve points.

Let us now define the elastic interface. The latter can be imaged by the evaluation of the change in the neighboring distances between successive atoms within an atomic line, along the direction of propagation. Let us denote by $d_{xx}(x, y)$ the x-y dependence of the nn distances, given by

$$d_{xx}(x, y) = \sqrt{(x(i + 1, j) - x(i, j))^2 + (y(i + 1, j) - y(i, j))^2}.$$
(9.32)

This quantity is equal to R_{LL} in the LS phase and R_{HH} in the HS phase and should change continuously in the interface region between these two values. The latter can be imaged by evaluating the change in the neighboring distances between successive atoms along the direction of propagation. Figure 9.31a shows the elastic interface derived from the analysis of the central atomic line corresponding to $j = L_Y/2 = 10$ at different instants during its propagation.

The elastic interface appears a little bit wider, due to the fact that the system needs several unit cells to accommodate the lattice parameter misfit between the LS and the HS phases. That is the excess of energy due to the presence of the interface, compared to that of the homogeneous system.

The 3D structure of the elastic interface, that is $d_{xx}(x, y)$, presented in Fig. 9.31b, resembles quite well to the experimental interface in Fig. 9.8. In the former figure, important fluctuations are easily remarked and are attributed to the stochastic nature of the MC simulation which allows existence of LS atoms at the border of the lattice, as clearly depicted in the snapshots of Fig. 9.30. The presence of these fluctuations specially at the border and the corner (and not in the center of the elastic lattice) are obviously due to the associated elastic energy. Indeed, the surface of the lattice permits a quick release of the extra cost in the elastic energy induced by the misfit in the lattice parameter between the HS and the LS states. At the border and the corner, the atoms have less neighboring sites, which favors their spin flip. An inspection of the simulated interface profiles of Fig. 9.31, are also a qualitative good agreement with optical microscopy observations [58, 59, 61].

We followed the interface position along $y = L_Y/2$, corresponding to the center of the lattice during the propagation process of Fig. 9.30 and results are summarized in Fig. 9.32. The propagation consists of two regimes: (i) a flow regime ([318–350] MCS) and (ii) an accelerated regime ([350–360] MCS). Such findings agree well with the available experimental data [57–59, 61–65]. The rather variational character of the velocity along the transformation process has to be discussed on the basis of the lattice shape. Indeed, the deterministic regime is attributed to the propagation of the interface along the longest edge of the rectangular lattice denoted here as L_X. Such regime is characterized by a constant velocity due to the uniformity of the central part of the rectangular lattice without geometrical fluctuations. While the drastic change in the position of the interface as a function of the MC time at the end of the relaxation process reveals an accelerated propagation and is assigned to the surface effect which allows an easy motion of the atoms unlike those in the central part of the lattice. These observations stress how the shape and the propagation velocity of the transformation

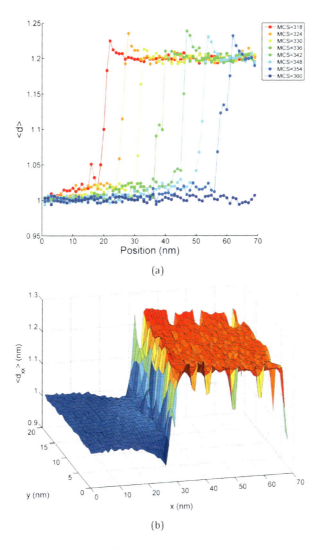

Figure 9.31 (a) Elastic profiles along the propagation direction (x axis) for the atomic line $j = L_Y/2 = 10$, plotted at different instants along the relaxation process. (b) 3D structure of the elastic HS–LS interface for $t = 330$ MCS. The blue area corresponds to the LS phase and the red part to the HS phase.

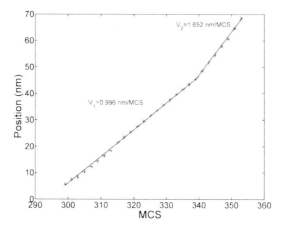

Figure 9.32 Time dependence of the HS:LS interface position along the propagation direction (x axis) for the reference atomic line $y = L_Y/2$ showing a constant velocity propagation in the flow regime. The acceleration observed at the end of the process is due to the surface effects.

front are critically dependent to the boundary conditions. Let us now discuss the relation between the interface length and shape and the elastic energy of the system. We have stated in the previous sections that at fixed HS fraction, the most stable interface shape is the one which minimizes the total elastic energy. This general behavior is, however, strongly dependent on the shape of the material, due to the long-range nature of the elastic interactions. It is then expected that in the case of a rectangular lattice, the maintaining of the interface length in the flow regime of propagation (that is the region where the interface is traveling at constant velocity) will result in an almost constant elastic energy. Figure 9.33 confirms very well this statement. Indeed, in the region of $n_{HS} \in [0.1, 0.9]$ the interface length (Fig. 9.33a) and the elastic energy (Fig. 9.33b) are almost constant up to fluctuations. In this region, the interface has more or less reached it stable shape which is maintained invariant in this HS fraction interval. We have as well checked that the elastic energy of the interface is proportional to the interface length. Such linear relation means that most of the elastic energy induced by the lattice parameter misfit is stored in a thick stripe around the interface region in very good agreement with the prior findings.

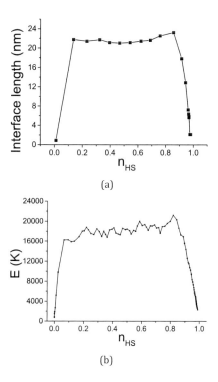

Figure 9.33 (a) Interface length Γ versus the HS fraction during the propagation process. (b) Associated elastic energy E_{elas} versus the HS fraction. The two curves show the existence of three regimes. In the first one, E_{elas} and Γ increasing with n_{HS} and correspond to the transverse propagation of the front. In the second regime E_{elas} and Γ are more or less constant, which corresponds to the flow regime (a constant velocity in Fig. 9.32). In third regime the interface accelerates due to border effects. Time is a hidden variable in this representation.

Consequently, at fixed HS fraction, the most stable interface is the shortest one which respects the boundary conditions, which are reflected here by a contact angle $\sim\pi/2$. So as in the previous case, the present problem is reduced to an interface optimization under the constraints imposed by the presence of a free surface (boundary). We should mention that the existence of different interface shapes is a signature of a nonmonotonous propagation velocity. It is interesting to recall that; usually experiments report that the propagation of the HS/LS interface takes place at constant velocity. However, the latter result is appropriate to plane waves or to circular waves that have grown sufficiently large to have a low curvature. For the initial development of radial interfaces, there is generally a high curvature. In this case, if the elastic interaction is too strong, the formation of an interface may fail as the perturbation will not reach the threshold value required for the electronic energy to overcome the elastic energy barrier. The relationship between the velocity of a planar interface wave, c_0, and that of a curved wave, c, can be approximated by the eikonal equation

$$c = c_0 - \frac{D}{r},\qquad(9.33)$$

where D is a kind of effective diffusion coefficient. If the curvature (resp. the radius) becomes too high (resp. too low), the interface velocity may vanish indicating a propagation failure. Then a critical radius for propagation initiation exists and is obtained by setting $c = 0$ in Eq. 9.33, giving $r_{crit} = D/c_0$.

9.8 Relation between the Interface Orientation and Lattice Symmetry

Up to now, we have only investigated the interface dynamic of lattices having a square symmetry associated with isotropic volume expansion upon the spin transition from LS to HS. For the sake of completeness, this section contains, therefore, a different investigation considering the effect of the lattice symmetry on the orientation of the interface. For that, the unit cell in the current study is set as a parallelogram of angle $\theta = \pi/3$. The lattice size is $L_x \times L_Y$ and the transition from LS to HS is accompanied with an expansion of the lattice parameters from $R_{LL} = 1$ nm to $R_{HH} = 1.2$ nm. Similarly as reported previously, the interactions between the molecules in the lattice are introduced through springs linking the sites. Each molecule has four nearest neighbors and four next nearest neighbors as illustrated in Fig. 9.34. Here also, we consider only lattice deformations inside the plane. The Hamiltonian is also written so as to maintain the symmetry between the LS and the HS

Relation between the Interface Orientation and Lattice Symmetry | 389

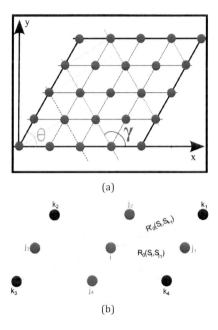

Figure 9.34 (a) Schematic representation of the lattice symmetry. (b) Topology of the bonds of the studied 2D lattice. Each site i interacts with its nearest neighbors (nn) j and next nearest neighbors (nnn) k. The angle θ between two nn sites can be either $\pi/3$ or $2\pi/3$, depending on the direction. The equilibrium bond length between nn (nnn) is denoted $R_0(S_i, S_j)$ ($R'_0(S_i, S_k)$) and depends on the spin states of the linked atoms.

states:

$$H = H_{\text{elec}} + H_{\text{elas}} \tag{9.34}$$

where H_{elec} is the electronic contribution:

$$H_{\text{elec}} = \sum_i \frac{1}{2} (\Delta - k_B T \ln(g)) S_i \tag{9.35}$$

and H_{elas} accounts for the elastic part:

$$H_{\text{elas}} = \sum_{i-j} \frac{A_{ij}(r_{ij})(r_{ij} - R_0(S_i, S_j))^2}{2}$$

$$+ \sum_{i-k} \frac{B_{ik}(r_{ik})(r_{ik} - R'_0(S_i, S_k))^2}{2}. \tag{9.36}$$

with

$$R_0(S_i, S_j) = \rho_0 + \rho_1(S_i + S_j) + \rho_2 S_i S_j \qquad (9.37)$$

and

$$R'_0(S_i, S_k) = R_0(S_i, S_k)cos(\theta), \qquad (9.38)$$

where θ is the angle connecting three nn sites like i, j_1, and j_2 in Fig. 9.34b.

The four nn sites are equivalent and their equilibrium distances $R_0(S_i, S_j)$ are equal to R_{LL} (R_{HH}) for LS–LS (HS–HS) configurations. For the HS–LS configuration, the equilibrium distance is $R_{HL} = \dfrac{R_{HH} + R_{LL}}{2}$. For the nnn sites, the equilibrium distances $R'_0(S_i, S_k)$ depend on the direction due to the parallelogram unit cell with two different diagonal length's. Indeed, in the present case, the equilibrium distances are longer in one direction than in the other one. In the case of a HS lattice for example, these distances are estimated to be $2R_{HH}$ and R_{HH}. Therefore, contrary to the square symmetry lattices, where all diagonal directions are equivalent, here the symmetry is broken along the two diagonals. As will be developed later such fact is of important consequence on the orientation of the interface. The other parameters are $R_{HH} = 1.2$ nm, $R_{LL} = 1$ nm, $R_{HL} = 1.1$ nm, and the harmonic elastic constants for nn (nnn) neighbors are $A_0 = 22000$ K ($B_0 = 10.0 \times A_0$), while the anharmonic contributions are $A_1 = 0.28 \times A_0$ for nn and $B_1 = 0.28 \times A_1$ for nnn neighbors. The calculations were performed using the MC method described above, performed on spin and lattice positions.

9.8.1 Thermal Properties and Isothermal Relaxation at Low Temperature

First, we investigated the thermal properties of the Hamiltonian 9.34. The corresponding results are illustrated in Fig. 9.35a and show a wide thermally induced hysteresis loop associated to a first-order transition. Figure 9.35b shows the snapshots of the spatial distribution of the HS fraction along the cooling and heating branches of the thermal hysteresis. First, in both processes the stochastic transformation starts around the acute angles of the

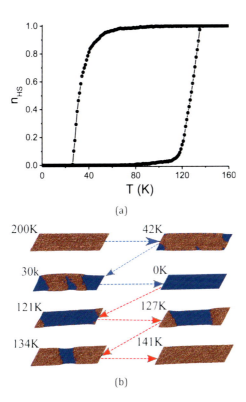

Figure 9.35 (a) Computed thermal dependence of the HS fraction on the lattice with the topology presented in Fig. 9.34. (b) Snapshots of the spatial distribution of the HS fraction at selected temperature along the cooling and the heating branches of the hysteresis loop. The blue (red) area represents the LS (HS) phase.

lattice and takes place through nucleation, growth and propagation of macroscopic domains accompanied with large deformations of the lattice. It is remarkable that all the HS–LS interfaces are oriented with an angle $\gamma = 2\pi/3$, which indicates that this orientation seems to be the stable one. To investigate more closely the dynamic of the interface in the present lattice and in order to better understand its orientation, it is preferable to reduce the thermal fluctuations through. For that, we propose to investigate the relaxation process of the metastable HS state trapped at low temperature. Such study is the main idea of the next subsection.

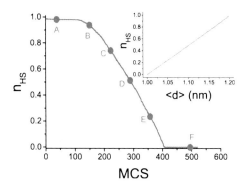

Figure 9.36 Time dependence of the HS fraction, n_{HS}, during the relaxation from HS to LS at 10 K. The sigmoidal shape of the curve clearly denotes the existence of an incubation period $t \in [0, 150]$ MCS, followed by a deterministic regime. Inset: correlation between the average nn distance, $\langle d \rangle$ and the HS fraction showing a linear relation between these two quantities.

9.8.2 Isothermal Relaxation of the Metastable HS Fraction at Low Temperature

Initially, the lattice was prepared in the full HS state, i.e., all spin values are equal to $+1$ and all nearest-neighbor distances are equal to R_{HH}. The relaxation is performed stochastically using the same MC procedure already presented in the previous sections. The goal of the study here is to confirm the influence of the lattice symmetry on the orientation of the propagation front.

The relaxation of the HS fraction is presented in Fig. 9.36 and shows a typical relaxation corresponding to a cooperative system. Along this relaxation, one can notice the presence of an incubation period resulting from the stochastic regime during which the HS fraction becomes invariant ($n_{HS} \simeq 1$) followed by a deterministic regime in which both spin states and atomic positions evolve in time.

The fact that the average value of the nn distances follow linearly the HS fraction (see inset of Fig. 9.36) means that for each spin configuration, the lattice was relaxed until reach the mechanical equilibrium. In this situation, the lattice is slaved to the electronic state, which means that the timescale of the lattice relaxation is much more shorter than that of the spin system. Of course, one

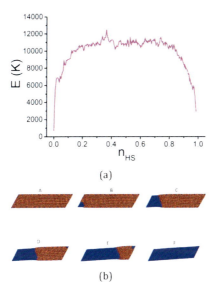

Figure 9.37 (a) Elastic energy during the relaxation process of Fig. 9.36 as function of the HS fraction. Three regimes can be identified (see text for more explanations). (b) Snapshot of the spatio-temporal configurations of the lattice during the thermal relaxation of the metastable HS state (point A) at 10 K of Fig. 9.36. Red (blue) color represents the HS (LS) phase. The LS state appears at the corner of the lattice (panel B) and then propagates with a stable interface (panels C-E), forming an angle of $\sim 2\pi/3$ with the direction of propagation.)

can also study the situation where the relaxation times of these two degrees of freedom interfere. This interesting question poses specific problems related to the creation of lattice defects and will be addressed elsewhere.

We should mention that the different regimes of relaxation could be as well be deduced from the dependence of the HS fraction n_{HS} to the total elastic energy of the system during the relaxation process (see Fig. 9.37a). Here, three regimes can be identified; $n_{HS} < 0.2$ corresponds to the development of the interface, and $0.2 < n_{HS} < 0.8$ is the flow regime of propagation at stable orientation, where the elastic energy is almost constant. From the latter observation, we then deduce that in this region the interface travels at constant velocity. And, the third regime corresponds to $n_{HS} > 0.8$, where the elastic energy decreases quickly due to the fast relaxation at the surface of the lattice, and consequently, the interface is expected to accelerate. The corresponding snapshots of the lattice during the relaxation process are illustrated in Fig. 9.37b and show that the nucleation starts preferentially around the acute angles where the cost of the interfacial energy is the lowest. At the beginning, the transformation front has a circular shape which switches then to straight one with a stable orientation. Indeed, the macroscopic shape of the interface has a crucial importance, since when the nucleation starts from one of the acute angles, it determines the future orientation of the interface.

Figure 9.38a shows the interface profile at different instants during the relaxation process. As stated in the previous sections, the elastic profile is obtained by monitoring the distance between successive sites along the line $j = L_y/2$. The fact that the profiles are regularly spaced proves that the interface travels at constant velocity, estimated here as 0.25 nm/MCS. The same figure shows also the three dimensional structure of the elastic interface calculated using the same procedure as in Section 9.7. Additionally to the interface orientation, one can remark its sharpness as well as the difference in the amplitude of fluctuations between the LS (blue area) and the HS (red) phase. The latter observation is a consequence of the difference in the elastic properties of the LS and HS phases. Indeed, the HS state constitutes the soft phase and also the metastable and it is then much easily subject to fluctuation than the LS state. By comparing the current results with those of the rectangular lattice reported earlier, one main difference arises regarding the orientation of the interface. Here, the interface is not oriented perpendicularly to the long edge of the lattice, which means that the lattice symmetry plays a significant role in the interface orientation. In the beginning the interface is circular and propagates toward the shortest border of the lattice, that is in the transverse direction. Once the interface has reached the lattice border it re-orient itself and becomes a straight line forming a more or less stable angle of $2\pi/3$ with the propagation direction (snapshots C-E in Fig. 9.37b). At the end, the system relaxes back to the LS state (snapshot F). Interestingly, the system chooses the dense atomic

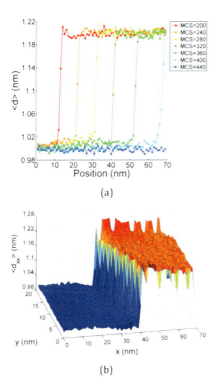

Figure 9.38 (a) 2D profile of the elastic interface at selected instants during the relaxation of Fig. 9.36. The figure shows a constant velocity travelling except at the end of the process. (b) Static view of the elastic interface corresponding to the point D on Fig. 9.36. $\langle d_{xx}\rangle$ is the distance between consecutive sites of the atomic chains along the x axis (the propagation direction). The blue area represents the LS state ($\langle d_{xx}\rangle \simeq 1$ nm) and the red one the HS state ($\langle d_{xx}\rangle \simeq 1.2$ nm).

plane as an interface, because it represents the best compromise for the interface length at fixed HS fraction. This optimization of the orientation front can be checked using a specific Monte Carlo simulation based on Kawasaki dynamics. For that, we prepare initially the system with a phase coexistence, with an interface oriented randomly and we allow atomic diffusion at fixed total spin state (conserved order parameter). The procedure aims to look for the best interface minimizing the total elastic energy of the system.

This method is very efficient to optimize orientation interfaces; unfortunately, its disadvantage is that it is very slow.

Let us now try to understand the origin of the present orientation of the interface. For that, we made different preparations of the system where the initial state was defined in a homogeneous structural state (HS) and biphasic spin state, so as to define straightforwardly the HS:LS interface. The MC simulations were performed on a lattice $L_x \times L_y = 70 \times 20$ at null temperature in a frozen spin configuration with free boundary conditions. The HS and LS states were defined on either side of a straight borderline passing through the center of the system with various orientations. The orientation angle is defined here by $\gamma = (\vec{L}_x, \vec{\Gamma})$, where $\vec{\Gamma}$ is a vector taken along the interface. For each of the orientations (i.e., γ value), we relaxed the elastic lattice until reach the stable mechanical state. Then we collected the total elastic energy and we draw it as a function of the orientation angle in Fig. 9.39. Here, for the sake of a comparison we show the results for the rectangular lattice ($\theta = \pi/2$) and for the present one ($\theta = \pi/3$). The curves of the elastic energy as a function of the orientation angle depict a single minimum in both cases. For the rectangular lattice this minimum is located at $\pi/2$ (see Fig. 9.39a), which corresponds to a most stable interface perpendicular to the long edge of the lattice. This result is in excellent agreement with the data reported above.

Figure 9.40 shows the elastic energy maps corresponding to some selected initial orientations of the electronic interface. Interestingly, long length interfaces show an important concentration of strain not only around the interface region but also inside the LS lattice. This behavior is normal since the initial state was an elastic HS state (all nn distances were equal to R_{HH}).

Moreover, for the lattice made of a parallelogram unit cell with an angle $\pi/3$, the stable position is obtained for $\gamma \simeq 2\pi/3$ (see Fig. 9.39b), which fairly agree with the simulation data, presented in Fig. 9.37b. It is then clear that this interface orientation is very stable and does not depend on the initial state. Let us now try to understand in a more deeper way why this orientation is the ones which minimizes the elastic energy of the lattice. Indeed, based on geometrical considerations of interface length, one can expect the existence of two equivalent interfaces whose angles are $\theta = \pi/3$ and

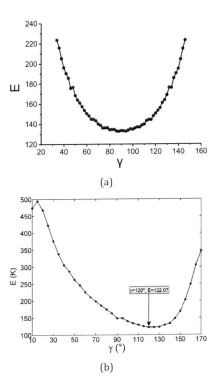

Figure 9.39 The relaxed elastic energy as a function of the interface orientation, after 100000 MCSs. (a) the case of the lattice with $\theta = \pi/2$ (rectangle) and (b) the case of $\theta = \pi/3$ (parallelogram). For the rectangular lattice the minimum of energy is for an interface oriented at $\pi/2$ with the long length, while this angle becomes $\sim 2\pi/3$ for the parallelogram.

$2\pi/3$ minimizing the elastic energy. Within the lattice symmetry of Fig. 9.34, these two orientations give the same interface length, and then, based on our discussion of Section 9.7, we expected to have the same elastic energy for these two orientations. However, Fig. 9.39b clearly reveals that these two situations are not symmetric. The origin of this asymmetry should be sought in the elastic energy stored in the system for these orientations.

Figure 9.41 shows the schematic view of the interface oriented with an angle $\gamma = \pi/3$. It is quite easy to calculate the total initial elastic energy stored in the interface region, due to the lattice

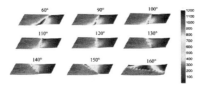

Figure 9.40 Spatial distribution of the elastic energy inside the mechanical relaxed lattice for different initial orientations of the electronic interface. Clearly for $\gamma = 120°$ we obtain an almost free stress interface.

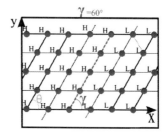

Figure 9.41 Schematic view of the electronic interface oriented with the angle $\gamma = \pi/3$. The lattice parameter is that of the HS state.

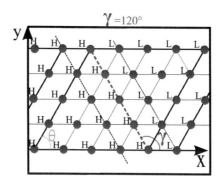

Figure 9.42 Schematic view of the electronic interface oriented with the angle $\gamma = 2\pi/3$. The lattice parameter is that of the HS state.

parameter misfit between the two phases. Such energy is written as

$$E_{\text{elas}}^{\pi/3} = \left[L_y A_0 + 5\left(L_y - 1\right) B_0\right] (R_{\text{HH}} - R_{\text{HL}})^2. \quad (9.39)$$

However, the energy corresponding to the interface oriented with an angle $\gamma = 2\pi/3$ (see Fig. 9.42) has the following expression:

$$E_{\text{elas}}^{2\pi/3} = \left[2 L_y A_0 + 4\left(L_y - 1\right) B_0\right] (R_{\text{HH}} - R_{\text{HL}})^2, \quad (9.40)$$

Thermal Behavior of Spin-Crossover Nanoparticles and Their Theoretical Description | 399

where A_0 and B_0 are, respectively, the elastic constants between the nearest and next nearest neighbors.

The energy difference between the two interface orientations is

$$\Delta E_{\text{elas}} = E_{\text{elas}}^{\pi/3} - E_{\text{elas}}^{2\pi/3} = L_y \left(R_{\text{HH}} - R_{\text{HL}} \right)^2 \left[A_0 - B_0 \right]. \quad (9.41)$$

We should mention that within the parameter values of the elastic constants used in our simulations, $A_0 = 22000$ K/nm^2 and $B_0 = 10 \times A_0$, the energy difference ΔE_{elas} in Eq. 9.41 is negative, and then the orientation $\gamma = 2\pi/3$ has the lower energy, which explains the origin of its stable orientation.

This study stress as well another important point that is the ratio between the elastic constants of the nearest and the next nearest neighbors. Such quantity has a crucial role and can govern the orientation of the interface. It is then expected that for $A_0 = B_0$, both orientations will be stable in the middle of the lattice. However, as stated before the shape of the lattice plays also an important role and once the interface started with some orientation, it is difficult from the energetic point of view to change it to the second stable orientation due to the existence of an energy barrier.

This result confirms our primary idea on the fact that the system optimizes its interface shape and length in order to minimize the total elastic energy. Obviously, if the expression of the elastic energy accounts for bending effects, it will affect the orientation of the interface and most likely its roughness. In addition, in systems with a quadratic symmetry experiencing anisotropic deformations along the x- and y-directions, two stable front orientations may appear as a result of the intrinsic symmetry of the lattice, as it was evidenced recently from both experimental and theoretical sides [60].

9.9 Thermal Behavior of Spin-Crossover Nanoparticles and Their Theoretical Description

Spin-crossover nanoparticles have attracted a lot of attention the past decade due to the development of new and efficient techniques of synthesis and preparation [117–122, 124–129]. From physical point of view, such nanoparticles constitute an excellent example of the effects of finite size on the occurrence of a first-order

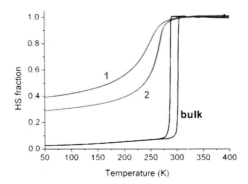

Figure 9.43 Experimental HS fraction of the bulk and nanoparticle samples of the SC compound Fe(pyrazine)[Pt(CN)$_4$]. Labels 1 and 2 stand for nanoparticle samples with average sizes 7 nm and 12 nm, respectively.

transition. However, since the things are never simple, it has been found that the thermal feature of the SC nanoparticles was not monotonously dependent on the size, such as a reduction in the size results a systematic decrease of the hysteresis loop, as expected from a simple point of view. Indeed, the experimental data are very rich: in some cases, the hysteresis loop survives until very small sizes, till 10 nm for the spin transition [117], while in the other cases it decreases monotonically; much more recently, it was reported a non-monotonous size dependence of the hysteresis loop of [Fe(hptrz)$_3$](OTs)$_2$ nanoparticles [130], but these results should be confirmed or reproduced by other experimental data, because they hit the common sense. The most usual results obtained for the SC nanoparticles are like those of Fe(pyrazine)[Pt(CN)$_4$] sample illustrated in Fig. 9.43, in which we show the thermal dependence of the HS fraction to the size of nanoparticle. Three important features can be deduced from these experimental results: (i) the size reduction of the hysteresis loop with decreasing the nanoparticle size, (ii) the enhancement of the residual HS fraction at low temperature, and (iii) the downward shift of the transition temperature upon reduction of size. Recent experimental investigations [128, 131] on SC nanoparticles inserted in a polymeric matrix lead also to interesting information about the type of the interaction between the nanoparticle and the matrix and confirmed the important role of the elastic properties of the region linking the nanoparticle to the matrix. Obviously, to the best of our knowledge, the experimental results reported in the literature on the thermal properties of SC nanoparticles concern average data collected on an ensemble of nanoparticles. So, one must keep in mind that the interactions between the nanoparticles, through their aggregation due to some electrostatic interaction for example, are definitely not excluded. Moreover, in isolated nanoparticles, surface effects are very important even though that the surface is the place of all possible defects. The discussion of the experimental results should then take into account these obvious facts. Of course, in the case of nanoparticles dispersed in a polymer, the situation is quite similar but then the interaction goes through the elastic properties of the polymer. In addition, one should mention that true mono-disperse nanoparticles are very difficult to produce (except in core–shell systems) and then another complication arises from the distribution of size.

These experimental facts make very hard any quantitative theoretical prediction of the thermodynamical properties of SC nanoparticles. Let us go further in the discussion of the simplifications of the description of this problem. Most of the experimental and theoretical studies (except rare cases [132]) forgot to discuss/mention that the lattice symmetry is impacted at very small sizes. This is an important point, since many previous studies on semiconductors have shown that the reduction of size of a material is accompanied by a change in the lattice symmetry. Indeed, the stability of the bulk structure for small particles is questionable. For example, it is quite known that small nanoparticles of Co-Pt bimetallic clusters have different stable structures, depending on their size. Thus, global structural optimization, performed using Monte Carlo simulations [132], based on a tight binding potential, has shown in this particular case that polyicosahedron-like, decahedron, and fcc structures are the most stable for nanoparticles with atom numbers <100, $100-400$ and $400-1000$, respectively. In addition, real nanoparticle materials involve surface relaxation and surface reconstructions, which may play the role of active points of defect for nucleation during the spin-crossover transition. At the present stage, there is no X-ray diffraction data on SC nanoparticles due to the high difficulty to perform experiments which remain a great challenge, mainly because of the presence of structural disorder at the surface and probably in the bulk too.

However, the role of the theory is to study simple and ideal objects. One of the most severe simplifications in theoretical models devoted to SC nanoparticles consists of assuming a 3D object as a 2D system. Accepting a set of sacrifices, and neglecting the elastic interactions between nanoparticles, as well as their size distribution and the variation of their shapes depending on their sizes, we can propose an electro-elastic model to describe thermal properties of SC nanoparticles. The choice was made here to discuss the electronic and elastic properties of circular SC nanoparticles embedded in a matrix. They are several relevant parameters controlling the physical properties. Hereafter, we quote only the physical parameters and we omit some chemical factors (like the effect of surfactants or special coordination properties of the surface). So, for the physicist, among the important parameters governing the nanoparticles behavior, we identify

- the size and the shape of the nanoparticle;
- the size of the matrix;
- the stiffness difference between the nanoparticle and the matrix;
- the elastic interaction between the nanoparticle and the matrix;
- the elastic misfit (or lattice parameter difference) between the SC nanoparticle and the matrix.

The "elastonic" Hamiltonian (Eq. 9.3) introduced earlier stands as a powerful tool to investigate the thermal properties of spin crossover nanoparticles while considering all the latter parameters. In doing so, two nanoparticles with different shapes (square and circular) were considered in a core–shell structure which consists on an inner spin crossover core and outer layer acting as a shell represented by nodes locked in the HS state. The usual Monte Carlo procedure was used to change the positions and spins of the core, and only the positions of the shell, whose the equilibrium parameter is denoted d_0. The nodes can move only inside the plane. A schematic

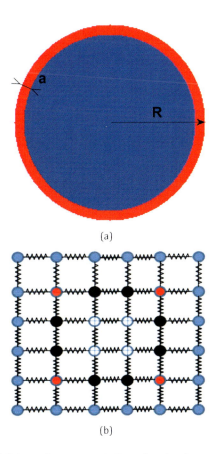

Figure 9.44 (a) Schematic representation of a circular core–shell system; R is the radius of the nanoparticle and a represents the thickness of the elastic shell (red) whose nodes are locked in the HS state unlike the core part (blue). (b) Elastic 2D lattice representing a square core–shell nanoparticle with one layer of fixed HS atoms. The core part is represented by red nodes having two fixed HS nearest neighbors, black nodes have one fixed HS neighbor, and white nodes are without fixed HS neighbor. For sake of clarity, the interactions with the next nearest neighbors are not presented here.

representation of the system is illustrated in Fig. 9.44. The choice of the HS shell here is justified by the assumption that the molecules located on the surface of the nanoparticles (the border) experience a weaker ligand-field. In a core–shell structure, the interactions between molecules are not the same depending on their positions. Indeed, the nodes in the core part experience the same kind of interactions with their four nearest neighbors unlike those located at the interface core/shell, which have less active spin crossover neighbors due to the inactive character of the shell. Let's start by the case of a circular particle (see Fig. 9.44a) whose R is the radius and a the thickness of the shell, we count $\pi(R-a)(R-a-2)$ nodes in the core part and $2\pi(R-a)\omega$ nodes on the core/shell interface, where ω is a unit distance required for dimensional homogeneity. The case of the square particle is more tricky due to the complexity of the geometry combining edges and corners. Indeed, in the latter case, the core/shell interface consists on molecules in the corner with two fixed HS nearest neighbors and molecules on the edge with one fixed HS neighbor. Accordingly, we have four molecules on the corners and those on the edge are size dependent and are estimated to $4(L-1)$, where L stands for the edge length.

9.9.1 Size and Shape Effects

The size dependence of the thermodynamic properties of SC nanoparticles is strongly related to their environment. For example, for a free particle, one has to consider the surface effects. Indeed, it is well known from the experimental literature that SC molecules at the surface experience a weaker ligand field than those of the bulk, due to their coordination to water or solvent molecules instead of nitrogen atoms. Thus, the molecules at the surface are usually in the HS state. So, the description of the SC nanoparticles as an object having an active core and an inactive thin shell maintained in the HS state is quite well admitted. Hereafter, we revisit some of the most important results arising from this description in the case of a 2D lattice with square symmetry. Obviously the results depend on the particle shape. Indeed, square, rectangular, circular, or elliptic shapes lead to different behaviors, not only because of the difference of their ratio surface/volume, but also because the elastic properties (like the electric properties in conductors) depend on the shape of the material. Of course, for large particle (active core) sizes the surface effect become negligible and all these shapes must lead to the same behaviors. Figure 9.45 illustrates the HS fraction

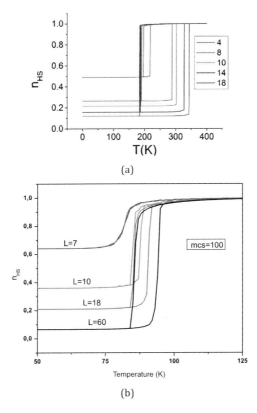

Figure 9.45 (a) Thermal dependence of n_{HS} in the case of circular (a) and square (b) core–shell nanoparticles for various active core sizes. In both cases, a net decrease of the thermal hysteresis width is obtained when decreasing the size (length or radius) of the active core. As a result, the transition temperature shifts and the low-temperature residual HS fraction increases.

dependence to the temperature for circular and rectangular shapes. In both cases, an increasing residual HS fraction is observed for small sizes (that is obvious), accompanied by a less trivial shift of the transition temperature to lower values. Concomitantly, the nature of the LS ↔ HS transformation changes from first-order transition to gradual spin-conversion by decreasing the size. These findings are in good agreement with the experimental observations [122, 123, 125]. At this stage we should mention that the tendencies described above are quite general and were obtained with both the elastic description [132, 133] and the Ising-like model [134] with specific surface effects.

In Fig. 9.45, we have monitored the equilibrium temperature for both circular and square nanoparticle system as a function of the size. It is interesting to mention that the size dependence of the equilibrium temperature is well described by a rational law with simple formula:

$$T_{eq}^{circle} = \frac{\Delta}{k_B \ln g} - \frac{m_c}{(R - a)} \tag{9.42}$$

and

$$T_{eq}^{square} = \frac{\Delta}{k_B \ln g} - \frac{m_s}{(L - 2)}, \tag{9.43}$$

where the factors m_c and m_s depend on the elastic interaction and the lattice parameter misfit. One can easily deduce that when the nanoparticle has a very large size its equilibrium temperature become similar to that one of the bulky spin crossover compounds ($T_{eq}^{bulk} = \Delta/k_B \ln g$) independently from the shape effect. However, and surprisingly, a difference could be noted between the circular and square particles is that in the former case the cooperative behavior of the transition appears to be less sensitive to the size modification, since the systems keeps the first-order character even at small particle sizes. This observation arises an important point that is about the shape effect on the thermal transition. Indeed, the shape effects can be easily investigated on 2D SC nanoparticles with the same number of particles but having different shapes: square, rectangle, circular, elliptic, or much more complicated shapes can be imagined. The most problematic aspect of this kind of study comes from the fact that for a fixed lattice symmetry, let's say square symmetry for example, the concentration of defects at the surface obviously depends on the chosen shape. Therefore, to study this specific problem, we recommend to use honeycomb structure which is one of the most stable and the efficient ways to pave a desired surface, thus minimizing the density of topological defects. This was a very general problem in mathematics, where Hales theorem [135] stated that the hexagonal lattice is the best way to divide a surface into regions of equal area with the smallest total perimeter.

For highly symmetric shapes well adapted to the lattice symmetry, such as squares and rectangles within the square lattice symmetry, the study is quite easy to perform. For example, at equal surfaces (or number of particles), the rectangle has always a bigger perimeter than the square. If the molecules at surface have a specific property (they are HS for example), then it is clear that the shape of the nanoparticle will play here an important role, and this will be free from the creation of any topological defects. This is not the case for a circular system with square lattice symmetry, where the discrete nature of the lattice gives rise to surface defects. So, it is expected to find different behaviors following the number of SC molecules at the surface. For example, this number is bigger in the rectangular-shaped nanoparticle than in the square one having the same total number of particles (or surface). To the best of our knowledge, there were no extensive studies on the role of the shape effect in SC literature, and one of the nice examples of the shape effect studies which remains to be realized is that of the particular case of the transformation of a lattice from an elliptic shape to a circular one by playing with the eccentricities, at constant volume (surface for a 2D case) using a hexagonal unit cell. This work is in progress and should be published prior to the present work.

9.9.2 Role of the Elastic Misfit between the Matrix and the SC Nanoparticle

One way to evaluate the effect of the lattice misfit between the shell or the matrix and the core of the nanoparticle is to study the thermal dependence of a nanoparticle inserted in a matrix whose equilibrium lattice parameter value d_0 is monitored from a value $d_0 < R_{LL}$ to a value $d_0 > R_{HH}$. When the equilibrium distance of the shell is equal to R_{HL}, it is expected a neutral elastic effect of the latter on the thermal properties of the nanoparticle. This prediction is checked in Fig. 9.47, which shows the temperature dependence of the HS fraction for various shell thickness when $d_0 = R_{HL}$. We see clearly that the hysteresis loop remain the same for the different shell thickness value, a result which contrast with the behavior of the thermal hysteresis when the shell lattice parameter is equal to that of the HS state ($d_0 < R_{HH}$) or the LS state $d_0 < R_{LL}$.

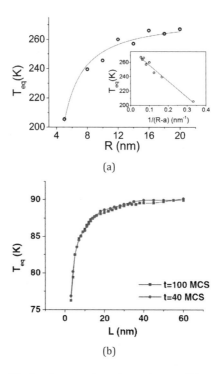

Figure 9.46 Equilibrium temperature dependence of the system for the size circularly (a) and square- (b) shaped core–shell nanoparticles.

Elastically active shell must present an elastic misfit with the nanoparticle core at the transition temperature. Indeed, at T_{eq}, the core has an average lattice parameter $R \simeq R_{HL}$. Therefore, shell with smaller or higher lattice parameter will affect the transition temperature of the core as well as the width of the related thermal hysteresis. Two limiting cases are considered here: that of the switchable nanoparticle surrounded by (i) a soft shell having the lattice parameters of the HS or a (ii) rigid shell having the elastic properties of the LS state.

We shall mention here that we did not consider in details the interaction between the nanoparticle and the matrix or the shell. Indeed, one can include in the total Hamiltonian of the system a specific interaction in the region linking the nanoparticle to the shell.

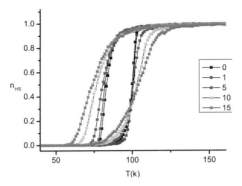

Figure 9.47 Thermal dependence of the HS fraction for a core–shell nanoparticle in the case of an elastically neutral shell. For that the lattice parameter of the shell is chosen as $d_0 = R_{HL}$. The results include different shell thickness. The hysteresis loop is almost independent of the shell thickness.

If we limit, in that border region, the interaction to the elastic one, then the elastic stiffness can be adapted according to that of the shell. In this case, we do not expect any new result compared to those presented here, except in the case where we consider the situation of broken links between the nanoparticle and the shell, as reported in [128, 133].

9.9.2.1 Effect of a soft shell

The effect of a soft shell, having the lattice parameter of the HS phase, on the first-order transition of a SC nanoparticle, is shown in Fig. 9.48. Obviously, the shell has no effect on the HS state. Actually, the soft shell decreases the transition temperature, by stabilizing the HS states, an effect which is enhanced when we increase the size (the number of layers) of the shell. To understand this behavior, one has to recall that the transition temperature for a free shell nanoparticle is given by $T_{eq}^0 = \dfrac{\Delta}{k_B T \ln g}$. Under the effect of the shell, the new transition temperature is the one which gives a null effective ligand field, i.e., $\langle h_i \rangle = \frac{1}{2}(\Delta - k_B T \ln g) - A_0 \rho_1 \sum_{j=1}^{z} (\langle r_{ij} \rangle - \rho_0) = 0$ that

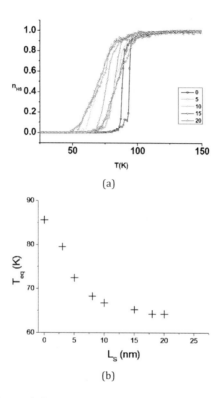

Figure 9.48 Thermal dependence of the HS fraction for square-shaped nanoparticles at constant core size $L = 10$ surrounded by a variable number of soft shell layers with a lattice parameter $d_0 = R_{HH}$. The hysteresis distorts and shifts downward as a result of the negative (and inhomogeneous) elastic pressure induces by the shell. The elastic stiffness between the chain layers is equal to that coupling two HS atoms in the core.

leads to

$$T_{eq}^{l_s} \simeq T_{eq}^0 - 4\frac{(\langle r_{ij} \rangle - \rho_0)}{k_B T \ln g}. \tag{9.44}$$

As a result of the expanded shell, the average value of lattice parameter, $\langle r_{ij} \rangle$, that is the distance between the nearest neighbors, is bigger than $\rho_0 = R_{HL}$ at $T = T_{eq}^0$, which decreases the transition temperature of the core, according to Eq. 9.44. The hysteresis loop is then shifted to the low-temperature region, thus stabilizing the HS phase. Here, we should mention that the shifted hysteresis is significantly distorted when we increase the number of shell layers, a behavior which can be attributed to the spatial fluctuation of the nn distances ($\langle r_{ij} \rangle$) which leads to disperse the values of the local effective ligand fields h_i (given in Eq. 9.16) acting on each site. In Fig. 9.48b, we present the size-shell dependence of the equilibrium temperature which behaves as $1/L_s$, a behavior which is also confirmed for circular core–shell nanoparticles.

9.9.2.2 Effect of a rigid shell

The case of a rigid shell is studied by considering a shell with a LS equilibrium lattice parameter. In this case the shell exerts a positive pressure on the core which increases the value of the local effective field acting on each site. Indeed, based on the same reasoning as in the previous section, the transition temperature increases (see Fig. 9.49a) when we increase the number of shell layers and the hysteresis width decreases monotonously but for large shell sizes. Moreover, it is remarked here that the transition temperature increases linearly with the number of shell layers (see Fig. 9.49b), without the presence of a significant deformation of the hysteresis's shapes. Interestingly, the same behavior is also obtained with circular core–shell nanoparticles. This result is compatible with the effect of a hydrostatic pressure on the core.

Let us discuss now the origin of the asymmetry in the behavior of the hysteresis loops for soft and rigid shells. The most important difference of the two situations comes from the fact that an expanded shell ($d_0 = R_{HH}$) acts mainly on the LS phase, which is situated in the low temperature region, while a compressed shell ($d_0 = R_{HH}$) acts on the HS phase, which is located in the high-temperature region. In the first case, the transition temperature is decreased and then kinetic effects on the relaxation of the elastic lattice are important, and enhanced with the number of shell layers. Therefore, for an expanded shell, nonrelaxed lattice positions are most probably at the origin of spreading of the thermal hysteresis loops. In contrast, for compressed shells, the transition temperature is shifted to higher temperatures, a region where the relaxation times of the elastic lattice are much shorter and then we can

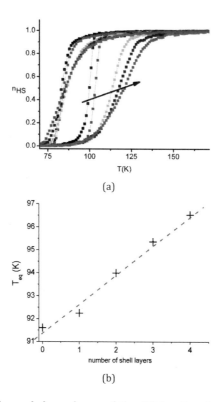

Figure 9.49 Thermal dependence of the HS fraction for square-shaped nanoparticles at constant core size $L = 10$ surrounded by a variable number of soft shell layers with a lattice parameter $d_0 = R_{HH}$. The hysteresis distorts and shifts downward as a result of the negative (and inhomogeneous) elastic pressure induced by the shell. The elastic stiffness between the chain layers is equal to that coupling two HS atoms in the core.

consider that the associated hysteresis loops are obtained with well relaxed mechanical systems.

9.10 Conclusion and Perspectives

The spatio-temporal aspects of spin-crossover single crystals undergoing first-order phase transitions is a fascinating subject that has recently received a considerable attention both experimentally and theoretically. From the experimental point of view, the observations of a macroscopic interface, separating the different spin states and propagating across the crystal, are possible in only few resilient systems where the molecular nature of the lattice allows the absorption of an important part of the volume change at the transition. Such systems can mechanically resist to ~30% of volume increase at the molecular scale which arises as ~3% of volume increase at the macroscopic scale. Therefore, among the essential prerequisite conditions for successful visualization of the transformation front (the interface), the synthesis of high quality single crystals since the defects may be source of dislocations and fractures upon the spin transition, leading to their breaking. Reduced volume change at the transition is also another important parameter to dampen the damage on the crystals as revealed by optical microscopy observations [65] on an incomplete spin transition resulting from two-steps spin crossover system. However, we should mention that the existence of a transformation front is a general problem in first-order solid-solid phase transitions in condensed matter. They have been reported in several other systems, like thermo-salient and jumping crystals [136–141] which also experience large volume changes with spectacular mechanical effects (jumping, rotation, breaking, explosions, etc.) at the transition. Although not mentioned here, similar effects are often observed in optical microscopy experiments on spin-crossover solids.

From the experimental point of view, several works remain to be done like the efficient control of the HS–LS interface. Since the latter has an elastic nature, it is expected that the access to a rapid control depends on the material size and shape. The existence of the interface itself at small sizes (nanometer scale) is an interesting issue, since it raises the question of the existence of a first-order transition at small scales. Recent and encouraging works devoted to the study of the spin crossover nanocrystals [142] and ultrathin films [143] have shown the existence of a spin transition at very small scale (~50 nm), which can be driven by an applied voltage through a Stark effect. These results open serious possibilities of applications or integrations of these systems in molecular electronics as nanoelectronic devices.

The width of the HS–LS interface reported in our optical microscopy is evaluated around ~4 μm via an experimental technique which has about ~0.3 μm as spatial resolution. The fundamental question that still arises concerns the width of this interface at the nanometric scale. To the best of our knowledge, there is no work done in this direction until now. Transmission electron microscopy (TEM) and atomic force microscopy (AFM) are highly desired, as well as other near-field techniques on systems with a static HS–LS interface. Since our work on the photo-thermal control of the interface propagation [61], this is now possible. Here, in addition to the determination of the true interface width, the goal is to clarify the dual nature (electronic and elastic) of the HS–LS interface by having access to the quantitative aspects of each of them. Moreover, snapshots of the moving interface, the relation between its width and its velocity as well its roughness are the essential information to understand the mechanism of the electro-structural relaxation.

From the theoretical point of view, we have seen that an electro-elastic model, equivalent to an Ising-like model on a deformable-lattice, allowed describing several important features of the spin transition. First-order and gradual transitions are reproduced and the conditions of their occurrence are well understood. In this model, however, we have not yet addressed the problem of two-steps spin transitions, which remains a challenging problem for the elastic models. Such a behavior requires the existence of competing (ferro- and antiferro-elastic) interactions in the system. One way to achieve this goal is to consider that the antiferro-interaction as electronic and then to induce an antiferro-magnetic-like interaction between neighboring spins. This appealing trivial solution is, however, a little physical and is only appropriate to binuclear molecules. However, for mononuclear systems [144, 145], the two steps transition results in the competition between ferro- and antiferro-elastic interactions. In the general case, the elastic interactions are anisotropic (directional) and they can change their sign from one direction to another.

Indeed, in real crystals, the elastic interactions are more complicated. Thus, for example, in weakly anisotropic cubic crystals there appears an interaction between local "defects" or "impurities"

Conclusion and Perspectives | 415

(HS atoms inside a LS matrix or inversely) which decays rather slowly, as $1/d_{ij}^3$, with a different sign in different directions [111, 146, 147]:

$$V\left(\vec{r}_j, \vec{r}_i\right) \sim -Dq_i q_j \frac{\Gamma(\vec{n}_{ij})}{\left|\vec{r}_j - \vec{r}_i\right|^3}, \qquad (9.45)$$

where $\vec{r}_j - \vec{r}_i = \left|\vec{r}_j - \vec{r}_i\right|.\vec{n}_{ij}$ and q_i and q_j are the change of the molecular volume (local sources of compression waves). The angular dependence of the interaction is determined by a function of the direction cosines of the vector $\vec{d}_{ij} = (\vec{r}_j - \vec{r}_i)$

$$\Gamma(\vec{n}_{ij}) = n_x^4 + n_y^4 + n_z^4 - \frac{3}{5}. \qquad (9.46)$$

The elastic anisotropy is characterized in this situation by the parameter $D = c_{11} - c_{12} - c_{44}$, where c_{ij} are the corresponding elastic moduli. One can see that for $D < 0$ (i.e., for $c_{44} > \dfrac{c_{11} - c_{22}}{2}$), the present interaction is repulsive (i.e., positive) along the directions [100], [010], and [001] and attractive along face and diagonals [110] and [111] and vice versa for $D > 0$. Thus it appears that the interaction between similar excitations may be repulsive in some directions and attractive along others. The situation becomes much more complicated if we consider the interaction between excitations of different types, i.e., HS and LS atoms, where the signs of q_i and q_j must be taken into account.

These anisotropies in the interactions have not yet been included in the atomistic elastic models describing the spin-transition. They can, however, be required in order to enable us to properly understand the situation of the existence of superstructures or cell doubling, and of course to provide the response to this particular organization of matter, with the associated symmetry breaking [148, 149] between the LS and the HS phases observed in some two-steps spin-crossover transitions.

On the other hand, it is well known that the characteristics of spin transition at the solid state are intimately related to the cooperativity inside the crystal lattice. In several examples, the spin transition occurs concomitantly with an order-disorder phase transition, where the competition between short- and long-range interactions gives rise to interesting phenomena. For example, the two-step transition in the compound $[Fe(2\text{-pic})_3]Cl_2 \cdot EtOH$ has been attributed to the existence of a superstructure in an intermediate phase coupled to two successive order-disorder phase transitions [144, 145, 150].

From the crystallographic point of view, it is worth to mention that the first evidence of nucleation and domain growth were evidenced on the strongly cooperative 2D system $Fe(btr)_2(NCS)_2 \cdot H_2O$ by S. Pillet et al. [151]. The same authors have also studied the interesting case of $Fe(abpt)_2[N(CN)_2]_2$ which crystallizes in two distinct phases. While the first phase [152] consists on one Fe site, the second one has shown an order-disorder phase transition at low-temperature induced by light or by a flash cooling. The resulting metastable phase evidenced an incommensurate structural modulation [153] which vanishes upon the relaxation from HS to LS. This modulation was attributed to the ordering of the ligands. In the same context, an aperiodic spin state ordering in a bistable spin-crossover solid and its photo-induced erasing was shown by Collet et al. [154].

These results demonstrate that the ligands, ignored in all the current models, can play an important role and can lead to new types of organizations. Their contributions can be also taken into account in the electro-elastic models which merit to be extended to include the order-disorder transitions resulting from ligand re-orientations and their influences on the spin-transition phenomenon.

Another important point concerns the theoretical description of SC thin films, which are now produced experimentally quite extensively using several techniques [155–163], to study their intrinsic properties as potential devices in molecular electronics. The theoretical description of SC thin films benefited of only one paper [94] in which the authors reported on the buckling and crumpling effects accompanying the relaxation of the HS fraction at low temperature. The idea was here to allow the atoms to move out of the plane. It is well known that systems constrained to distort in the plane (transverse and longitudinal deformations) as well as out of the plane, create large and well organized deformation structures at the surface. This phenomenon should be fully investigated from the experimental side and should be extended to include the case of thermal SC transitions from the theoretical side. In addition, the thermo-mechanical behavior of switchable SC thin films (made of multi-layers) as function of the number of the atomic layers, is an interesting question which merits further research. We believe that this fundamental work will help pave the way for the understanding of the future experimental results of the behavior of SC thin films.

The fact that the electro-elastic model can be mapped under the form of an Ising like model, with effective interaction parameters depending on the lattice misfit between the LS and HS phases, opens interesting perspectives. Indeed, we can exploit this analogy to describe the spatio-temporal dynamics of spin-crossover solids using reaction diffusion equations. Actually, the propagation of the front resembles quite well to the propagation of infectious diseases described in the past by Fisher and Kolmogorov [164, 165]. It is then possible to build up a description in this direction by using the free energy as a driving force. Since, we are dealing with an Ising-like description, the free energy can be calculated in mean-field theory and its expansion about the power of its gradients [166] leads to the following general equation of motion, of the HS fraction, n:

$$\frac{dn}{dt} = -\Gamma_0 \frac{\partial F}{\partial n} + D\nabla^2 n + \xi\,(\vec{r},\,t)\,, \tag{9.47}$$

with,

$$F = \frac{1}{2}J\,m^2 - k_B T \, \ln 2 \cosh\left[\beta\,(J\,m + h(\vec{r}))\right], \tag{9.48}$$

where m is the average local magnetization ($n = \dfrac{1+m}{2}$). D a diffusion coefficient, Γ_0 is a frequency factor, ∇^2 is the Laplacian operator, and ξ is a fluctuating force, whose time and spatial averages are equal to zero. The noise term obeys the fluctuation-dissipation theorem:

$$\langle \xi(t) \rangle_{\vec{r}} = 0 \ \text{ and } \ \left\langle \xi(\vec{r},\,t)\xi(\vec{r}',\,t') \right\rangle = 2\Gamma_0 \delta\left(\vec{r} - \vec{r}'\right) \delta\left(t - t'\right).$$
$$\tag{9.49}$$

Equation 9.47 expresses a Langevin-type dynamics and suggests a new kind of description of the spatio-temporal dynamics of SC solids, accounting for the stochastic process of nucleation and growth. Compared to the usual homogeneous mean-field-type dynamical equations of motion [75, 167–169], it brings the additional information on the spatial dependence of the order parameter, but compared to the presented MC simulations on the electro-elastic models, Eq. (9.47) has the major drawback of not including the volume change accompanying the variation of the order parameter as well as any information on the lattice symmetry. One simple solution is assuming that the elastic lattice is slaved to the motion of the HS fraction. Then, at each node, the local order parameter will follow the equation of motion (9.47) in which the free energy depends on the lattice positions through the ligand field (see Eq. 9.13), and the lattice positions will evolve in time using MD simulations, based on the expression of the potential energy given in Eq. 9.5, leading to the equations of motion

$$\frac{d^2\vec{r}}{dt^2} = -\vec{\nabla} V\left[m\left(\vec{r}\right),\vec{r}\right] - \gamma \frac{d\vec{r}}{dt}, \tag{9.50}$$

where γ is a damping factor. The elastic potential V, in which it is sufficient to consider only the harmonic part, has the expression

$$V\left(m,\vec{r}\right) = \frac{A_0}{2} \sum_{\vec{r}_i} \sum_{\vec{r}_j} \left[|\vec{r}_j - \vec{r}_i| - R_0\left(m_i, m_j\right)\right]^2, \tag{9.51}$$

where the equilibrium distance $R_0\left(m_i, m_j\right)$ writes in the local mean-field approach as follows:

$$R_0(m_i, m_j) = \rho_0 + \rho_1(m_i + m_j) + \rho_2 m_i m_j. \tag{9.52}$$

The effective ligand-field,

$$h_i = \frac{1}{2}\left(\Delta - k_B T \ln g\right) - A_0\rho_1 \sum_{j=1}^{z} \left(r_{ij} - \rho_0\right), \tag{9.53}$$

appearing in the equation of motion of local HS fractions (Eq. 9.47) clearly connects the value of the effective field h_i acting on the magnetization at this site on the positions of the neighbors. Then Eqs. 9.47 and 9.50 represent a set of closed equations in which the evolution of the local magnetizations changes the equilibrium distances between the spin states which itself act on the magnetization. The set of equations (9.48–9.53) represent a $h_i = \frac{1}{2}\left(\Delta - k_B T \ln g\right) - A_0\rho_1 \sum_{j=1}^{z} \left(r_{ij} - \rho_0\right)$ is the average magnetization at site i (j).

So, this simple model is equivalent to a set of coupled mechanical reaction diffusion equations in which the spin and the elastic subsystems are closely interlinked. The resolution of this partial derivative equations can be performed using several numerical methods (finite difference, finite elements method, etc.). As well as a gain in simplicity, the savings in time calculations is remarkable within this approach. Moreover, in this procedure the timescale can be adapted to that of the experiment, contrary to the MC procedure, in which the Monte Carlo steps are hardly comparable to that of the experimental kinetics. We also believe that this approach can allow the calculation of the local strains, displacements, and deformation fields, which govern the transition mechanism. In addition, the use of MD simulations to treat the lattice relaxation, should allow the observation of the acoustic shock waves developing upon the relaxation or the thermal transition of these bistable molecular solids. This may constitute a powerful incentive to pursue the future experimental studies in this direction.

Acknowledgments and Appendixes

The present work was supported by the University of Versailles Saint-Quentin, Centre National de la Recherche Scientifique (CNRS), Groupement de Recherche International (GDRI) France-Japan, "Agence Nationale de la Recherche" (ANRproject BISTA-MAT:grant number ANR-12-BS07-0030-01). We are also indebted to Prof. S. Miyashita, Dr. M. Nishino, Dr. C. Enachescu, Prof. S. Kaizaki, and Prof. P. Naumov for fruitful and helpful discussions on the topics.

References

1. P. Gütlich, H. A. Goodwin (2004). *Topics in Current Chemistry, Spin Crossover in Transition Metal Compounds I-III*, Springer, Heidelberg, Berlin, pp. 233–235.
2. P. Gütlich (1981). Spin crossover in iron(II)-complexes, *Struct. Bonding (Berlin)*, 44, 83.
3. P. Gütlich, A. Hauser, and H. Spiering (1994). Thermal and optical switching of iron(II) complexes, *Angew. Chem., Int. Ed. Engl.*, 33, 2024.

4. E. König, (1992). Bonding, Nature and dynamics of the spin-state interconversion in metal-complexes, *Struct. Bonding (Berlin)*, 76, 51.

5. O. Kahn and C. J. Martinez (1998). Spin-transition polymers: From molecular materials toward memory devices, *Science*, 279, 44.

6. A. Bousseksou, K. Boukheddaden, M. Goiran, C. Consejo, M.-L. Boillot, and J.-P. Tuchagues (2002). Dynamic response of the spin-crossover solid Co(H-2(fsa)(2)en)(py)(2) to a pulsed magnetic field, *Phys. Rev. B*, 65, 172412.

7. W. Kosaka, K. Nomura, K. Hashimoto, and S. Ohkoshi (2005). Observation of an Fe(II) spin-crossover in a cesium iron hexacyanochromate, *J. Am. Chem. Soc.*, 127, 8590.

8. D. Papanikolaou, W. Kosaka, S. Margadonna, H. Kagi, S. Ohkoshi, and K. Prassides (2007). Piezomagnetic behavior of the spin crossover Prussian blue analogue CsFe[Cr(CN)6], *J. Phys. Chem. C*, 111, 8086.

9. F. Prins, M. Monrabal-Capilla, E. A. Osorio, E. Coronado, H. S. J. van der Zant (2011). Room-temperature electrical addressing of a bistable spin-crossover molecular system, *Adv. Mater.*, 23, 1545–1549.

10. A. Bousseksou, F. Varret, M. Goiran, K. Boukheddaden, and J.-P. Tuchaguess (2004). The spin crossover phenomenon under high magnetic field, *Top. Curr. Chem.*, 235, 65–84.

11. L. Cambi, and L. Szego (1931). Über die magnetische Susceptibilität der komplexen Verbindungen, *Chem. Ber. Dtsch. Ges.*, 64(10), 2591.

12. C. D. Coryell, F. Stitt, and L. Pauling (1937). The magnetic properties and structure of ferrihemoglobin (methemoglobin) and some of its compounds, *J. Am. Chem. Soc.*, 59, 633–642.

13. L. Orgel (1956) Ion compression and the colour of ruby, *Nature*, 179, 1348–1348.

14. W. A. Baker Jr. and H. M. Bobonich (1964). Magnetic properties of some high-spin complexes of iron(II), *Inorg. Chem.*, 3, 1184.

15. P. G. Sim and E. Sinn (1981). 1st Manganese(iii) spin crossover and 1st d4 crossover—comment on cytochrome-oxidase, *J. Am. Chem. Soc.*, 103, 241.

16. D. Cozak, F. Gauvin, and J. Demers (1986). Direct observation of low-spin high-spin electronic ground-states and crossover exchange in manganocene derivatives, (ETA5-C5H4H)2MN, (ETA5-C5H4CH3)2MN, (ETA5-C5H4C2H5)2MN by paramagnetic nuclear-magnetic-resonance, *Can. J. Chem.*, 64, 71.

17. D. M. Halepoto, D. G. L. Holt, L. F. Larkworthy, G. L. Leigh, D. C. Povey, and G. W. Smith (1989). Spin crossover in chromium(II)

complexes and the crystal and molecular-structure of the high-spin form of BIS[1,2-bis(diethylphosphino)ethane]di-iodochromium(II), *J. Chem. Soc., Chem. Comm.*, 1322.

18. A. K. Hughes, V. J. Murphy, and D. O'Hare (1994). Synthesis-X-ray structure and spin-crossover in the triple-decker complex [(ETA(5)-C(5)ME(5))CR(MU(2)ETA(5)-P5)CR(ETA(5)-C(5)ME(5))]+[A]- (A=PF6, SBF6), *J. Chem. Soc., Chem. Comm.*, 2, 163.

19. W. Kläui (1979) high spin-low spin equilibrium in a 6-co-ordinate cobalt(III) complex, *J. Chem. Soc., Chem. Comm.*, 16, 700.

20. M. G. Simmons and L. J. Wilson (1977). Magnetic and spin lifetime studies in solution of a deltas=1 spin-equilibrium process for some 6-coordinate bis(N-R-2,6-pyridinedicarboxaldimine)cobalt(II) complexes, *Inorg. Chem.*, 16(1), 126–130.

21. L. G. Marzilli and P. A. Marzilli (1972). Magnetic spin equilibria in some new 5-coordinate schiff-base complexes of cobalt(II), *Inorg. Chem.*, 11, 457.

22. J. Zarembowitch and O. Kahn (1984). Magnetic-properties of some spin-crossover, high-spin, and low-spin cobalt(II) complexes with Schiff-bases derived from 3-formylsalicylic acid, *Inorg. Chem.*, 23, 589.

23. L. Sacconi (1972). Influence of geometry and donor-atom set on spin state of 5-coordinate cobalt (II) and nickel (II) complexes, *Coord. Chem. Rev.*, 8, 351.

24. L. Sacconi (1971). Conformational and spin state interconversions in transition metal complexes, *Pure Appl. Chem.*, 27, 161.

25. W. Kläui, K. Schmidt, A. Bockmann, P. Hofmann, H. R. Schmidt, and P. Staufert (1985). Conformational and spin state interconversions in transition metal complexes, *J. Organomet. Chem.*, 286, 407.

26. H. Werner, B. Ulrich, U. Schubert, P. Hofmann, and B. Zimmer-Gasser (1985). Heterometallic binuclear complexes in the triple-decker sandwich complexes [NI2(C5H4R)3]BF4 and cyclopen-tadienylmetal-bisthiolates—crystal-structure and spin-crossover behavior of [(C5H5)2MO(MU-SBU-TERT)2NI(C5H4R)]BF4, *J. Organomet. Chem.*, 297, 27.

27. D. M. Halepoto, D. G. L. Holt, L. F. Larkworthy, D. C. Povey, G. W. Smith, G. J. Leigh, and D. C. Povey (1989). Spin crossover in chromium(II) complexes, *Polyhedron*, 8, 1821.

28. M. Sorai, Y. Yumoto, D. M. Halepoto, and L. F. Larkworthy (1993). Calorimetric study on the spin-crossover phenomenon between (3)T(1) and (5)E in trans-bis[1,2-bis(diethylphosphino)

ethane]diiodochromium(II), [CRI2(DEPE)2], *J. Phys. Chem. Solids*, 54(4), 421.

29. H. Sitzmann, M. Schar, E. Dormann, and M. Kelemen (1997). Octaisopropylchromocene: The first chromocene with gradual low spin high spin-transition, *Z. Anorg. Allg. Chem.*, 623, 1850.

30. J. H. Ammeter, R. Bucher, and N. Oswald (1974). High-spin-low-spin equilibrium of manganocene and dimethylmanganocene, *J. Am. Chem. Soc.*, 96, 7833.

31. M. E. Switzer, R. Wang, M. F. Rettig, and A. H. Maki (1974). Electronic ground-states of manganocene and 1,1'-dimethylmanganocene, *J. Am. Chem. Soc.*, 96, 7669.

32. J. H. Ammeter (1978). EPR of orbitally degenerate sandwich compounds, *J. Mag. Reson.*, 30, 299.

33. J. H. Ammeter, L. Zoller, J. Bachmann, E. Blatzer, E. Gamp, R. Bucher, and E. Deiss (1981). The influence of molecular host lattices on electronic-properties of orbitally (near-) degenerate transition-metal complexes, *Helv. Chim. Acta.*, 64, 1063.

34. D. Cozak and F. Gauvin (1987). Electronic ground-states and isotropic proton nmr shifts of manganocene and its derivatives, *Organometallics*, 6, 1912.

35. N. Hebedanz, F. H. Kohler, G. Muller, and J. Riede (1986). Electron-spin adjustment in manganocenes: Preparative, paramagnetic NMR, and X-ray study on substituent and solvent effects, *J. Am. Chem. Soc.*, 108, 3281.

36. Y. Garcia, O. Kahn, J.-P. Ader, A. Buzdin, Y. Meurdesoif, and M. Guillot (2000). The effect of a magnetic field on the inversion temperature of a spin crossover compound revisited, *Phys. Lett. A*, 271, 145.

37. H. J. Krokoszinski, C. Santandrea, E. Gmelin, and K. Barner (1982). Specific-heat anomaly connected with a high-spin low-spin transition in metallic MNAS1-XPX crystals, *Phys. Status Solidi B*, 113, 185.

38. L. Kaustov, M. E. Tal, A. I. Shames, and Z. Gross (1997). Spin transition in a manganese(III) porphyrin cation radical, its transformation to a dichloromanganese(IV) porphyrin, and chlorination of hydrocarbons by the latter, *Inorg. Chem.*, 36, 3503.

39. S. Decurtins, P. Gütlich, C. P. Kohler, H. Spiering, and A. Hauser (1984). Light-induced excited spin state trapping in a transition-metal complex: The hexa-1-propyltetrazole-iron (II) tetrafluoroborate spin-crossover system, *Chem. Phys. Lett.*, 105, 1.

40. G. D'Avino, A. Painelli, and K. Boukheddaden (2011). Vibronic model for spin crossover complexes, *Phys. Rev. B*, 84, 104119.

41. S. W. Biernacki and B. Clerjaud (2005). Thermally driven low-spin/high-spin phase transitions in solids, *Phys. Rev. B*, 72, 024406.

42. C. P. Kohler, R. Jakobi, E. Meissner, L. Wiehl, and H. Spiering (1990). Nature of the phase-transition in spin crossover compounds, *J. Phys. Chem. Solids*, 51, 239.

43. R. Jakobi, H. Spiering, and P. Gütlich (1992). Thermodynamics of the spin transition in [FEXZN1-X(2-PIC)3]CL2.ETOH, *J. Phys. Chem. Solids*, 53, 267.

44. K. Boukheddaden, J. Linares, E. Codjovi, F. Varret, V. Niel, and J. A. Real (2003). Dynamical Ising-like model for the two-step spin-crossover systems, *J. Appl. Phys.*, 93, 7103.

45. A. Bousseksou, J. Nasser, J. Linares, K. Boukheddaden, and F. Varret (1992). Ising-like model for the 2-step spin-crossover, *J. Physique*, 2, 1381–1403.

46. C. P. Slichter and H. G. J. Drickamer (1972). Pressure-induced electronic changes in compounds of iron, *Chem. Phys.*, 56, 2142.

47. J. Wajnflasz and R. Pick (1971). Transitions low spin—high spin dan les complexes de Fe^{2+}, *J. Phys. (Paris). Colloq*, 32, 91–92.

48. S. Doniach (1978). Thermodynamic fluctuations in phospholipid bilayers, *J. Chem. Phys.*, 68, 4912.

49. S. Miyashita, Y. Konishi, M. Nishino, H. Tokoro, and P. A. Rikvold (2008). Realization of the mean-field universality class in spin-crossover materials, *Phys. Rev. B*, 77, 014105.

50. M. Nishino, K. Boukheddaden, Y. Konishi, and S. Miyashita (2007). Simple two-dimensional model for the elastic origin of cooperativity among spin states of spin-crossover complexes, *Phys. Rev. Lett.*, 98, 247203.

51. M. Nishino, C. Enachescu, S. Miyashita, K. Boukheddaden, and F. Varret (2010). Intrinsic effects of the boundary condition on switching processes in effective long-range interactions originating from local structural change, *Phys. Rev. B, Rapid Comm.*, 82, 020409.

52. C. Enachescu, L. Stoleriu, and A. Stancu (2009). Model for elastic relaxation phenomena in finite 2D hexagonal molecular lattices, *Phys. Rev. Lett.*, 102, 257204.

53. M. Nishino, C. Enachescu, S. Miyashita, P. A. Rikvold, K. Boukheddaden, and F. Varret (2011). Macroscopic nucleation phenomena in continuum media with long-range interactions, *Sci. Rep.*, 1, 162.

54. L. Stoleriu, P. Chakraborty, A. Hauser, A. Stancu, and C. Enachescu (2011). Thermal hysteresis in spin-crossover compounds studied within the mechanoelastic model and its potential application to nanoparticles, *Phys. Rev. B*, 84, 134102.

55. W. Nicolazzi and S. Pillet (2012). Structural aspects of the relaxation process in spin crossover solids: Phase separation, mapping of lattice strain, and domain wall structure, *Phys. Rev. B*, 85, 094101.

56. A. Slimani, K. Boukheddaden, F. Varret, H. Oubouchou, M. Nishino, and S. Miyashita (2013). Microscopic spin-distortion model for switchable molecular solids: Spatiotemporal study of the deformation field and local stress at the thermal spin transition, *Phys. Rev. B*, 87, 014111.

57. F. Varret, A. Slimani, K. Boukheddaden, C. Chong, H. Mishra, E. Collet, J. Haasnoot, and S. Pillet (2011). The propagation of the thermal spin transition of [Fe(btr)2(NCS)2].H2O single crystals, observed by optical microscopy, *New J. Chem.*, 35, 2333.

58. A. Slimani, F. Varret, K. Boukheddaden, C. Chong, H. Mishra, J. Haasnoot, and S. Pillet (2011). Visualization and quantitative analysis of spatiotemporal behavior in a first-order thermal spin transition: A stress-driven multiscale process, *Phys. Rev. B*, 84, 094442.

59. C. Chong, A. Slimani, F. Varret, K. Boukheddaden, E. Collet, J. C. Ameline, R. Bronisz, and A. Hauser (2011). The kinetics features of a thermal spin transition characterized by optical microscopy on the example of [Fe(bbtr)3](ClO4)2 single crystals: Size effect and mechanical instability, *Chem. Phys. Lett.*, 504, 29.

60. M. Sy, F. Varret, K. Boukheddaden, G. Bouchez, J. Marrot, S. Kawata, and S. Kaizaki (2014). Structure-driven orientation of the high-spin–low-spin interface in a spin-crossover single crystal, *Angew. Chem.*, 126, 1–5.

61. A. Slimani, F. Varret, K. Boukheddaden, D. Garrot, H. Oubouchou, and S. Kaizaki (2013). Velocity of the high-spin low-spin interface inside the thermal hysteresis loop of a spin-crossover crystal, via photothermal control of the interface motion, *Phys. Rev. Lett.*, 110, 087208.

62. F. Varret, C. Chong, A. Slimani, D. Garrot, Y. Garcia, and D. Naik Anil (2013). Real-time observation of spin-transitions by optical microscopy, in *Spin-Crossover Materials: Properties and Applications* (ed. M. A. Halcrow), 1st ed., John Wiley and Sons. Ltd.

63. S. Bedoui, G. Molnár, S. Bonnet, C. Quintero, H. J. Shepherd, W. Nicolazzi, L. Salmon, and A. Bousseksou (2010). Raman spectroscopic and optical

imaging of high spin/low spin domains in a spin crossover complex, *Chem. Phys. Lett.*, 499, 94.

64. S. Bedoui, M. Lopes, W. Nicolazzi, S. Bonnet, S. Zheng, G. Molnár, and A. Bousseksou (2012). Triggering a phase transition by a spatially localized laser pulse: Role of strain, *Phys. Rev. Lett.*, 109, 13.

65. S. Bonnet, G. Molnár, J. S. Costa, M. A. Siegler, A. L. Spek, A. Bousseksou, W. Fu, P. Gamez, and J. Reedijk (2009). Influence of sample preparation, temperature, light, and pressure on the two-step spin crossover mononuclear compound [Fe(bapbpy)(NCS)(2)], *Chem. Mater.*, 21, 1123.

66. K. Boukheddaden, M. Nishino, and S. Miyashita (2007). Molecular dynamics and transfer integral investigations of an elastic anharmonic model for phonon-induced spin crossover, *Phys. Rev. Lett.*, 100, 177206.

67. J. Jeftic, F. Varret, A. Hauser, O. Roubeau, M. Matsarski, and J.-P. Riviera (1999). Patterns during photoexcitation and high-spin → low-spin relaxation in [Fe(Ptz)(6)](BF4)(2) spin transition crystal, *Mol. Cryst. Liq. Cryst.*, 335, 511–520.

68. Y. Ogawa, S. Koshihara, K. Koshino, T. Ogawa, C. Urano, and H. Tagaki (2000). Dynamical aspects of the photoinduced phase transition in spin-crossover complexes, *Phys. Rev. Lett.*, 84, 3181–3184.

69. C. Chong, H. Mishra, K. Boukheddaden, S. Denise, G. Bouchez, F. Varret, Y. Garcia, E. Collet, and J.-C. Ameline (2010). Electronic and structural aspects of spin transitions observed by optical microscopy. The case of [Fe(ptz)6](BF4)2, *J. Phys. Chem. B*, 114, 1975–1984.

70. K. Nakano, N. Suemura, S. Kawata, A. Fuyuhiro, T. Yagi, S. Nasu, S. Morimoto, and S. Kaizaki (2005). Substituent effect of the coordinated pyridine in a series of pyrazolato bridged dinuclear diiron(II) complexes on the spin-crossover behavior, *Dalton Trans.*, 4, 740.

71. C. J. Schneider, J. D. Cashion, B. Moubaraki, S. M. Neville, S. R. Batten, D. R. Turner, and K. S. Murray (2007). The magnetic and structural elucidation of 3,5-bis(2-pyridyl)-1,2,4-triazolate-bridged dinuclear iron(II) spin crossover compounds, *Polyhedron*, 26, 1764.

72. M. Avrami (1939). Kinetics of phase change I: General theory, *J. Chem. Phys.*, 7, 1103.

73. O. Fouche, J. Degert, G. Jonusauskas, N. Daro, J.-F Létard, and E. Freysz (2009). Mechanism for optical switching of the spin crossover [Fe(NH2-trz)3](Br)2[middle dot]3H2O compound at room temperature, *Phys. Chem. Chem. Phys.*, 12, 3044–3052.

74. A. Hauser (2004). Light-induced spin crossover and the high-spin $\rightarrow$ low-spin relaxation, *Top. Curr. Chem.*, 234, 155–198.

75. K. Boukheddaden, I. Shteto, B. Hôo, and F. Varret (2000). Dynamical model for spin-crossover solids. I. Relaxation effects in the mean-field approach, *Phys. Rev. B*, 62, 14796.

76. K. Boukheddaden, I. Shteto, B. Hôo, and F. Varret (2000). Dynamical model for spin-crossover solids. II. Static and dynamic effects of light in the mean-field approach, *Phys. Rev. B*, 62, 14806.

77. F. Varret, K. Boukheddaden, C. Chong, A. Goujon, B. Gillon, J. Jeftic, and A. Hauser (2007). Light-induced phase separation in the [Fe(ptz)(6)](BF4)(2) spin-crossover single crystal, *EPL*, 77, 30007.

78. K. Koshino and T. Ogawa (1998). Domino effects in photoinduced structural change in one-dimensional systems, *J. Phys. Soc. Jpn.*, 67, 2174–2177.

79. K. Yonemitsu and K. Nasu (2008). Theory of photoinduced phase transitions in itinerant electron systems, *Phys. Reports*, 465, 1.

80. C. Enachescu, M. Nishino, and S. Miyashita (2013). Theoretical descriptions of spin-transitions in bulk lattices, in *Spin-Crossover Materials: Properties and Applications* (ed. M. A. Halcrow), 1st ed., John Wiley and Sons. Ltd., pp. 454–474.

81. K. Ishida and K. Nasu (2008). Nonlinearity in the dynamics of photoinduced nucleation process, *Phys. Rev. Lett.*, 100, 116403.

82. E. Collet, M. H. L. Cailleau, M. B. Le Cointe, H. Cailleau, M. Wulff, T. Luty, S. Koshihara, M. Meyer, L. Toupet, P. Rabillerand, and S. Techert (2003). Laser-induced ferroelectric structural order in an organic charge-transfer crystal, *Science*, 300, 612.

83. N. Willenbacher and H. Spiering (1988). The elastic interaction of high-spin and low-spin complex-molecules in spin-crossover compounds, *J. Phys. C: Solid State Phys.*, 21, 1423.

84. H. Spiering and N. Willenbacher (1989). Elastic interaction of high-spin and low-spin complex-molecules in spin-crossover compounds. 2, *J. Phys.: Condens. Matter*, 1, 10089.

85. H. Spiering (2004). Elastic interaction in spin crossover compounds, *Topics Curr. Chem.*, 235, 171.

86. W. Nicolazzi, S. Pillet, and C. Lecomte (2008). Two-variable anharmonic model for spin-crossover solids: A like-spin domains interpretation, *Phys. Rev. B.*, 78, 174401.

87. T. Nakada, P. A. Rikvold, T. Mori, M. Nishino, and S. Miyashita (2011). Crossover between a short-range and a long-range Ising model, *Phys. Rev. B*, 84, 054433.

88. C. Enachescu, M. Nishino, S. Miyashita, L. Stoleriu, A. Stancu, and A. Hauser (2010). Cluster evolution in spin crossover systems observed in the frame of a mechano-elastic model, *Eur. Phys. Lett.*, 91, 27003.

89. C. Enachescu, M. Nishino, S. Miyashita, L. Stoleriu, and A. Stancu (2012). Monte Carlo Metropolis study of cluster evolution in spin-crossover solids within the framework of a mechanoelastic model, *Phys. Rev. B.*, 86, 054114.

90. S. Galam (1987). Plastic crystals, melting, and random-fields, *Phys. Lett. A*, 122, 271.

91. S. Galam and P. Depondt (1988). Rotational states and self-dilution vs translational disorder in plastic neopentane, *Europhys. Lett.*, 5, 43.

92. A. Bousseksou, J. J. McGarvery, F. Varret, J. A. Real, J. P. Tuchagues, A. C. Dennis, and M. L. Boillot (2000). Raman spectroscopy of the high- and low-spin states of the spin crossover complex Fe(phen)(2)(NCS)(2): An initial approach to estimation of vibrational contributions to the associated entropy change, *Chem. Phys. Lett.*, 318, 409–416.

93. G. Brehm, M. Reiher, and S. Shneider (2002). Estimation of the vibrational contribution to the entropy change associated with the low- to high-spin transition in Fe(phen)(2)(NCS)(2) complexes: Results obtained by IR and Raman spectroscopy and DFT calculations, *J. Phys. Chem. A*, 106, 12024–12034.

94. K. Boukheddaden and A. Bailly-Reyre (2013). Towards the elastic properties of 3D spin-crossover thin films: Evidence of buckling effects, *Eur. Phys. Lett.*, 103, 26005.

95. S. Miyashita, Y. Konishi, H. Tokoro, M. Nishino, K. Boukheddaden, and F. Varret (2005). Structures of metastable states in phase transitions with a high-spin low-spin degree of freedom, *Prog. Theor. Phys.*, 114, 719.

96. K. Boukheddaden (2004). Anharmonic model for phonon-induced first-order transition in 1-D spin-crossover solids, *Progr. Theor. Phys.*, 112, 205.

97. D. J. Bergman and B. Halperin (1976). Critical behavior of an Ising-model on a cubic compressible lattice, *Phys. Rev. B.*, 13, 2145.

98. M. Sorai (2004). Spin crossover in transition metal compounds III, *Top. Curr. Chem.*, 235, 153.

99. B. Gallois, J.-A. Real, C. Hauw, and J. Zarembowitch (1990). Structural-changes associated with the spin transition in Fe(Phen)2(NCS)2: A single-crystal X-ray-investigation, *Inorg. Chem.*, 29, 1152.

100. J. Jung, F. Bruchhäuser, R. Feile, H. Spiering, and P. Gütlich (1996). The cooperative spin transition in [FexZn1-x(ptz)(6)](BF4)(2). 1. Elastic properties: An oriented sample rotation study by Brillouin spectroscopy, *Z. Phys. B: Condens. Matter*, 100, 517.

101. H. Spiering, K. Boukheddaden, J. Linares, and F. Varret (2004). Total free energy of a spin-crossover molecular system, *Phys. Rev. B.*, 70, 184106.

102. T. Nakada, T. Mori, S. Miyashita, M. Nishino, S. Todo, W. Nicolazzi, and P. A. Rikvold (2012). Critical temperature and correlation length of an elastic interaction model for spin-crossover materials, *Phys. Rev. B*, 85, 054408.

103. K. Frankzrahe, P. Nielaba, and S. Sengupta (2010). Coarse-graining microscopic strains in a harmonic, two-dimensional solid: Elasticity, nonlocal susceptibilities, and nonaffine noise, *Phys. Rev. B.*, 82, 016112.

104. H. Wagner and H. Horner (1974). Elastic interaction and phase-transition in coherent metal-hydrogen systems, *Adv. Phys.*, 23, 587.

105. R. A. Cowley (1976). Acoustic phonon instabilities and structural phase-transitions, *Phys. Rev. B.*, 13, 4877.

106. E. Courtens, R. Gammon, and S. Alexander (1979). KH2PO4 in a field-transition without critical microscopic fluctuations, *Phys. Rev. Lett.*, 43, 1026.

107. H. Wagner and J. Swift (1970). Elasticity of a magnetic lattice near magnetic critical point, *Z. Phys.*, 239, 182.

108. H. Wagner (1970). Phase transition in a compressible Ising ferromag-net, *Phys. Rev. Lett.*, 25, 31.

109. H. Wagner (1970). Correction, *Phys. Rev. Lett.*, 25, 261(E).

110. E. R. Grannan, M. Randeria, and J. P. Sethna (1990). Low-temperature properties of a model glass. 1. Elastic dipole model, *Phys. Rev. B.*, 41, 7784.

111. J. D. Eshelby (1956). The continuum theory of lattice defects, *Sol. Stat. Phys.*, 3, 79.

112. J. F. Létard, S. Asthana, H. J. Shepherd, P. Guionneau, A. E. Goeta, N. Suemura, R. Ishikawa, and S. Kaizaki (2012). Photomagnetism of a sym-cis-dithiocyanato iron(II) complex with a tetradentate N,N'-Bis(2-pyridylmethyl)1,2-ethanediamine ligand, *Chem. Eur. J.*, 18, 5924–5934.

113. C. Chong, M. Itoi, K. Boukheddaden, E. Codjovi, A. Rotaru, F. Varret, F. A. Frye, D. R. Talham, I. Maurin, D. Chernyshov, and M. Castro (2011). Metastable state of the photomagnetic Prussian blue analog $K_{0.3}Co[Fe(CN)_6]_{0.77}\cdot 3.6H_2O$ investigated by various techniques, *Phys. Rev. B*, 84, 144102.

114. The Neumann boundary condition is one of the three types of boundary conditions commonly encountered in the solution of partial differential equations. It specifies the normal derivative of the function, U, for example, on a surface. It writes, $\vec{n}.\vec{\nabla}U = f(\vec{r}, t)$, where $\vec{n}$ is the normal vector to the surface at $\vec{r}$.

115. M.-Y. Im, L. Bocklage, G. Meier, and P. Fischer (2012). Magnetic soft X-ray microscopy of the domain wall depinning process in permalloy magnetic nanowires, *J. Phys. Condens. Matter*, 24, 024203.

116. S. Lemerle, F. J. Ferré, C. Chappert, V. Mathlet, T. Giamarchi, and P. Ledoussal (1998). Domain wall creep in an Ising ultrathin magnetic film, *Phys. Rev. Lett.*, 80, 849.

117. M. Nishino, T. Nakada, C. Enachescu, K. Boukheddaden, and S. Miyashita (2013). Crossover of the roughness exponent for interface growth in systems with long-range interactions due to lattice distortion, *Phys. Rev. B.*, 88, 094303.

118. E. Coronado, J. R. Galán-Mascarós, M. Monrabal-Capilla, J. García-Martínez, and P. Pardo-Ibáñez (2007). Bistable spin-crossover nanoparticles showing magnetic thermal hysteresis near room temperature, *Adv. Mater.*, 19, 1359–1361.

119. F. Prins, M. Monrabal-Capilla, E. A. Osorio, E. Coronado, and H. S. J. van der Zant (2011). Room-temperature electrical addressing of a bistable spin-crossover molecular system, *Adv. Mater.*, 23, 1545–1549.

120. J. R. Galán-Mascarós, E. Coronado, A. Forment-Aliaga, M. Monrabal Capilla, E. Pinilla-Cienfuegos, and M. Ceolin (2010). Tuning size and thermal hysteresis in bistable spin crossover nanoparticles, *Inorg. Chem.*, 49, 5706–5714.

121. A. Rotaru, F. Varret, A. Gindulescu, J. Linares, A. Stancu, J. F. Létard, T. Forestier, and C. Etrillard (2011). Size effect in spin-crossover systems investigated by FORC measurements, for surfaced [Fe(NH2-trz)3](Br)2·3H2O nanoparticles: Reversible contributions and critical size, *Eur. Phys. J. B*, 84, 439.

122. T. Forestier, S. Mornet, N. Daro, T. Nishihara, S. Mouri, K. Tanaka, O. Fouche, E. Freysz, and J.-F. Létard (2008). Nanoparticles of iron(II) spin-crossover, *Chem. Commun.*, 36, 4327.

123. T. Forestier, A. Kaiba, S. Pechev, D. Denux, P. Guionneau, C. Etrillard, N. Daro, E. Freysz, and J.-F. Létard (2009). Nanoparticles of [Fe(NH2-trz)(3)]Br-2 center dot 3H(2)O (NH2-trz=2-Amino-1,2,4-triazole) prepared by the reverse Micelle technique: Influence of particle and coherent domain sizes on spin-crossover properties, *Chem. Eur. J.*, 15, 6122.

124. V. Martinez, I. Boldog, A. B. Gaspar, V. Ksenofontov, A. Bhattacharjee, P. Gütlich, and J. A. Real, Spin crossover phenomenon in nanocrystals and nanoparticles of [Fe(3-Fpy)2M(CN)4] (MII = Ni, Pd, Pt) two-dimensional coordination polymers, *Chem. Mater.*, 22, 4271–4281.

125. L. Salmon, G. Molnar, D. Zitouni, C. Quintero, C. Bergaud, J.-C. Micheau, and A. Bousseksou (2010). A novel approach for fluorescent thermometry and thermal imaging purposes using spin crossover nanoparticles, *J. Mater. Chem.*, 20, 5499.

126. I. Boldog, A. B. Gaspar, V. Martinez, P. Pardo-Ibanez, V. Ksenofontov, A. Bhattacharjee, P. Gütlich, and J. A. Real (2008). Spin-crossover nanocrystals with magnetic, optical, and structural bistability near room temperature, *Angew. Chem., Int. Ed.*, 47, 6433–6437.

127. C. Bartual-Murgui, N. A. Ortega-Villar, H. J. Shepherd, M. C. Munoz, L. Salmon, G. Molnar, A. Bousseksou, and J.-A. Real (2011). Enhanced porosity in a new 3D Hofmann-like network exhibiting humidity sensitive cooperative spin transitions at room temperature, *J. Mater. Chem.*, 21, 7217.

128. J. Larionova, L. Salmon, Y. Guari, A. Tokarev, K. Molvinger, G. Molnar, and A. Bousseksou (2008). Towards the ultimate size limit of the memory effect in spin-crossover solids, *Angew. Chem., Int. Ed.*, 47, 8236.

129. A. Tissot, C. Enachescu, and M.-L. Boillot (2012). Control of the thermal hysteresis of the prototypal spin-transition FeII(phen)2(NCS)2 compound via the microcrystallites environment: Experiments and mechanoelastic model, *J. Mater. Chem.*, 22, 20451–20457.

130. F. Volatron, L. Catala, E. Rivière, A. Gloter, O. Stéphan, and T. Mallah (2008). Spin-crossover coordination nanoparticles, *Inorg. Chem.*, 47, 6584.

131. I. A. Gural'skiy, B. G. Molnár, I. O. Fritsky, L. Salmon, and A. Bousseksou (2012). Synthesis of [Fe(hptrz)3](OTs)2 spin crossover nanoparticles in microemulsion, *Polyhedron*, 38(1), 245–250.

132. I. A. Gural'skiy, C. M. Quintero, G. Molnár, I. O. Fritsky, L. Salmon, and A. Bousseksou (2012). Synthesis of spin-crossover nano- and micro-objects in homogeneous media, *Chem. Eur. J.*, 18, 9946–9954.

133. H. Oubouchou, A. Slimani, and K. Boukheddaden (2013). Interplay between elastic interactions in a core-shell model for spin-crossover nanoparticles, *Phys. Rev. B*, 87, 104104.

134. A. Slimani, K. Boukheddaden, and K. Yamashita (2014). Thermal spin transition of circularly shaped nanoparticles in a core-shell structure investigated with an electroelastic model, *Phys. Rev. B*, 89, 214109.

135. A. Muraoka, K. Boukheddaden, J. Linares, and F. Varret (2011). Two-dimensional Ising-like model with specific edge effects for spin-crossover nanoparticles: A Monte Carlo study, *Phys. Rev. B*, 84, 054119.

136. T. C. Hales (2001). The honeycomb conjecture, *Disc. Comp. Geom.*, 25, 1–22.

137. R. Medishetty, A. Husain, Z. Bai, T. Runcevski, R. E. Dinnebier, P. Naumov, and J. J. Vittal (2014). Single crystals popping under UV light: A photosalient effect triggered by a [2+2] cycloaddition reaction, *Angew. Chem. Int. Ed.*, 53, 5907.

138. A. S. C. Sahoo, N. K. Nath, L. Zhang, M. H. Semreen, T. H. Al-Tel, and P. Naumov (2014). Actuation based on thermo/photosalient effect: A biogenic smart hybrid driven by light and heat, *RSC Adv.*, 4, 7640.

139. N. K. Nath, L. Pejov, S. Nichols, C. Hu, N. Saleh, B. Kahr, and P. Naumov (2014). Model for photoinduced bending of slender molecular crystals, *J. Amer. Chem. Soc.*, 136(7), 2757.

140. J.-H. Ko, K.-S. Lee, S. C. Sahoo, and P. Naumov (2013). Isomorphous phase transition of 1,2,4,5-tetrabromobenzene jumping crystals studied by Brillouin light scattering, *Solid State Commun.*, 173, 46.

141. S. C. Sahoo, S. B. Sinha, M. S. R. N. Kiran, U. Ramamurty, A. F. Dericioglu, C. M. Reddy, and P. Naumov (2013). Kinematic and mechanical profile of the self-actuation of thermosalient crystal twins of 1,2,4,5-tetrabromobenzene: A molecular crystalline analogue of a bimetallic strip, *J. Amer. Chem. Soc.*, 135, 13843.

142. P. Naumov, S. C. Sahoo, B. A. Zakharov, and E. V. Boldyreva (2013). Dynamic single crystals: Kinematic analysis of photoinduced crystal jumping (the photosalient effect), *Angew. Chem. Int. Ed.*, 52, 9990.

143. R. M. van der Veen, O.-H. Kwon, A. Tissot, A. Hauser, and A. H. Zewail (2013). Single-nanoparticle phase transitions visualized by four-dimensional electron microscopy, *Nat. Chem.*, 5, 395.

144. M. Gruber, V. Davesne, M. Bowen, S. Boukari, E. Beaurepaire, W. Wulfhekel, and T. Miyamachi (2014). Spin state of spin-crossover complexes: From single molecules to ultrathin films, *Phys. Rev. B*, 89, 195415.

145. K. W. Törnroos, M. Hostettler, D. Chernyshov, B. Vangdal, and H.-B. Bürgi (2006). Interplay of spin conversion and structural phase transformations: Re-entrant phase transitions in the 2-propanol solvate of tris(2picolylamine)iron(II) dichloride, *Chemistry*, 12, 6207.

146. M. Hostettler, K. W. Törnroo, Chernyshov, B. Vangdal, and H.-B. Bürgi (2004). Challenges in engineering spin crossover: Structures and magnetic properties of six alcohol solvates of iron(II) tris(2-picolylamine) dichloride, *Angew. Chem. Int. Ed. Engl.*, 6, 43(35) 4589.

147. D. I. Khomskii and F. V. Kusmartsev (2004). Intersite elastic coupling and the Invar effect, *Phys. Rev. B*, 70, 012413.

148. D. I. Khomskii and U. Low (2004). Superstructures at low spin-high spin transitions, *Phys. Rev. B*, 69, 184401.

149. D. Chernyshov, H.-B. Bürgi, M. Hostettler, and K. W. Törnroos (2004). Landau theory for spin transition and ordering phenomena in Fe(II) compounds, *Phys. Rev. B*, 70, 094116.

150. D. Chernyshov, N. Klinduhov, K. W. Törnroos, M. Hostettler, B. Vangdal, and H.-B. Bürgi (2007). Coupling between spin conversion and solvent disorder in spin crossover solids, *Phys. Rev. B*, 76, 014406.

151. M. Marchivie, P. Guionneau, J. F. Létard, and D. Chasseau (2003). Towards direct correlations between spin-crossover and structural features in iron(II) complexes, *Acta Cryst. B*, 59, 479–486.

152. S. Pillet, J. Hubsch, and C. Lecomte (2004). Single crystal diffraction analysis of the thermal spin conversion in [Fe(btr)(2)(NCS)(2)](H2O): Evidence for spin-like domain formation, *Eur. Phys. J. B*, 38, 541–552.

153. N. Moliner, A. B. Gaspar, M. C. Munoz, V. Niel, J. Cano, and J. A. Real (2001). Light- and thermal-induced spin crossover in {Fe(abpt)(2)[N(CN)(2)](2)}. Synthesis, structure, magnetic properties, and high-spin ↔ low-spin relaxation studies, *Inorg. Chem.*, 40, 3986.

154. S. Pillet, E. E. Bendeif, S. Bonnet, H. J. Shepherd, and P. Guionneau (2012). Multimetastability, phototrapping, and thermal trapping of a metastable commensurate superstructure in a Fe^{II} spin-crossover compound, *Phys. Rev. B*, 86, 064106.

155. E. Collet, H. Watanabe, N. Bréfuel, L. Palatinus, L. Roudaut, L. Toupet, K. Tanaka, J.-P. Tuchagues, P. Fertey, S. Ravy, B. Toudic, and H. Cailleau (2012). Aperiodic spin state ordering of bistable molecules and its photoinduced erasing, *Phys. Rev. Lett.*, 109, 257206.

156. T. Miyamachi, M. Gruber, V. Davesne, M. Bowen, S. Boukari, L. Joly, F. Scheurer, G. Rogez, T. K. Yamada, P. Ohresser, E. Beaurepaire, and

W. Wulfhekel (2012). Robust spin crossover and memristance across a single molecule, *Nat. Commun.*, 3, 938.

157. T. G. Gopakumar, F. Matino, H. Naggert, A. Bannwarth, F. Tuczek, and R. Berndt (2012). Electron-induced spin crossover of single molecules in a bilayer on gold, *Angew. Chem., Int. Ed.*, 51, 6262.

158. T. Mahfoud, G. Molnár, S. Cobo, L. Salmon, C. Thibault, C. Vieu, P. Demont, and A. Bousseksou (2011). Electrical properties and non-volatile memory effect of the [Fe(HB(pz)3)2] spin crossover complex integrated in a microelectrode device, *Appl. Phys. Lett.*, 99, 053307.

159. S. Shi, G. Schmerber, J. Arabski, J. B. Beaufrand, D. J. Kim, S. Boukari, M. Bowen, N. T. Kemp, N. Viart, and G. Rogez (2009). Study of molecular spin-crossover complex Fe(phen)(2)(NCS)(2) thin films, *Appl. Phys. Lett.*, 95, 043303.

160. H. Naggert, A. Bannwarth, S. Chemnitz, T. vonHofe, E. Quandt, and F. Tuczek (2011). First observation of light-induced spin change in vacuum deposited thin films of iron spin crossover complexes, *Dalton Trans.*, 40, 6364.

161. T. Palamarciuc, J. C. Oberg, F. E. Hallak, C. F. Hirjibehedin, M. Serri, S. Heutz, J.-F. Létard, and P. Rosa (2012). Spin crossover materials evaporated under clean high vacuum and ultra-high vacuum conditions: From thin films to single molecules, *J. Mater. Chem.*, 22, 9690.

162. A. Pronschinske, Y. Chen, G. F. Lewis, D. A. Shultz, A. Calzolari, M. Buongiorno Nardelli, and D. B. Dougherty (2013). Modification of molecular spin crossover in ultrathin films, *Nano Lett.*, 13, 1429.

163. M. Bernien, D. Wiedemann, C. F. Hermanns, A. Krüger, D. Rolf, W. Kroener, P. Müller, A. Grohmann, and W. Kuch (2012). Spin crossover in a vacuum-deposited submonolayer of a molecular iron(II) complex, *J. Phys. Chem. Lett.*, 3, 3431.

164. B. Warner, J. C. Oberg, T. G. Gill, F. El Hallak, C. F. Hirjibehedin, M. Serri, S. Heutz, M.-A. Arrio, P. Sainctavit, M. Mannini, G. Poneti, R. Sessoli, and P. Rosa (2013). Temperature- and light-induced spin crossover observed by X-ray spectroscopy on isolated Fe(II) complexes on gold, *J. Phys. Chem. Lett.*, 4, 1546.

165. R. A. Fisher (1937). The wave of advance of advantageous genes, *Ann. Eug.*, 7, 355.

166. A. N. Kolmogorov, I. G. Petrovsky, and N. S. Piskunov (1937). Investigation of the equation of diffusion combined with increasing of

the substance and its application to a biology problem, *Bull. Moscow State Univ. Ser. A: Math. Mech.*, 1(6), 1.

167. M. Paez-Espejo, M. Sy, F. Varret, and K. Boukheddaden (2014). Quantitative macroscopic treatment of the spatiotemporal properties of spin crossover solids based on a reaction diffusion equation, *Phys. Rev. B.*, 89, 024306.

168. H. Spiering, E. Meissner, H. Kopen, E. W. Müller, and P. Gütlich (1982). The effect of the lattice expansion on high-spin reversible low-spin transitions, *Chem. Phys.*, 68, 65.

169. A. Hauser, P. Gütlich, and H. Spiering (1986). High-spin-]low-spin relaxation kinetics and cooperative effects in the [Fe(1-propylte-trazole)6](BF4)2 AND [Zn1-XFeX(1-propyltetrazole)6](BF4)2 spin-crossover systems, *Inorg. Chem.*, 25, 4245.

170. A. Hauser (1991). Intersystem crossing in the [Fe(ptz)6](BF4)2 spin crossover system (ptz = 1-propyltetrazole), *J. Chem. Phys.*, 94, 2741.

Index

antidot lattices 258, 287, 302, 307–308
antidots 310, 319–320
antivortex motion 54
apoferritin 262–264, 273
atomic stiffness 20–22, 26
azimuthal modes 136–138, 143, 147–148
  second-order 142, 144
azimuthal number 134, 137–139, 142–143

backward volume magnetostatic wave geometry (BVMW) 290, 302, 311, 313–314, 319
bandgap, directional 268–269
Bloch functions 266, 301, 304
Bloch theorem 283–284, 300, 303, 312
Brillouin light scattering 32, 50–51
Brillouin zone (BZ) 138, 266, 268, 270, 293, 301–302, 311–314, 316
BVMW, *see* backward volume magnetostatic wave geometry
BZ, *see* Brillouin zone

chevrons 35, 40, 43
cobalt nanodots 197, 199, 201
  self-organized 203

core–shell nanoparticles 405, 408–409
  circular 411
core–shell structure 274, 402–403
crystalline symmetry 24, 26–27

Debye temperatures 338, 354
demagnetizing field 3, 46, 62, 66–67, 72, 228, 230, 234, 236, 243, 265, 293, 297–299, 302–304, 309
  macroscopic 242, 244
  static 289, 303, 313, 315
density functional theory (DFT) 17–18, 61, 352
deterministic fractals 164, 168–169, 172–173, 180–181, 187, 190
deterministic Sierpinski fractals 160, 164, 176, 178
DFT, *see* density functional theory
dipolar effects 3, 5–6, 43, 59–60
dipolar forces 28, 30, 43
dipolar interactions 3–4, 6–7, 44–45, 48, 50, 60, 62–66, 72, 82–84, 123–125, 141, 143, 200–201, 244, 259
  long-ranged 1, 83
dipole-exchange spin waves 290, 310, 315
dipole interactions 292, 310, 319–320, 322
  long-range 285, 319, 321

dipole–dipole interactions 29, 101–104, 106, 120

dynamical matrix method 127, 129, 131, 153

EAM, *see* embedded atom model

eigenfrequencies 57, 168

eigenmodes 114, 167, 169, 171, 190, 304

eigenvalues 63, 167, 304–305, 322

electrodeposition 222–223, 225

electrons 18, 31, 60, 283, 294, 335

ELM, *see* empty lattice model

embedded atom model (EAM) 22

empty lattice model (ELM) 310–315

equations of motion 107, 130–131, 418

exchange Hamiltonian, local 64–65

exchange interactions
short-range 319, 321
weak 141, 149

exchange stiffness constant 227, 268, 289

fast Fourier transform (FFT) 42, 106–107

FEMM, *see* finite element method magnetics

ferritin 262–263, 275

ferromagnetic films 284, 289, 300, 306, 308–309, 312

ferromagnetic materials 259, 284, 307–309, 320

ferromagnetic matrix 264, 266, 273–274, 307

ferromagnetic phase 173, 188, 190

FFT, *see* fast Fourier transform

finite element method magnetics (FEMM) 244

first Brillouin zone 265–267, 293, 301, 303, 310–311, 321

fractal dimension 21–23, 163–165, 168–169, 173, 175, 178–180, 184–186

fractal nanostructures 175, 185

fractals, phase transitions on 173, 175, 177, 179, 181, 183, 185, 187, 189

gold 14, 17

Green function matrices 80

gyrotropic motion 52, 54–55, 57, 78, 102, 126–127, 138, 152

gyrotropic vortex motion 44, 46, 52, 78, 83

Heisenberg spins 43, 45, 48–49

high-aspect-ratio nanoparticles 213–248
magnetism of 227–239

high-resolution electron microscopy (HREM) 11, 13, 219

hotoemission electron microscopy (PEEM) 37

HREM, *see* high-resolution electron microscopy

hybridization 146–148

Ising model 173, 175, 178–179, 181, 363

Ising spins 43–44, 48

Kerr effect 4, 13, 35

Landau state 125–126, 151–152

Landau–Lifshitz equation 101, 104, 106, 108, 300, 321

Langevin simulation 4, 44–45, 54–55, 58, 101–102, 104, 106, 108, 110, 112, 114, 116, 118, 120

LEED, *see* low-energy electronic diffraction

LEMS, *see* low-energy magnetic structures

like spin domains (LSDs) 339

liquid crystals 7, 84

low-energy electronic diffraction (LEED) 11

low-energy magnetic structures (LEMS) 5

low-spin domains 357, 365, 367, 377, 381–382

low-spin states 335, 342, 347–348, 352, 354, 378, 380–381, 384, 393–395, 407–408

LSDs, *see* like spin domains

magnetic anisotropy 193–194, 196, 198–200, 202, 204–206, 294

magnetic anisotropy energy 200–202

magnetic carriers 60, 83

magnetic excitations 7, 31–32, 51–52, 59–60, 82–85, 171

magnetic force microscopy (MFM) 40–41

magnetic interactions 3, 7, 18, 60, 83, 101–102

magnetic moments, elementary 127–128, 134, 136

magnetic resonance 50–51

magnetization, circular 125–126, 128, 150–151

magnetization reversal 50–51, 193–194, 196, 198, 200–204, 206, 232–233, 249

magneto-crystalline energy 227, 230

magneto-optical Kerr effect (MOKE) 35–36

magnetocrystalline anisotropy 3, 206, 215, 223, 228–229, 231, 238–239, 241, 243, 246, 284–285, 289, 293, 318, 321

magnetocrystalline anisotropy constants 227

magnetoferritin-based magnonic crystals 266–267, 269, 271, 273

magnetoferritins 260, 262–263, 275

magnetostatic interactions 265, 269, 272, 292, 305

magnetostatic self-energy 297

magnonic band structure 287, 293, 305, 308, 310–313, 315–316, 322

magnonic bandgap 259–261, 266–270, 273–274

magnonic bands 311, 317

magnonic crystals (MCs) 258–261, 264, 266–267, 269, 273–274, 283–286, 288–290, 292–294, 296–302, 304–321, 358, 371, 375, 381–382, 384–385

magnonic crystals, three-dimensional 257

magnonic gap 267, 270, 273–274, 317, 319

magnonics 82, 259, 284–287, 295, 301, 318

magnons 5, 43, 54, 114, 124–125, 167

Maxwell equation 62, 294, 298, 302

MCs, *see* magnonic crystals

MD, *see* molecular dynamics

MFM, *see* magnetic force microscopy

mFT crystals 261, 267, 269, 273–274
mFT magnetic 272–273
micromagnetic simulations 36, 102, 127, 139, 143, 148, 152, 234, 314
MOKE, *see* magneto-optical Kerr effect
molecular dynamics (MD) 17–18, 42
Monte Carlo simulation 20, 149, 180, 182, 188–190, 351, 383, 395

nanodots 15–16, 124, 126–128, 130, 132, 134, 136, 138, 140, 142, 144, 146, 148, 150, 259
nanomagnets 194, 196, 200–202, 204, 206
nanoparticles
  alloyed 10, 194
  biomimetic 258, 261, 263
  biomimetic magnetic 257
  cobalt 197, 199
  CoPt 198, 206
  elongated 214, 221, 226, 231
  magnetic 1–2, 5–8, 13, 16–17, 29–30, 41–44, 47–52, 57, 82–83, 85, 197, 202, 229–230, 233, 257
  magnetic excitations of 50, 53
  magnetic structure of 31, 42
  magnetoferritin 257–258, 260–264, 266, 268, 270, 272, 274–275
  mFT 263–264, 266–267, 269, 271–274
  spin-crossover 334, 399–402, 404, 406, 409
  square-shaped 410, 412
  three-dimensional 102, 204

nanowires 213–214, 217, 219, 223, 226
  cobalt 232–233, 239, 246
  elongated aggregates of 244
  magnetic 245–247, 249
  metal 222
Néel–Brown Model 200–202
NIDOS, *see* normalized integrated density of states
normalized integrated density of states (NIDOS) 168–170, 172
nuclear spins 7, 83
nucleation 51, 337, 342, 344, 347, 361, 365, 367, 371–372, 379, 381–382, 391, 394, 401, 416–417
  single domain 372
nucleation modes 132, 142

optical microscopy 338–339, 341, 343, 345, 347, 349, 365, 367, 414
organometallic chemistry 219, 238, 248

PEEM, *see* hotoemission electron microscopy
phase transitions
  order-disorder 415–416
  second-order 173, 177–178, 182
photonic crystals 258, 260, 269, 284, 318
physical vapor deposition 202, 206
planar magnonic crystals 307, 310
plane wave method (PWM) 261, 264–266, 287, 301, 304, 316, 321
polyol process 215–217, 219–220, 226, 238–239
porous systems 12–13

Potts model 181–182, 184–185, 187–188
  14-state 188–190
protein cages 261–262, 275
PWM, *see* plane wave method

quasicrystalline structures 26
quasicrystals 6, 84

random discrete scale invariance 186–187, 190
Ruderman–Kittel–Kasuya–Yosida interaction 3, 60, 62

S-shaped structures 76–77
S-shaped vortex 74–75, 77–78, 83
SC, *see* spin-crossover
scanning electron microscopy with polarized analysis (SEMPA) 36
second-order transitions 165, 184–188
self-organized magnetic arrays (SOMA) 198
self-organized surfaces 194–195, 198
SEMPA, *see* scanning electron microscopy with polarized analysis
Sierpinski fractals 160–161, 163
SMSs, *see* switchable molecular solids
solid-state physics 284, 310
SOMA, *see* self-organized magnetic arrays
spin-crossover (SC) 334, 336–338, 341–342, 348, 354, 379, 400, 412, 417
spin-crossover materials 338
spin-crossover molecules 351, 404, 407

spin-crossover problem 351, 353, 355, 357, 359
spin-crossover solids 333–336, 338, 340, 342, 344, 346, 348, 350, 352, 354, 356, 358, 360, 412–414, 416–418
spin-crossover transition 335–336, 401
spin wave dynamics 285–286, 288, 295, 297, 305, 321
spin wave propagation 287, 290, 305, 309, 320–321
spin waves 124–125, 127, 138, 140–141, 143, 258, 283–288, 290–293, 295–297, 301, 303, 308–309, 311–312, 314–316, 321
  exchange-dominated 291
spintronics 2, 285–286
sputtering methods 11–12, 14, 17
square nanoparticle system 406
STM, *see* surface tunneling microscopy
Stoner–Wohlfarth model 230–232, 234, 237, 241
superstructures 415–416
surface magnetocrystalline anisotropy 300, 309
surface tunneling microscopy (STM) 11, 13–14, 195, 203
switchable molecular solids (SMSs) 336, 339, 361

thermodecomposition 215, 221–222
three-dimensional magnetic nanodots, vortex lines in 101–120

ultrathin films 10, 13, 35–36, 68, 124, 132, 202, 204, 413

X-ray magnetic circular dichroism (XMCD) 37, 39
X-rays 39, 41

XMCD, *see* X-ray magnetic circular dichroism

zero-frequency modes 132, 136, 140–141